THE GERM OF AN IDEA

The Germ of an Idea

Contagionism, Religion, and Society in Britain, 1660-1730

Margaret DeLacy

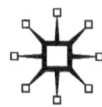

First published 2016 by
PALGRAVE MACMILLAN

The author has asserted her right to be identified as the author of this work
in accordance with the Copyright, Designs and Patents Act 1988.

Palgrave Macmillan in the UK is an imprint of Macmillan Publishers
Limited, registered in England, company number 785998, of Houndmills,
Basingstoke, Hampshire, RG21 6XS.

Palgrave Macmillan in the US is a division of Nature America, Inc., One
New York Plaza, Suite 4500, New York, NY 10004-1562.

Palgrave Macmillan is the global academic imprint of the above companies
and has companies and representatives throughout the world.

Hardback ISBN: 978-1-137-57527-2
E-PUB ISBN: 978-1-137-57528-9
E-PDF ISBN: 978-1-137-57529-6
DOI: 10.1057/9781137575296

Distribution in the UK, Europe and the rest of the world is by Palgrave
Macmillan®, a division of Macmillan Publishers Limited, registered in
England, company number 785998, of Houndmills, Basingstoke,
Hampshire RG21 6XS.

Library of Congress Cataloging-in-Publication Data

Names: DeLacy, Margaret, author.
Title: The germ of an idea: contagionism, religion and society in Britain,
    1660-1730 / Margaret DeLacy.
Description: New York, NY: Palgrave Macmillan, [2015]
Identifiers: LCCN 2015023117 | ISBN 9781137575272 (hardback: alk. paper)
Subjects: LCSH: Medicine—Great Britain—History—17th century. |
    Medicine—Great Britain—History—18th century. | Epidemics—Great
    Britain—History. | Plague—Great Britain—History. |
    Medicine—Philosophy. | Diseases—Philosophy. | Medicine—Religious aspects.
Classification: LCC R486 .D43 2015 | DDC 610.941—dc23 LC
    record available at http://lccn.loc.gov/2015023117

A catalogue record for the book is available from the British Library.

This book is dedicated to the memory of my brother, John Calvert Eisenstein (1953–1974)

CONTENTS

Preface

The Eighteenth-Century Slump?[1]

Sometimes a single question or comment will trigger a long research program. My current project began with the tantalizing remark of Charles-Edward Amory Winslow:

> By 1700 there was available theoretical and observational evidence which should have made possible the formulation of our modern germ-theory of disease. Kircher had advanced the concept of a *contagium animatum* . . . Redi had presented convincing evidence that living things . . . were not spontaneously produced . . . Leeuwenhoek had actually described . . . protozoa and bacteria . . . If an open-minded and imaginative observer had put the work of these three pioneers together, the germ-theory of disease could have been developed in the seventeenth century instead of the nineteenth. Why [were] medical thinkers . . . diverted?[2]

Winslow's frustration with early-eighteenth-century medicine echoed the bleak view of his predecessor, the medical historian Fielding H. Garrison, who had argued that the eighteenth century was a "lull" characterized by "exaggerated sobriety and apparent content with the old order of things," "tedious and platitudinous philosophizing," and a "mania for sterile, dry-as-dust classifications."[3]

These claims left me wondering how early modern doctors actually approached the problem of explaining disease. How close had they come to developing a theory that some diseases might result from living contagions—that is, a theory resembling what Winslow loosely termed the "modern germ-theory of disease"?[4] Was there an abrupt break in this pattern of thought after 1700? Did the "medical revolution" that reached a peak in Harvey's discovery of the circulation of the blood in the seventeenth century simply fizzle out? Finally, apart from the tidiness of being able to assign "precursor-hood" appropriately, why should we care? In what ways might the ideas of Restoration medical thinkers matter to us? Being new to this field, I thought it would be easy to find a book that would answer these questions.

When I turned to eighteenth-century medical works to learn about the conceptualization of disease during the period, one reason many historians have avoided this subject became evident. I could not understand what I read. When they discussed philosophy, politics, and even religion, eighteenth-century authors used language that at least seems to be clear and comprehensible, but when they wrote about medicine they drew on a vocabulary and an underlying conceptual framework that is almost impenetrable today.

Medical ideas from this time cannot be mapped isomorphically onto any corresponding set of terms and ideas of our own. Events and conditions that began the century as strings of adjectives such as "bilious," "putrid," "low," "pestilential," and "intermittent" were being discarded or rearranged into distinct diseases. The names for these illnesses were becoming more consistent in both English and Latin, enabling physicians to communicate more fruitfully with each other. This process began well before 1700 and ended long after 1900, but it was gathering steam in this period, when completely different systems for conceptualizing disease competed with each other. It is not surprising that this period of complex transitions seemed inchoate.

Many years later, I have reached some conclusions. First, claims that this period saw an abrupt break with the past or future are exaggerated or misplaced. Second, the interest of authors in ideas about contagion was related to their status as outsiders in England or their experience as religious dissenters (broadly defined).[5] Moreover, this relationship persisted from before the Restoration right through this period. Third, what was important about their idea of contagion was not whether it accurately anticipated Pasteur's (or Bassi's or Henle's, or twentieth-century) ideas but its contribution to the shift from a physiological to an ontological view of acute diseases.[6] In other words, contagionism partly replaced the conceptualization of illness as an imbalance or dysfunction with the conceptualization of separate disease entities with their own characteristic patterns.[7] The theory that the *contagium* itself might be alive sometimes played a role in this transformation, but the idea of contagion itself affected disease theory even when the nature of the agent was seen as inanimate or was not specified.

The rest of this book will discuss the second two conclusions in greater detail, but the hypothesis of an eighteenth-century "lull" deserves further consideration here. In the paragraph before the one quoted above, Winslow had acknowledged that "Benjamin Marten in 1720, three years before Leeuwenhoek's death, assumes that tuberculosis is caused by invisible 'animalculae,' like those discovered by Leeuwenhoek . . ."[8] This led me back to Clifford Dobell's biography of Leeuwenhoek and to William Bulloch's *History of Bacteriology*.[9] The latter contained a more detailed discussion of Marten's work, including quotations from *Marten's New Theory of Consumptions* (1720), but it ended with what would become a familiar assessment: "Marten's book . . . was soon relegated to oblivion. Marten could never have dreamt that his views would turn out to be so correct. How different were they from those of his day and long after."[10] After very brief and dismissive references to two other eighteenth-century authors (Carl Linnaeus and Marc Anton von Plenciz), Bulloch, like many of his colleagues, turned with relief to his real subject, the enormous strides that were made in bacteriology in the nineteenth century. Clearly, Marten was an interesting but inconvenient snag complicating the story of eighteenth-century nescience.

Bulloch's work in turn led me to the articles by Charles Singer on Marten himself. In 1910, Singer, then a young physician, stumbled on a copy of Marten's *New Theory* when browsing through some used books and purchased it for a few pennies.[11] This beguiling work changed Singer's life forever. He quickly published an article on Marten and then, with help from his new wife, the scholar Dorothea Waley Singer, researched and self-published a book on the idea of living contagion from 1500–1750.[12] He conclusively demonstrated that many earlier European authors had expressed an interest in ideas about pathogenic "worms" or "animalcules," a fact that was known to eighteenth- and nineteenth-century writers but still seems to surprise most people today. Even though his title implied that there was something worthy of discussion up to 1750, he concluded that after 1720 the story of the idea of *contagium vivum* was one of "progressive degradation" throughout the eighteenth century.[13]

This view was still common in the middle of the twentieth century. In fact, with a few important exceptions (such as the work of Mabel Buer), historians have not revised this gloomy picture even as they have rewritten our view of other aspects of eighteenth-century thought.[14] In addition to Winslow, Richard Shryock had entitled two chapters of his widely read textbook on medical history (1936) "The Partial Failure of Physical Science, 1700–1800" and "Social Factors in [the] Medical Lag after 1700."[15] Although Peter Gay did argue in 1967 that the Enlightenment stress on "experience, clinical study, and experimentation revolutionized medicine," he subverted this claim by adding that, nevertheless, a "sick man who did not consult a physician had a better chance of surviving than one who did."[16]

The final years of the twentieth century saw a renaissance in the study of early modern natural philosophy, early modern medicine, and related sciences (especially chemistry) in the wake of groundbreaking publications by Walter Pagel, Charles Webster, and Allen Debus that connected new conceptualizations about disease in the sixteenth and early seventeenth centuries to both religious heterodoxy and Paracelsian and (later) Helmontian medicine.[17] Of special relevance here are studies of plague and syphilis. Jon Arrizabalaga, John Henderson, and Roger French have traced transformations in the way physicians characterized and understood syphilis from its first appearance in Europe to the publication of Sennert's synthetic account in 1641.[18] By the seventeenth century, belief in its contagiousness was widespread. Studies of the plague and especially the Great Plague of London have amply documented the belief of many early modern writers that the plague was contagious, although their ideas about the nature of this contagion varied.[19] Thus, I have not belabored the development by 1700 of a belief in the contagiousness of plague and syphilis because it has already been well established.

Most works in early modern medical history, however, terminated before 1700. Margaret Pelling even commented that Charles Webster deliberately avoided the eighteenth century because it saw the defeat and eclipse of the utopian radicals he had lovingly profiled in *The Great Instauration*.[20] Shryock did have a twinge of remorse because he revisited the subject of the early eighteenth century in 1972 as the social history of medicine was becoming popular.[21] Confessing that as "a mere impressionist" he had no computer data on how many men had written or thought about pathogenic microorganisms between 1700 and 1860, he admitted that there was "almost a flurry" of references to the idea between 1705 and 1730. Moreover, the authors of these works listed others with similar views. Perhaps, he suggested, a computer could retrieve these references, but his "hunch" was that the effort would not justify the cost and effort required.[22]

In 1971, William Le Fanu, who had been Garrison's assistant, delivered the annual Fielding H. Garrison lecture to the American Association for the History of Medicine titled "The Lost Half-Century in English Medicine, 1700–1750."[23] His title accurately described his conclusions. He commented, "How widely the concept of contagion was accepted is very uncertain; it was not the general belief of physicians."[24] Le Fanu's article remains influential, despite Adrian Wilson's observation that "It is strange to reflect that it took Le Fanu some thousands of words merely to summarize the many non-events, as he saw them, of this 'lost half-century.'"[25]

Just three years after Le Fanu, sociologist Nicholas Jewson published an explanation for this sterility, which he extended from fifty years to an entire century.[26] He claimed that in the period between the late seventeenth century and the "decades after 1780," English medicine was characterized by the dependence of physicians on their patrons and the dependence of the other branches of the profession on physicians. This led to a competitive atmosphere that discouraged real innovation, experimentation, and collegiality. It underpinned a conservative medical ideology that classified diseases by symptoms, not causes, and relied on speculative theories, not empirical evidence.[27]

More recently, David Wootton has also complained that eighteenth-century medicine failed to develop a "germ theory of disease," commenting,

> Conventional histories of medical knowledge . . . are incapable of giving a coherent account of eighteenth-century biology and medicine because the story cannot be told as a story of progress; it is rather a story of squandered opportunities, of wasted effort, of intellectual dead ends. It is a history of failure.[28]

He added: "Doctors successfully pushed the microscope, and all the questions that it generated, out of medicine in the 1690s, and kept it out until the 1830s." J. N. Hays's popular and well-reviewed survey *The Burdens of Disease* concluded that contagionism "withered" in the eighteenth century.[29] Other medical history surveys literally wrote off eighteenth-century medicine either by beginning in the nineteenth century or skipping over the eighteenth century after barely mentioning a minimal number of figures.[30] This surely led readers to conclude that eighteenth-century medical thought was not worth studying, despite revived interest in eighteenth-century science, philosophy, and natural history.[31]

The many works of the historian Roy Porter, although they have contributed greatly to our understanding of medicine and related sciences during
this period, did not challenge this overall view. Porter, who often quoted
the negative comments of Garrison, Gay, and others with relish, proclaimed
both his ambivalence about the "optimistic" picture of eighteenth-century
medicine and his belief that in any case this was the wrong way to approach
the period.[32]

Porter complained that "there is a Whiggism ingrained in traditional medical history which instinctively portrays the story of medicine as the progressive triumph of light over darkness . . . And yet there are critical objections
to the . . . view that the Enlightenment spawned medical progress, blooming
health and population growth."[33] He argued that advances in medical philanthropy were due to "evangelical Christianity not Enlightenment beliefs"
and that "inoculation was certainly no triumph of new scientific medicine:
it was an empiric folk-remedy imported from the Near East," which was
readily accepted in England because it was patronized by "court physicians
such as [Hans] Sloane and [Richard] Mead."[34] These comments slide over
the important questions of why the Dissenters he mentions initiated medical
philanthropy and why Sloane and Mead put their reputations on the line in
support of inoculation.[35]

Instead of focusing on the ideas of physicians and the collection of medical data, Porter argued, historians should recognize that "the locus of the
English medical Enlightenment lay with the private, the individual, the local,
the personal, the voluntary," and they should turn their attentions to the
much broader cultural changes taking place among the English laity.[36] The
fundamental task of the medical historian, he stated, did not lie in analyzing
"scientific medicine, the professionalization of doctors, nor in a state medical police" but instead was "to grasp contemporaries' consciousness of the
dimensions of living" and their changing views of "life and death, health
and sickness, the normal and the deviant, the stages of life, . . . the family,
childhood, madness, sexuality, demography."[37] Is the field so limited that it
cannot include both approaches?

Porter certainly acted on his argument. Although he was so prolific that
he did occasionally raise the sort of topics he had discountenanced, most of
his work addressed such questions as the relationship between medicine and
culture, patients' medical knowledge, medical communication, quackery,
venereal disease and sexual maladies, mental illness, philanthropy, literature
and medicine, and medical competition in the marketplace.[38] In addition
to Porter's own influence, a determination to avoid the historian's foibles
of Whiggery, naïveté, "precursor-itis," and presentism has deterred younger
medical historians from pursuing older questions. A shifting of interest away
from English history, eighteenth-century history, and intellectual history
toward colonial and world history, nineteenth- and twentieth-century history, and social and cultural history has also played a role.[39]

Porter's colleagues and successors have produced a growing and solid
body of work, but it gives little attention to early eighteenth-century theories

of disease and contagionism.[40] James Riley's book on disease prevention in the eighteenth century offered a detailed and serious engagement with eighteenth-century authors but concluded that "more and more, eighteenth-century epidemiology was a medicine of the environment, rather than of diseases."[41] One dissident, Adrian Wilson, argued in separate essays both in favor of analyzing the development of ideas about individual diseases and in defense of early eighteenth-century medicine, but Wilson's own work has focused on topics other than contagion.[42]

The monographs published since 1970 on more detailed aspects of eighteenth-century medicine include evidence about the skill, learning, charity, and good judgment of many eighteenth-century doctors, but the reader will find very little discussion of ideas about contagion in these works, let alone any effort to connect this theory with other contemporary developments.[43] Thus, no recent historian seems to have revisited the claim that contagionism was all but dead in the early eighteenth century.

Our picture of the other end of the century, on the other hand, is still colored by Erwin Ackerknecht's seminal article "Anticontagionism between 1821 and 1867."[44] Ackerknecht's article reviewed the rise of opposition to contagionism and quarantines in the early nineteenth century, but he included a sketch of the status quo that his subjects were seeking to transform. Focusing on four major diseases—yellow fever, cholera, the plague, and typhus—and working backward from the time and place he knew best (nineteenth-century Paris), he argued that the brief heyday of anticontagionism coincided with the revolutions in America and France, where anticontagionism was already especially popular. He claimed that anticontagionists were "liberals" opposed to quarantines that hurt free trade and thus suggested that their enemies, the contagionists, were traditionalists. He located the center of this old-fashioned contagionism in the conservative London College of Physicians.[45] Not only did this argument suggest that eighteenth-century reformers or proto-liberals would have been anticontagionists, but it focused the interest and attention of medical historians on cholera, contagionism, and the state in the nineteenth century, where it has largely remained.[46]

Margaret Pelling's book on changing views of cholera in nineteenth-century Britain offered a major revision of Ackerknecht's theories, but Pelling agreed with him that a majority of the conservative English medical profession at the turn of the century was contagionist.[47] Many of the works on nineteenth-century epidemics that followed also assumed that contagionism was widespread by the beginning of the nineteenth century. They worked backward from Victorian debates about contagion, infection, and miasma, not forward from early modern views, and did not see contagionism as new or puzzling.[48]

Ackerknecht's work thus posed two problems. First, if ideas about contagionism in Britain peaked about 1700 and then immediately plummeted, as Winslow and his colleagues had thought, how did it come about that so many members of the profession were contagionists by 1800 as Ackerknecht claimed? Was Ackerknecht simply wrong about this or was part of the eighteenth-century story missing? Second, if anyone supported contagion-ism during this period, it was physicians who were also Dissenters and/or Scottish Presbyterians.[49] These men were reformers, even radicals at times, not the conservative English loyalists that Ackerknecht had described.[50]

Moreover, as noted above, a growing literature on early modern science and medicine had uncovered a close connection between religious hetero-doxy and Paracelsian or Helmontian ideas in medicine in the sixteenth and early seventeenth century. However, there didn't seem to be any way to con-nect the Dissenting medical reformers of the end of the eighteenth century to the heterodox authors who had pioneered ideas of contagion and *conta-gium vivum* at the end of the seventeenth century.

While I was pondering the problem of the missing links between reli-gion, early modern contagionism, and contagionism in the early nineteenth century, historians of related domains published work that demanded a broader reappraisal of eighteenth-century medical thought. For example, Lise Wilkinson, in a series of articles beginning in 1974, traced the establish-ment of veterinary medicine and the development of what she termed "the virus concept," showing how this affected the understanding of cattle plague (rinderpest) in the eighteenth century.[51]

Catherine Wilson's book on early microscopy included a discussion of Marten and other early believers in living contagion.[52] Echoing Winslow's conclusions but moving the timeline, she concluded that "a germ theory of animate contagion can thus be said to have emerged between 1650 and 1720, supported however, only by circumstantial and not by direct evi-dence."[53] Wilson did not extend her narrative on through the eighteenth century, but Mark Ratcliff's study of microscopy in the Enlightenment con-tinued the story to 1800.[54] Although he did not analyze changes in ideas about contagion and disease, he discussed some of the themes that I felt were also important for the medical history of this period. These included scien-tific communication and scientific networks, the relationship between natural history and theoretical science, and the interplay between what he calls "sys-tematical" (i.e., taxonomic and linguistic) and "practical" activities.[55]

Ulrich Tröhler demonstrated that quantitative medicine did not begin in nineteenth-century Paris but in eighteenth-century Britain.[56] The James Lind Library, to which Tröhler is the main contributor, continues to provide articles and resources on eighteenth-century quantification.[57] A. J. Cain, M. M. Slaughter, Rhodri Lewis, and John Wilkins discussed the evolution and importance of early modern systematics and taxonomy, dispelling the view that the eighteenth-century interest in classification and nosology was ster-ile.[58] Frank Egerton, in a long-running series of articles in the *Bulletin of the Ecological Society of America*, traced the eighteenth-century contribution to ecology, whose importance had been unrecognized by older histories of sci-ence. The term appeared only in the nineteenth century, but the discipline itself has its roots in earlier natural philosophy.[59] Eighteenth-century science was more creative than earlier historians had believed—did this also extend to eighteenth-century medicine?

Meanwhile the "lost half-century" in medicine seemed to be shrinking. Some scholars had already moved its beginning forward from 1700 to 1720 to accommodate the publication of Marten's *New Theory*. But the appearance of this work was followed by the publication of Marten's second edition in 1722, the work of Leeuwenhoek up to his death in 1723, the publications of Richard Bradley, and the continuing discussions about the plague following the Plague of Marseilles. The following decade saw a parody of contagionism in the *Grub Street Journal* (1730) and the publication of John Andree's English translation of Desault's treatise on venereal disease (1738) in addition to a continuing stream of other books, new editions, and articles on inoculation, including Sloane's "Account of Inoculation," which was written in 1736 although it was published posthumously in 1756.[60] The half-century was becoming a decade, or perhaps merely a lustrum.[61]

In the 1730s, both Richard Mead and Hans Sloane, who had supported quarantines for plague as well as smallpox inoculation, were still alive and very active. Sloane resigned as President of the Royal Society in 1741 and retired from practice in 1746; Mead published his final work in 1751. Two men who would become pivotal figures in the second half of the century, William Cullen and Carl Linnaeus, were paying extended visits to London, in about 1731 and 1736 respectively; a third, John Fothergill, studied in London from 1736 to 1738 and began his own London practice in 1740. These years were coming to seem more like a changing of the guard than a full-scale retreat.

After changing my mind (and title) several times, I decided that it made the most sense to end this part of the story in 1730, which does seem to mark a period of transition from an occasional reference to *contagium vivum* to a growing belief in contagion from an unspecified agent as the critical factor in the spread of many acute diseases. The decade between 1730 and 1740 also saw the death of the Leyden professor Hermann Boerhaave (1738) and the beginning of a new wave of physicians mostly educated in Scotland. If I could not offer an account of the rise and fall of contagionism, at least I could provide a more detailed and coherent examination of the discussion of this idea as it developed from the Restoration into the early eighteenth century and try to situate it more fully in its religious, professional, and social context.

Many other events and transitions that occurred in the eighteenth century should be interleaved with the story of contagionism, but this is unlikely to happen as long as historians in other domains consider eighteenth-century medical thought tedious, sterile, or unapproachable. We still lack a full account of the ways such individual diseases as measles, scarlet fever, whooping cough, pneumonia, and poliomyelitis were understood during the century.[62] There are few detailed studies of eighteenth-century ideas about disease and contagion in other Western countries.[63] It is time to situate contagionism and the history of medicine in general more securely in its religious, political, philosophical, professional, and social context.

This book begins with the philosophical and religious developments underlying the cultural landscape of early modern British medicine: in particular the classical four-element theory of matter, the nature of life, and the role of the soul. It briefly traces the development of ideas about the nature of contagion within this tradition and concludes that the monist philosophy underlying contagionism was more compatible with a heterodox religious outlook than with either a traditional Catholic or a traditional (magisterial) Protestant approach to natural philosophy.

Chapter 2 discusses the medical views of sectarians during the English Civil War and the effort to stamp out sectarianism, which continued to shape the medical profession throughout the century. It describes William Harvey's vitalist theory of contagion and the pervasive influence of the Belgian philosopher and chemist J. B. Van Helmont on what might be termed "alternative medicine" during the Restoration. It then depicts the fractious, cosmopolitan, multicultural world of early modern London, which undermined efforts to maintain an orthodox and homogeneous profession, creating fissures that allowed contagionist ideas to seep in.

Chapter 3 focuses on three unorthodox medical populists: Marchamont Nedham, the opportunistic journalist who admired Helmont's work and published the first English work on living pathogens; Gideon Harvey, who thought that venereal diseases arose from living particles; and the "Chymical Physitian" Everard Maywaringe, who attributed diseases to "seminal agents," provided a clear description of the mechanism of contagion, and argued that "Qualities cannot be Diseases."

Chapter 4 discusses the effort of more established Restoration authors to find a middle position between Galenism and Helmontianism that was informed by a new empiricism and incorporated the "natural history" approach of the parliamentarian physician Thomas Sydenham and his associates. It briefly considers discussions by Fellows of the Royal Society about the sources of the plague epidemic of 1665. It then addresses their work on taxonomy and the biological (or generative) conceptualization of species that led John Wilkins to publish a classification of individual diseases in his work on artificial languages.

Chapter 5 reviews European developments in biology and microscopy, especially debates over spontaneous generation and living contagion, and the reception of these ideas by members of the Royal Society. It considers several works that attributed diseases to parasites or "worms." It concludes with the international rinderpest epizootic, which inspired a widespread discussion of the mechanism of contagion.

Chapter 6 explores the connection between contagionist authors and the President of the Royal Society, Hans Sloane: an Irishman trained in chemistry and botany. Sloane led the effort to introduce smallpox inoculation in England and was also one of three physicians commissioned by the British government to develop a plan to prevent or contain plague. The chapter depicts Sloane's "natural history" or Baconian approach to science and relates it to his approach to classification.

Chapter 7 focuses on Benjamin Marten's *New Theory of Consumptions*: the first full presentation of a theory of *contagium vivum* in Britain. It explores

Marten's milieu and his links to "fringe" practitioners associated with Sloane and evaluates the likely impact of the *New Theory*. It also reviews several related theories of pathogenesis, the appropriation of *contagium vivum* by a con man in Paris, and the survival of the idea of parasitic pathogens after 1730.

Chapter 8 provides a political and professional context for the introduction of smallpox inoculation into England that includes the involvement of the Royal Society and the inaction of the College of Physicians. It shows how the Puritan minister Cotton Mather was influenced by his reading of Van Helmont and Marten to introduce inoculation in Boston. It then summarizes the impact of inoculation on ideas about disease transmission.

Chapter 9 turns to another terrifying disease that shaped British thought during this period: the plague and the British response to outbreaks in Europe. A lively discussion emerged about the value of quarantines and other preventive measures based on competing theories about how the plague spread. The chapter focuses on the 1720s but extends through the century.

The conclusion points out that although many British authors before 1730 entertained ideas about contagion, it remained an unorthodox idea with roots in a suspect Helmontian philosophy. The infrastructure necessary to investigate, sustain, and institutionalize contagionism, or to deploy it systematically as an aid in conceptualizing, classifying, and investigating acute diseases, did not exist. Only after the rise of Scottish medical education at midcentury would a research community possess the numbers, status, infrastructure, intellectual background, and geographic reach to adopt contagionism as a shared ideology.

ACKNOWLEDGMENTS

This book draws on a reference base of thousands of books and articles. Many of them were borrowed for me by the interlibrary loan department of the Multnomah County Public Library in Portland, Oregon. I thank the library itself and the dozens of libraries that generously made these works available to me through interlibrary loan, in most cases free of charge. This book would have been impossible without their help.

Many libraries also made materials available online or allowed walk-in access to their holdings. They include the Wellcome Library for the History of Medicine, the British Library, and the Library of the Royal College of Physicians in London; the Bodleian Library in Oxford; the National Library of Medicine in Bethesda, Maryland, which also provided and created microfilms; and in Portland: the Family History Center of the Church of the Latter Day Saints, the Oregon Health Sciences University, Portland State University, and Reed College libraries. Libraries and librarians who assisted with specific problems are named in the notes, but I would like to express overall gratitude to Harold Cook, Geoffrey Davenport of the Royal College of Physicians Library, Gina Douglas of the Linnean Society, Stephen Greenburg of the National Library of Medicine, David Harley, Jeff and Liz McBride, and David Zuck and to acknowledge debts to the late Arthur J. Cain, James Cassedy, Worth Estes, and John Symons.

The Humanities, Science, and Technology program of the National Endowment for the Humanities program provided a three-year grant for 1989–1992 that initiated this project. An earlier fellowship from the American Council for Learned Societies first enabled me to study the history of medicine.

Family members, including my mother, Elizabeth Eisenstein, and my brother, Edward Eisenstein, read and commented on early drafts. My husband, John DeLacy, not only put up with this seemingly endless project but provided outstanding in-house IT service.

Abbreviations and Short Titles

AIM25:   Archives in London and the M 25 area, www.aim25
                .ac.uk

DNB: *Dictionary of National Biography, 1885–1996*

ECCO: Eighteenth-Century Collections Online, part 1 and 2,
                created by Gale Digital Collections from digitally
                scanned microfilms of books published in Great Britain
                during the eighteenth century

EEBO: Early English Books Online. Online image files created
                from microfilms of books published in England before
                1700 included in the English Short-title Catalogue or
                the Thomason Tracts

FRCP: Fellow of the Royal College of Physicians of London

FRS: Fellow of the Royal Society

Google: Book database at books.google.com

Hathi Trust: The Hathi Trust Digital Library, www.hathitrust.org

JHMAS: *Journal of the History of Medicine and Applied Sciences*

LRCP: Licentiate of the Royal College of Physicians of London

MB: Bachelor of Medicine (including Oxford BM)

MD: Doctor of Medicine (including Oxford DM)

Munk's *Roll:* William Munk, *The Roll of the Royal College of Physicians
                of London*

N&R: *Notes and Records of the Royal Society of London*

ODNB: *Oxford Dictionary of National Biography* (2004–)
                www.oxforddnb.com

Wallis: *Medics.* Peter Wallis, Ruth Wallis, and T. D. Whittet, *Eighteenth
                Century Medics: Subscriptions, Licenses, Apprenticeships*
                (Newcastle upon Tyne: 1988)

C H A P T E R   1

INTRODUCTION: MEDICAL THEORY IN EARLY MODERN EUROPE

INTRODUCTION

Contagionism (literally "touching together") is the idea that diseases are transmitted by the transfer of a pathogenic substance from one person to another in a chain of contacts. In the period between the Restoration and the 1720s, new ideas in natural philosophy and developments in technology coincided with serious epidemics to produce a new interest in theories of contagion, including theories of animate contagion. Contagionism, in turn, encouraged medical authors to reconsider the nature of disease itself and the way that acute diseases were categorized and understood, transforming medical theory and practice during the succeeding centuries.

This book will consider the development of contagionism in the context of the religious and professional environment of early modern Britain, focusing on the role played by physicians and medical authors who were not members of the Church of England or who held ideas that were slightly outside the professional mainstream.

Medical philosophy has always concerned itself with such religious issues as the relationship between the soul and the body and the nature of life itself. Early modern religious views constrained medical inquiry and activity: the ideas physicians held about disease, the ways they sought knowledge, and the sorts of questions they asked. Even if they were not connected in strict logic, certain ideas about contagion challenged traditional medical and philosophical theories and by doing so became associated with religious dissent and social marginalization.

In the early modern period, most educated medical authors based their theories on a "physiological theory" of disease originally derived from Galenic medicine and classical natural philosophy. They thought that disease was the result of a malfunction that took place within the body's own systems for maintaining health, often triggered by an imbalance in the patient's environment or mode of living (for example, diet, sleep, or exercise). They viewed disease as a dis-order—as an imbalance of fluids (humors) or flows in the body. They thought that most epidemic diseases (those affecting a large number of people at once) resulted from environmental factors that poisoned or altered the air or water and caused harmful changes in a patient's humoral balance or led the humors to become corrupted. They saw the environmental events that sometimes precipitated humoral malfunctions as both difficult to control and less relevant than the steps the body took to return itself to balance, aided by a physician who understood each patient as an individual.

In this perspective, every disease was tied to a particular person and his own humoral balance or to a particular place and its characteristic environment. Whenever many people fell ill in the same place, they in effect had the "same" illness, although each person's temperament would cause different symptoms to appear in response. On the other hand, diseases that appeared in different places or different times had no actual physical relationship to each other. Even if the scattered patients had very similar symptoms, these merely reflected underlying similarities in the patients' physiological state or environment. Such diseases might have the same *kind* of cause, but they did not have the *same* cause.

Contagionism is not the same thing as an ontological theory of disease. Contagionism is a theory of disease transmission, not a description of its nature or physical cause. Nevertheless, there is a close relationship between contagionism and the depiction of diseases as entities, which in turn facilitates a useful classification of acute diseases. Just as we may differentiate between species of plants and animals by their ability to produce fertile seeds or offspring, we can differentiate between species of contagious diseases by asking whether the same pattern of symptoms is spreading from one person to another and yet another. Even though the offspring of a pair of plants or animals is not identical to its parents in every respect, there is enough of a likeness between them to enable us to say that they are all members of the same species.

Members of a particular species of life are related to each other by a physical chain of causation that reaches back through time. Contagious diseases also maintain their identities across place and time. If two patients contract measles, they acquired their disease through a chain of contacts to a common index (source) case that was also measles. Every new episode has the same material cause.

One of the problems that confronted medical authors during this period was that they were trying to understand the epidemiology of a group of acute diseases, some of which spread through strict contagion—direct contact from person to person—and others that did not. How were they related to each other (if they were)? Medical writers wanted to understand diseases in terms of one kind of cause, but every time they tried they were frustrated by conflicting evidence, technical problems, and the fragmentary nature of early modern communications and research communities.

In London there was a small but growing group of well-educated physicians whose status was circumscribed by their nationality, education, and/or religious loyalties. These men were more likely to have a broad patient base than the Anglican physicians who usually treated the elite—primarily wealthy adults who often suffered from chronic illnesses. These outsiders were more likely to have worked or trained as apothecaries, surgeons, or military doctors before becoming physicians; to have lived or worked in other countries and retained cosmopolitan connections; to treat poor or middle-class patients, servants, and children; and to see patients suffering from acute and epidemic illnesses. In addition, their religious heritage, which stressed individual freedom of conscience, made them more sympathetic to earlier authors who had criticized traditional medical philosophy and Galenism. Antiestablishment sympathies spilled over from religion to politics and from politics to medicine.

By the early modern period, most doctors agreed that a limited number of diseases—including syphilis, inflamed eyes, and the itch, and, less often, consumption, plague, and dysentery—were often contagious.[1] They attributed most other acute diseases to changes in the weather or individual errors in regimen. Most learned physicians, however, thought that advising their clients on ways to preserve their health by living moderately and restoring patients' humoral balance through intervention when they fell ill were more important than musing about the possible and presumably uncontrollable causes of epidemic diseases. Their patients filled their time and consumed most of their energy.

By the late seventeenth century, some doctors were questioning the value of the classical humoral tradition and seeking new answers through new methods of gathering knowledge. In England, increased access to advanced instruction, intense competition, and barriers to professional advancement created a group of highly literate doctors with time on their hands. Medical publishing became a path to a larger practice or an alternative career—a way for a doctor to advertise his competence without undermining his status (at least if the publication was well received). As changes in the medical market after the Restoration encouraged "outsider" doctors to publish medical works, changes in ideology, broadly defined, encouraged them to challenge authority.

All across Europe, botanists, zoologists, and physiologists were exploring the processes and boundaries of living organisms using new analytic techniques and new tools, including the microscope. They were frantically trying to organize, study, and assimilate a torrent of new minerals, plants, and animals that not only poured in through the rapid expansion of trade and commerce but had also emerged through new ways to see and understand what had been there all along. This work was eagerly followed by members of the Royal Society, such as Hans Sloane, William Petty, Nehemiah Grew, Frederick Slare, and Edward Tyson. Their interest in microscopic life led several of them to entertain ideas of living pathogens—a hypothesis that was not new but gained a different meaning after the idea of spontaneous generation fell into abeyance.

Interest in venomous and pathogenic organisms and theories of disease transmission crossed paths frequently during the period between the Restoration and the 1720s, especially in the Royal Society, where physicians, botanists, chemists, and biologists mingled constantly. Different authors combined these two disparate fields of thought in different ways: some, such as Richard Bradley, believed that pathogenic insects or pathogenic animated particles were everywhere; others, such as the microscopist Antoni van Leeuwenhoek, believed that animated matter was harmless or appeared as a consequence, not as a cause, of illness. Still others thought of *contagiums* as catalysts or ferments—but these weren't well understood either. Perhaps the majority of the medical profession, like some lay writers, saw these ideas as purely speculative, a waste of valuable time, without practical benefits, and even as risible or absurd. The doctors who were interested in the idea lacked the ongoing and interactive community that might have winnowed and refined these ideas, brought a wide range of clinical observations into the conversation, and produced a consensus that was solid enough to withstand neglect, criticism, and ridicule.

At the beginning of the eighteenth century, an epidemic of an animal disease, rinderpest, strengthened belief in contagion, as did two factors that increased interest in human epidemics and their etiology. One factor was the outbreak of plague in Marseilles, which occasioned a revival of earlier contagionist treatises and the publication of many additional works. In the hope of preventing another devastating epidemic in England, the government commissioned a report from Richard Mead, a cosmopolitan physician with deep roots in the Dissenting community. Mead's report summarized the contagionist view of plague and led to a maritime quarantine policy that survived in some form for the next two centuries. The quarantine policy in and of itself ensured that arguments about its content and value would continue to take place.

The other factor was information trickling in during the early eighteenth century from credible witnesses about a folk practice: smallpox inoculation. This practice may have been widespread, but diplomats, traders, and physicians sent the first clear and detailed written descriptions from the Near East. Smallpox inoculation was introduced to England by two courageous women, Lady Mary Wortley Montagu and Queen Caroline, and by a group of physicians led by Hans Sloane, who was, like his friend Richard Mead, a cosmopolitan figure with close ties to many Dissenters. Both the research on inoculation and the campaign to promote its use centered on the Royal Society of London.

The introduction of smallpox inoculation revealed the existence of a specific disease-causing substance that could be carried around in a box and cause the "same" disease in every patient in succession. Smallpox became the model of a contagious disease. For the rest of the century, medical authors sought to understand how many other illnesses were transmitted in the same way. Was smallpox the exception or the rule? Did severe epidemic diseases always or often travel from patient to patient by contact or did they emerge from some general poisoning of the air in a particular place? Did acute diseases have rules? Did different species of disease exist, and, if so, what were they? Did different rules apply to different species of illness? To what extent were the diseases of plants or animals related to those of humans? Did the processes of fermentation, putrefaction, or even generation offer useful analogies? Were they simply analogies, or were they in fact the same processes caused by the same entities?

One English physician in the early eighteenth century, Benjamin Marten, published a completely worked-out account of *contagium vivum* and its relationship to contagious diseases. Perhaps the close resemblance of his work to the germ theory of disease that emerged in the nineteenth century was due more to coincidence than any special endowments of his own. By the standards of his time, Marten was adequately or even well educated; his book was moderate and learned, but his marginal position in the medical profession and the absence of a community of like-minded colleagues doomed his work to marginal status. It was a suggestion, not a school.

After a surge in interest in the 1720s, British interest in contagionism apparently declined during the 1730s during a period of relative peace and good health when there were few severe epidemics prompting an anxious public to demand medical information or intervention.

THE GALENIC HERITAGE

Although no one has traced its origins, it is likely that some sort of contagionism has always been a part of popular ideas about disease.[2] A few nonmedical authors even related certain diseases to "particulate agents of infection and insect carriers."[3] Classical historians beginning with Thucydides recognized that epidemics spread from the sick to the well. Some writers, such as Lucretius and Varro, even identified "infection with 'seeds' which 'fly about and cause disease and death,' or 'animalculae which cannot be seen with the eyes and which we breathe through the nose and mouth.'"[4]

Potentially contagionist views of a few diseases including "leprosy" appear in the Bible.[5] In his classic history of contagionism, historian Charles-Edward Amory Winslow concluded that the Old Testament offers "the first clean-cut conception of contagion and—built upon this conception—a definite and well-conceived program of differential diagnosis, isolation, quarantine and disinfection."[6] Later scholars have questioned whether non-Western societies shared Western contagionist views and have suggested that even in the West contagion was often understood in a less concrete way, as an evil intention or emanation.[7]

Although a few classical medical works mentioned contagion, most focused on environmental or physiological factors. Early modern medical education was dominated by the surviving works of Hippocrates and Galen, but neither gave much attention to contagion. Galen had adopted Aristotle's theory that the universe was composed of just four elements: earth, air, fire, and water. Every natural substance contained these elements but in different intensities or degrees. The four liquids or humors in the body—blood, yellow bile (choler), black bile, and phlegm—were also composed of these elements, and their relative balance determined a person's "temperament."[8] Thus diseases were not separate things; they were imbalances (dyscrasias) or abnormalities.[9]

Galen did not reject contagion entirely, but he minimized its role. The seeds of disease, if they existed, would not grow unless the humors provided the proper climate. The physician's duty was to focus on the patient, not the disease, and to restore the balance of the humors through "treatment by opposites." For example, the excessive heat of many fevers was blamed on a problem with blood, which was hot and moist. The cure was to draw blood, to provide medicines containing the right combination of cooling elements, and to impose a cooling regimen. Galen had acknowledged that diseases had different "seats" in the body, but orthodox physicians avoided "specifics" (remedies aimed at a particular symptom) in prescribing, instead combining ingredients with different degrees of each quality so the remedy would counteract the humoral imbalance. The use of these compound medicines with dozens of ingredients continued into the eighteenth century.[10]

Galen retained some older theories about disease transmission. Hippocrates had discussed the effect of different climates, noting, for example, that certain diseases such as agues (intermittent fevers) were more common in the summer while others, such as respiratory diseases, prevailed in the winter because the atmosphere varied in its composition and in its temperature and moisture.[11] Most dangerous of all was warm air that had been poisoned or "corrupted" by proximity to marshes or unburied corpses. Only this pervasive corruption of the air could explain the simultaneous outbreak of many illnesses. Even here, however, fevers were more about quantities than about kinds: they were grouped by their duration and by whether the fever was constant or came and went. By the early modern period, the term "Hippocratic" had become more a slogan than a specific set of medical practices, usually representing the ideal opposite of anything an author disliked about the medicine of his own day.[12] Historian Vivian Nutton has shown that in three tracts Galen did refer to diseases caused by the presence of minute "seeds" from outside the body. He conceded that *psora* (itch), *opthalmia* (inflamed eyes), *phthisis* (tuberculosis or consumption), and plague might be contagious, but most of his work downplayed the idea because he associated it with Lucretian atomism, which he opposed.[13]

Galen's work appealed to medieval authors because he was a monotheist who believed in a life-giving spirit that was drawn in with each breath. Christians interpreted this "spirit" as the human soul. Galen, however, had been a materialist who believed that the individual soul perished with the body, a view that was unacceptable to the Christian Church, which held that the soul was immortal. By the early modern period the Church backed the theory of Thomas Aquinas that the soul and the body were entirely different.[14] The dualism of body and soul became essential to certain central Catholic teachings, but the need to reconcile Galen's material and mortal "spirits" with the immaterial and eternal Christian "soul" caused many difficulties for medical writers.

Early modern contagionists challenged Galenism and religious orthodoxy by developing alternative theories about the relationship between matter and spirit that relied on monism (or vitalism) and/or atomism. Paracelsus, the alchemist and medical author who rejected Galenism, was a monist. Girolamo Fracastoro, author of the first contagionist medical treatise (1546), was an atomist.

MONISM: QUALITIES ARE NOT SEPARATE FROM THINGS

Monism returned to a conception of movement as inherent within matter: no spiritual entity was needed to form and direct it.[15] In denying the divide between material and incorporeal matter, monism also implicitly denied the distinction between body and soul. This led to mortalism: if souls were inherent in the organization of bodies, then they must perish with the dissolution of the body. Some monists retained the immortal soul by emphasizing the unity of God with his creation: if God was omnipresent, the soul was an expression of his immanence, and, since God was eternal, the soul was too. Many believed that souls "slept" with the body until it was resurrected. Orthodox Catholics and Protestants both rejected monism, but it was compatible with the views held by sectaries outside the state Reformed churches and with certain approaches within the established churches, such as pietism and mysticism.[16]

Monism was especially attractive to physicians.[17] Unlike many other natural philosophers, physicians could not isolate certain phenomena for separate investigation: they always confronted a whole patient, body and soul together.[18]

PARACELSUS (CA. 1493–1541): DISEASES ARE LIVING THINGS

The author known as Paracelsus saw body and soul as a single integrated unit and believed the universe was teeming with life forms; indeed, almost everything was animated, including ideas and minerals.[19] People were not the result of a mixture of humors but the unique outcomes of a generative process. The fact that each disease also had its own nature and organizing principle explained why different diseases appeared in people near one another and similar diseases appeared in those who were distant.[20] These individual diseases could be named and classified.[21] Although Paracelsus emphasized the separate nature of every disease, he was not a contagionist. For example, he believed that plague resulted from human imagination, which created a "*semina*": the seed or germ of a disease. This rose to heaven, where it angered the cosmic bodies and rained down again. It disrupted both the universe and the individual body, but it did not pass from person to person.[22]

ATOMISM: THINGS COME FROM SEEDS

Atomists thought that everything, including the soul, consisted of atoms and perished with the body, thus undermining a belief in divine judgment. Qualities such as hot and cold were not inherent in matter but were merely the impression left by certain atomic patterns. Instead of humoral imbalances, they blamed diseases on hidden particles or "seeds" that might rise up from putrefying ground or rain down from the heavens; a mass of them could putrefy the air itself. People became ill either by breathing in the poison directly or by ingesting food or water that contained the poisonous particles.[23]

The Lucretian idea that everything came from seeds contributed to a "generative conception of species": rejecting Aristotle's theory of spontaneous generation, atomists argued that the same fruit always came from the same trees.[24] When applied to illness, this metaphor suggested that species of diseases came from species of seeds.[25] Atomism did not always entail acceptance of a belief in contagious diseases: a seed theory of disease only entails contagionism if it is first agreed that the seeds cannot be spontaneously generated or persist without a host.[26] Poisonous particles could cause illness without being transmitted from person to person. Some writers thought they were omnipresent, raining down from the planets or rising from subterranean vents and caves.[27] "Seed" theories fostered contagionism in a way that physiological explanations did not, but determining whether a given disease was contagious still required careful clinical observation. Conversely, contagionists could believe that a pathogen had passed from person to person without believing that it consisted of particles or living entities.

Belief in contagion evolved in tandem with changing conceptualizations about disease categories. The belief that diseases were entities, the characteristics of the lines that could be drawn around diseases, and the decision about where the lines could be drawn, all emerged together and also responded to changing ideas about the meaning of "species" in natural philosophy as a whole.

GIROLAMO FRACASTORO (1478-1553): CONTAGIOUS SEEDS

Girolamo Fracastoro, the father of modern contagionism, was educated at Padua, the center of Italian free thought, medical heresy, and doctrines of contagion from the Renaissance into the eighteenth century.[28] He retired in about 1530, about the time he published his poem *Syphilis*.[29] In 1545, just one year before the publication of his work *On Contagion*, he was appointed as a physician to the Council of Trent, and it was on his advice that the Council was transferred to Bologna in 1547 to avoid an outbreak of typhus.[30]

Fracastoro was not the first early modern physician to discuss contagion, but his work provided the most systematic discussion of the idea. He was the first to definitively describe the three categories of infection: by contact, by fomites, and at a distance, a framework that lasted for centuries.[31] He emphasized the idea that the disease was defined by its cause, and in contagious diseases that cause was the seed:

> For the humors in some individual may be in a normal condition . . . and yet the plague may be contracted from another person. Therefore there must be present some other principle . . . of that contagion . . . The principles of contagions per se are the germs themselves . . . endowed with a special nature and method of causing putrefaction . . . They have not only a material but also a spiritual antipathy to the natural heat of the body and to the soul itself."[32]

The contagion *was* the disease: neither the symptoms, nor "obstruction or plethora or a malignant condition, be it quantitative or qualitative of the humors," nor the process of the disease, such as a putrefaction of the heart.[33] A seed of a particular plant can only reproduce that plant; Fracastoro's use of the word "seed" does not imply that he believed that the pathogen was alive, but it does imply specificity. Moreover, for Fracastoro these seeds were no mere metaphor: he insisted that disease particles reproduced themselves. It may seem that Fracastoro was very close to a doctrine of *contagium vivum*, but in fact his conception of the nature of the disease-causing matter was closer to that of an enzyme: some substance that could cause a chemical recombination of the elements within matter. Indeed, he eschewed vitalism. Germs were carried through the body by the movement of its own "spirits," "for we must not say . . . that poisons and contagions try above all to make for . . . the heart, like an enemy, as though they possessed cognition and will."[34]

To explain the mechanism of contagion, Fracastoro drew an analogy with fermentation. Ferments, unlike poisons, generated substances that resembled themselves.[35] There were two kinds of fermentation. In the first, normal putrefaction, the particles lost their usual combination of heat and moisture, broke down, and became smelly and offensive. In the second, instead of merely disassociating, the heat and moisture were regenerated into a new substance, like vinegar, and did not become disagreeable. This two-stage process explained how some diseases, such as rabies, could lie hidden for some time.

The seeds of disease could be spontaneously generated either in the atmosphere or in the body when terrestrial or celestial influences changed the heat and moisture of the air; the air itself served only as a medium for transporting them. Like plants, however, different diseases had different seeds. Once it began, a given disease would retain its identity, not grow into some other disease, so each disease had its own natural history. Each kind of seed required "its own special and peculiar treatment" according to the nature of the disease, not the patient's constitution.[36] Bloodletting was ineffective because the seeds were dispersed through the body.[37] From its first systematic formulation, therefore, contagionism was closely associated with therapeutic reform, the use of specifics, and opposition to venesection.

Because physicians must recognize different varieties of disease to select the best remedies, Fracastoro described individual diseases that he regarded as contagious. Among these he included plague, poxes, measles/scarlet fever, "sweating sickness," "lenticulae" or spotted fever, phthisis, rabies, syphilis, scabies, some forms of dysentery, some scrofulous tumors, elephantiasis and leprosy, herpes/shingles, and some other skin infections that are difficult to translate into modern terms.[38] Fracastoro did not create these disease groupings but borrowed them either from the medical literature or from popular terminology.

According to some earlier historians, Fracastoro's theories had little influence in his own day.[39] However, Fracastoro's work was still well known to educated physicians who read Latin in the seventeenth and eighteenth centuries.[40] Moreover, Singer found at least ten contemporary authors who echoed or simply copied his work, including the neo-Platonist physician Jerome Cardan (1501–1576), who developed a theory of matter based on "seminaria" that were alive and reproduced in a manner similar to minute animals.[41] Among Cardan's followers was Gabriele Fallopius (1523–1562), Professor of Anatomy and Surgery at Padua, who believed that phthisis and syphilis were caused by living entities exhaled by those afflicted by these diseases.[42] Fallopius was the first author to recommend the use of a sort of condom to intercept the transmission of syphilis.[43] He claimed that he had tried it on 1100 men, none of whom became infected—an early clinical trial of a medical device and a demonstration of the idea that a pathogen was a necessary cause of a contagious disease.[44]

JEAN FERNEL (1497?–1558) AND FELIX PLATER (1536–1614): DISEASES FROM HIDDEN POISONS

Two years after Fracastoro published *De Contagione*, the Parisian physician Jean Fernel published a treatise entitled *On the Hidden Causes of Things (De Abditis)*. Fernel had delayed its publication for years, apparently fearing that it would be considered too radical. Although he claimed to be writing within the Galenic tradition, he found that traditional explanations did not adequately explain some illnesses that seemed to attack patients without regard to their temperaments. In addition, some illnesses could not be assigned to corruption in a specific part of the body but instead attacked the body as a whole.[45]

Fernel described these as diseases of the whole substance of the body: a concept that he seems to have originated. Contagious diseases included both diseases due to manifest putrefaction (such as putrid fevers, ulcers, phthisis, scabies, leprosy, and itching) and diseases due to hidden properties.[46] The hidden properties caused illness in people when they inhaled poisonous seeds in the air, ingested or touched a mineral poison, or had contact with a venomous animal or a person with a contagious disease.[47] Malign planetary conjunctions unleashed streams of occult poisons into the air and caused widespread pestilences. As we found with Fracastoro, Fernel's hypothesis that different external causes resulted in different diseases led to the idea of individual disease entities; Fernel's *Medicina*, first published in 1554, has been described as the "first serious attempt to develop a comprehensive approach to the classification of disease."[48]

Fernel's work attracted many European readers, including Felix Plater, a Basel physician.[49] His book *On Fevers* (1597) attributed pestilential fevers to an inflammatory poison from the stings and bites of venomous bodies and a hidden pestilent poison that probably came from an infected person near the patient.[50] Unlike Fernel, who thought the poisonous seeds of disease either emanated from the stars or were generated in the body, Plater argued that the hidden pestilential poison probably originated in humans at the beginning of the world; some people could carry this poison without being ill themselves. Thus, it always survived in some part of the world, as did other kinds of poison, and from thence it infected others.[51]

Historian David Wootton has argued on the basis of this passage that Plater "is thus a proper germ theorist, the earliest known to me . . . all his language implies that he is thinking in terms of animate contagion."[52] Seeds, however, were not necessarily viewed as animate in this period; many authors, including Fernel, had referred to them in the sense of seminal ideas or seminal principles: the unknown forces that gave form to undifferentiated matter. Animals, plants, and minerals, the examples Plater (and Fernel) had used, had innate poisons that were not animate. Plater's work was translated into English by Nicholas Culpeper and Abdiah Cole in 1662 as part of a series of translations of standard medical works intended as a sort of desk reference or *vade mecum* for ship's surgeons and others who needed a compact medical library.[53] Wootton notes that a substantial extract reappeared in the early eighteenth century in a note entitled "Of Contagion, the Chief Cause of a Plague," appended to Thomas Creech's widely read translation of Lucretius in 1714. The anonymous annotator summarized Plater's views in detail.[54]

DANIEL SENNERT (1572–1637): CONTAGIOUS DISEASES FROM HIDDEN SEEDS

Daniel Sennert, Professor of Medicine at Wittenberg, was a student of Johannes Jessenius, who had studied in Padua and brought Italian ideas to Wittenberg.[55] Sennert also borrowed from Fracastoro and from the heretic Giordano Bruno, who spent two years in Wittenberg between 1586 and 1588 giving private lectures.[56] His books, republished in many editions, became a major source for contagionist English writers during the Restoration, including Marchamont Nedham (see below).[57] Several were translated into English; his translators include the populist medical reformers Nicholas Culpeper and Henry Care.[58]

Sennert discussed contagious diseases in his early textbook *Institutionum Medicinae Libri V* (Wittenberg: 1611), which was translated into English in 1656 by "N. D. B. P." as *The Institutions or Fundamentals of the Whole Art, both of Physick and Chirurgery, Divided into Five Books*.[59] In Book 2, Part 1:

"Of Diseases," Sennert divides diseases into different categories, including diseases of intemperature, diseases of the whole substance, organic diseases, diseases of conformities, diseases of number, and diseases of magnitude. His Galenic discussion of diseases of intemperature attributes them to heating substances, such as hot air, warm drinks, too much stirring of the humors, costiveness, and retention of the hot steam generated in the body.[60] In chapter 12, however, he turns briefly to diseases of the whole substance.[61] Like diseases of intemperature, these affect the whole body, but they are caused by "occult" factors known only through their effects. Contagion, defined as an effect communicated from one body to another either by touch or at a distance, was one of many occult factors he lists.[62]

Sennert did not expand on this comment in this work, but he returned to it in later works. A book he published on fevers in 1619 included a chapter on contagion arguing that although the "pestilent poison" of the plague may be generated in the air or in a human body, yet often "neither the ayr, nor evil diet, nor any of the rest of these causes have stirred up the pestilence, but otherwise from elsewhere being brought . . . by contagion, and afterwards by contagion also it is diffused into more places."[63]

About a decade after this, he included another chapter on the hidden causes and differences of diseases.[64] The hidden causes included vapors that corrupted the air and also contact with contagious individuals, as in scurvy, elephantiasis, dog bite (rabies), the pox (i.e., syphilis), and the like. Poison from plants, minerals, or venomous animals was also a hidden cause, as was witchcraft.[65] These hidden causes produced diseases such as contagious but nonfatal fevers accompanied with a cough, the French pox, leprosy, and malignant fevers that occurred with or without the plague. Hidden poisons in water could cause diseases such as dysentery, dropsy, and scurvy, but it might be corrected by boiling or quenching steel, stone, or iron in it, or, as a last resort, by straining. Physicians could determine whether these hidden causes had been bred within the patient or had come from outside by taking a history and asking about any encounters with people suffering from the same disease.[66]

The source of a contagious disease from hidden causes could be an animal as well as a man:

> That is only contagious that can breed any thing in it self, which being sent to another of the same kind, produceth the like disease. When that Contagion passeth to another body, with which it hath some likeness, the passage is by infection or seed, in which there is force to act by the quality that flows from the seed . . . This quality and form are in as small a body as an Atome, and . . . the infection of diseases is multiplied by little bodies, that like seeds, comprehend the whole essence of the disease.[67]

This infection was not putrefaction because putrefaction operated gradually whereas contagion infected immediately and often multiplied, rendering the infected body contagious and killing it before there was any putrefaction.

This "miasma or contagion" was spread through the pores of the skin as well as the breath and sometimes by matter coming out of ulcers. Sometimes the contagious "atomes" flew through the air, and sometimes they were conveyed by "fuel" (i.e., fomites) such as porous cloth, feathers, and skins. Soiled stones and metals could carry contagious particles, but clean ones were less common as carriers. The body that was infected must bear some resemblance to the contagious body because infected clothing did not become diseased. Nor did every contagion affect all animal bodies, for sows contracted a plague that did not affect men or oxen, and the plague that affected men did not affect hogs.[68]

Among contagious diseases Sennert lists catarrhs, malignant pestilent fevers, sore eyes, consumption, dysenteries, scurvy, scabs, itch, scald heads, "Arabian Leprosie," rabies, and syphilis. Sennert's debt to Fracastoro is evident, but he adds "catarrhs" (that is, colds).[69] By "malignant pestilent fevers" in *Practical Physick*, he probably meant the plague, but his *Institutions* had included other fevers and mild fevers that accompanied coughs.[70] This goes beyond the traditional collection of skin diseases plus the plague (syphilis can be considered a skin disease because its first symptom is a chancre). However, unless he saw them as mild fevers, this list omits several illnesses mentioned by Fracastoro, including measles, scarlet fever, and spotted fever.[71]

Sennert's "atoms" were packets of matter and form combined together. There is no reason to assume that he understood his "seeds" that "bred" infection as living entities, but he does not disavow this possibility.[72] It was certainly possible for some of Sennert's legions of readers to infer this from his work. Like Fracastoro, Sennert's later work insists that the seeds of contagious diseases are specific entities that cause "like" diseases or the "same" disease to be transmitted from person to person. In other words, he also emphasizes disease specificity.

ATHANASIUS KIRCHER (1602–1680) AND PATHOGENIC ANIMALCULES

The eccentric Jesuit Athanasius Kircher thought that the contagion of the plague arose from the spontaneous generation of poisonous particles, either animate or inanimate, in the putrefying bodies of plague victims.[73] Kircher's *Scrutinium Pestis* (1658) drew on his own extensive experience with treating victims of the disease in Rome in 1656.[74] He had volunteered to work in the city hospitals, including the enormous Christ's Hospital managed by his friend, the Englishman James Alban Gibbs, where he worked closely with the medical staff.[75]

Kircher thought that the plague was caused by "hidden seeds of a deadly nature" in its victims. When they exhaled, they emanated poisonous corpuscles that acquired life from the warmth of the air.[76] He insisted on the physical nature of the infection and warned that medical attendants or domestic animals might carry the seeds of disease on their instruments, hands, or bodies.[77] Kircher defended his assertion that the effluvia might be animate by referring to his own microscopic observations, claiming to have seen little "worms" in the blood of plague victims.[78]

Kircher listed 12 ways that contagion might spread: (1) contact, (2) fomites, (3) healthy animal carriers, (4) by inert substances such as wine, fats, and metals, (5) animals sick of the same disease, although he denied this, (6) sight, sound, and smell, (7) letters and merchandise, (8) physician carriers, (9) winds, (10) sexual intercourse, (11) certain fomites that could contain the disease for many years, and (12) through the air to a distance of not more than five or six feet.[79]

Kircher also examined pus from the pustules of smallpox patients and saw "animalcules and vermicules" that he blamed for causing smallpox and other putrid diseases. He suggested that conjunctions of Mars and Saturn gave off exhalations of putrescent air that created a favorable atmosphere for the generation of baneful animalcules. His theory had been anticipated in 1650 by August Hauptmann, who attributed malignant fevers and contagious diseases to microscopic animalcules that were either worms or their eggs generated from certain sorts of corruption.[80] Kircher's claim was also enthusiastically supported by Christian Lange, who believed he had seen organisms in the sputum of patients with phthisis.[81] Although Kircher minimized the role of the imagination as a cause of disease, he believed in spontaneous generation and thought that sensation could cause plague.[82] His work was not translated into English, but Marchamont Nedham would introduce his ideas to English readers.[83]

CONCLUSION

By the mid-seventeenth century, the belief that diseases spread through contagion had a long and respectable ancestry. It would be further developed and refined in succeeding decades. Ideas about living agents of disease were less common, but they had at least been raised. What had evidently not been suggested was the idea of different species of living agents as the *exclusive* causes of some kinds of diseases. Such a claim presumed a conceptualization of "life" as the exclusive product of a process of generation—an idea that was not yet available.

An ontological conceptualization of disease underlies many other medical innovations. The only way to decide whether any medical or public health intervention works to prevent or heal a given illness is to compare like with like—and the concept of separate species of diseases often defines what "like" means for acute illnesses. It is impossible to determine whether quarantines prevent the plague if no one has sorted out the difference between plague and food poisoning or typhus. Once the authorities have established a *cordon sanitaire*, or lighted bonfires, or administered their favorite remedies, and people are still dying, is it because they have failed to cure the plague or is some entirely different ailment killing them? The difficulty of establishing truly comparable patient groups—that is, groups suffering from the same disease—helps explain the lag between progress in surgery and improvements in medical therapy during the early modern period.

The transformation from a theory of acute disease that emphasized the body's physiological response to a theory that emphasized the role of an initial invasion by disease-causing substances whose own differences accounted for different patterns among contagious diseases took place very gradually during the early modern period. Its roots lay in monist ideas about the nature of matter, the relationship between matter and life, and the nature of the soul. These were unsettling to orthodox religious establishments, both Catholic and Protestant, and outside orthodox medical training, ensuring that contagionism was always an embattled idea. Over the following centuries, contagion and classification would evolve together, affected by other transformations in ideology, religious authority, education, professional organization, and communication.

Chapter 2

Restoration Medicine and the Dissenters

Introduction: Medical Sectarianism during the Civil War Period

During the Civil War and Interregnum, many monist sects, including the Familists, Quakers, Muggletonians, Levellers, "Ranters," and Seekers, emerged. Some of their adherents were rationalists who believed that churches should consist only of freely consenting members, others were evangelicals who stressed the need for a personal conversion experience, and still others believed that the Bible could only be understood through the inspiration of God within each believer.[1] Their belief in the direct presence of God in the heart of each believer justified resistance to the authority wielded by traditional political elites and to traditional Reformed churches, including the Church of England and Calvinist Presbyterianism.[2]

Reformers who challenged the political and religious legitimacy of the English state also attacked both the university-based learning of the Fellows of the College of Physicians and their claim to regulate medical practice. The growing belief that all freeborn Englishmen (and sometimes women) should rule their own affairs and follow their own conscience without interference by state-appointed bishops extended to the view that patients should also be free to choose their own healers without interference from state-sanctioned professional monopolies.

Published in 1628, William Harvey's *De Motu Cordis* had also launched a devastating attack on Galenic physiology by demonstrating that blood circulated throughout the body and that the Galenic distinction between arterial and venous blood was unfounded. Although Harvey was a Royalist, his work only gained acceptance in Britain in the new intellectual environment of the Interregnum: it was a further half-century before it was reinterpreted by mechanists and conquered the Galenic strongholds on the Continent.[3]

Harvey's monism led him to develop an ontological view of disease that was very similar to that of his contemporary Joan Baptista van Helmont (see below). He believed that disease was due to the growth of a living entity, a *contagium*, which was separate from that of its host. It developed its own purposes and plan and behaved like a parasite, robbing the host of its nutrition.[4] Harvey saw the contagium as a generative, not a putrefactive, entity capable of independent life, able to multiply and create similar diseases in other organisms.[5] The circulation of the blood implied that pathogens were carried by the blood and dispersed throughout the body, invalidating earlier claims that the pooling of blood in some places caused it to stagnate and putrefy like brackish water. Unlike van Helmont, Harvey accepted the traditional belief that people might suffer from an excess of blood, a "plethora," which could lead to disease and might be cured by venesection. Nevertheless, he saw blood as not merely a carrier of heat and life but as alive in itself.[6]

In the mid-seventeenth century, the London College of Physicians held the legal authority to regulate physicians within the city and for seven miles around it. Although most College doctors accepted Harvey's work on circulation, and many kept an open mind about the value of chemistry, the College itself was still a Galenic institution.[7] Not only was its Fellowship limited to graduates of Oxford and Cambridge, but they had to pass Latin examinations on texts of Hippocrates and Galen before becoming Candidates awaiting a vacancy among the Fellows.[8] The College's authority on paper vastly outstripped its actual power. During the Interregnum, London's lucrative medical marketplace gave rise to a panoply of interlopers selling medicines, amulets, and charms, booksellers offering self-help advice (including thinly disguised advertisements for the remedies also on sale), astrologers, quacks, bonesetters, magicians, and faith healers.[9]

Some of these practitioners were well educated but lacked the credentials or English birth required of College physicians. Their desire to make a living often coincided with a sincere belief that traditional medical theories were wrong and that corporate censorship or regulation was unjustified. If traditional learning was to be discredited, however, some other explanatory system was needed. Astrology offered one alternative. It had prospered side by side with Church doctrine and university scholarship for centuries, and it saw a major revival during this period despite the crumbling of Ptolemaic astronomy.[10] Astrologers, however, were so well entrenched that most of them felt no need to challenge the legitimacy or authority of Church and state, and astrology did not offer a coherent rationale for rebellion.[11]

As philosophers eroded traditional learning from the inside, sectaries outside sapped its authority by appealing to a higher power: either the direct inspiration of God or the study of God as revealed in nature. The monist views of the Paracelsians matched their own outlook. Paracelsian chemistry also appealed to some adepts because it was only accessible to those schooled in its language and practices. This solved the problem of demolishing existing monopolies without literally enabling every man to become his own physician: God might dwell in every heart, but only some would possess the knowledge to harness the natural forces hidden in every substance, ensuring a market for new products.

The Interregnum coalescence of Paracelsian alchemy, sectarianism, political radicalism, and occultism meant that after 1660 the newly restored Anglican Church discouraged alchemy in addition to proscribing Dissenters and "enthusiasts."[12] The Church maintained that it should monopolize any effort to harness spiritual or unseen powers. In the end, those who wanted to build a reputation or participate in public life were forced to trumpet their conformity and minimize any interest in the occult "principles" that had fascinated the chemists.[13]

Sectarian medical writers were especially attracted to the monist works of the Belgian Catholic Joan Baptista van Helmont (1577–1644), a follower of Paracelsus and a pioneering chemist.[14] He was born just one year before Harvey, but his many difficulties with the Spanish Inquisition delayed publication of his works. A collected edition appeared posthumously in Latin in 1648.[15] Although Dr. Walter Charleton published a selection in English as *A Ternary of Paradoxes* and *Deliramenti Catarrhi* in 1650, a complete English version only appeared after the Restoration as *Oriatrike* (1662), translated by the sectary John Chandler.[16]

Van Helmont was in no sense more "materialistic" than Paracelsus, but he was more naturalistic, rejecting astrology and the Paracelsian belief in a connection between the microcosm and the macrocosm.[17] He believed every entity contained an *"Archaeus"* that mingled matter and spirit in a single unit.[18] There were only one or two elements: water and (possibly) air. Matter was simply water, unformed and homogenous. Within it were vital seeds originally created by God.[19] Within each seed was an immaterial guiding principle or *Archeus*. Through a process of fermentation, these seeds reorganized matter into discrete objects and substances: animal, vegetable, and mineral. The body of an animal or human contained separate local *Archei* that not only gave form to its various parts but recognized and assimilated or repelled invaders.[20] What was transferred was not corporeal: it was the image, or the impression, of the disease that impinged upon the *Archaeus*.[21] Because the properties of bodies did not result from the balance of their elements, there could not be "degrees" of activity or treatment by opposites: each substance acted in accord with its own unique nature, nor could diseases come from elemental imbalances.[22]

Like Paracelsus, Helmont believed that diseases were individual entities: one disease could not transmute into another. Disease was not merely an abstract idea; it was also generated from an external seed and possessed a more powerful *Archaeus* than did its host.[23] Diseases could be classified into species according to their cause, and they could reproduce and cause the same species of disease in another person.

Van Helmont's belief that disease was a "thing" that one either "had" or did not led him to propose an innovative clinical trial. In support of his opposition to bloodletting, he challenged the Galenists: a few hundred poor patients would be selected at random, half to be bled and half not.[24] His

English followers borrowed this idea. For example, George Starkey, a member of the London Society of Chemical Physicians, challenged the Galenists to a trial of cures in 1657. Patients were to be divided into groups of ten each by one of the parties, with the opponent choosing five patients in each group to be cured within a stated number of days—four days for continued fevers, fluxes, and pleurisies and thirty or at most forty for chronic diseases.[25] In 1666, George Thomson, another member of the Chemical Physicians, also invited members of the London College of Physicians to compete to see whether the Helmontians or the Galenists had the most effective method of treatment.[26] Their challenge was ignored: to Galenic physicians, for whom disease was a matter of degree, not of kind, such a trial would have been pointless: as every disease was the unique response of an individual, grouping cases for clinical study was useless.[27]

Van Helmont saw respiration as a chemical process in which the volatile alkaline salts in venous blood combined with an aerial ferment in the lungs and were exhaled into the air, thus changing a "constituent in the blood for a catalyst from the air."[28] The English Helmontians later identified the "vital spirits" with these volatile salts.[29] Mixed with the air, salts such as "aereal nitre" had vitalizing qualities.[30] Dissolved in water, they explained the healing qualities of mineral waters, but they could also cause explosions within the earth or in the atmosphere.[31] Many chemists believed that these volatile salts could also cause illness when they were exhaled or emitted by people or by animals. This idea could support miasmatism. It could also explain contagion as the result of a chemical process, not an animate entity—or embrace both contagion and miasmatism. In any case, the early chemists did not draw a line between chemical and vital entities. To them, all vital processes were ultimately chemical. By the end of the century, volatile acids were replacing Helmont's alkaline salts. Newton, for example, believed that acids accounted for fermentations that rearranged matter, causing both putrefaction and the creation of new organic entities.[32]

ORTHODOXY, HETERODOXY, AND DIVERSITY: THE RESTORATION RELIGIOUS SETTLEMENT AND THE OUTSIDERS

In 1661, Parliament, determined to prevent future revolts, passed the first of a set of statutes collectively known as the Clarendon Code.[33] These laws reduced religious liberty and determined the framework of medical education and elite practice for the succeeding century and a half. Many Presbyterians had expected to be included in the religious settlement; to their dismay they were pushed outside the pale with their enemies, the radical sectaries, leaving them with the painful choice of conforming as Anglicans or joining the ranks of proscribed Dissenters unwilling to subscribe to Anglican doctrine. Most Englishmen chose to conform.

The Code offered most religious Nonconformists limited toleration at the cost of participation in civic life. In effect, it made them aliens in their own land. In the second half of the century, they would be joined by other aliens: political and religious refugees from elsewhere in Europe, forming the

substrate of a multicultural medical community. Nearly two thousand ministers and schoolmasters were ejected from their parishes or lost their posts.[34]
Dissenters could not serve in municipal government, assemble in groups
of five or more for worship, serve as schoolmasters or tutors, or remain in
the pulpit. This prevented ejected ministers from turning to teaching and
ensured that for years to come efforts to create a distinctive Nonconformist
educational system would be discontinuous and dangerous.[35]

The Five Mile Act of 1665 prohibited ejected ministers and preachers
from going within five miles of their former parishes or five miles of any city
or town. This led many Nonconformists to settle in unincorporated urban
areas that were growing fast enough to make immigration possible: most of
these were in the North of England. Over the next century, this influx would
contribute to the growth and prosperity of these communities, including the
development of many provincial medical institutions.

For a time, the Crown tried to avoid enforcing the religious laws. Royal
Declarations of Indulgence in 1662 and 1672 offered dispensations from
several provisions of the Acts. In 1679, however, Parliament tried to exclude
James, an avowed Catholic, from succeeding his brother Charles as king.
After Charles defeated this effort, his opponent, the Earl of Shaftesbury,
was forced into exile in Holland along with his advisor, the physician John
Locke, and the government began to enforce the Clarendon Code more
vigorously.[36] The parties later known as "Whig" and "Tory" coalesced during the late seventeenth century.[37] The Tories favored James's claim to the
crown and emphasized the importance of nonresistance to an anointed monarch, however detested. The Whigs favored greater indulgence toward the
Dissenters and tried to secure a Protestant succession to the throne. Most
Dissenters supported the Whigs as the lesser of two evils, but most Whigs
were Latitudinarian Anglicans who opposed excessive formalism and legal-
ism in Church ritual but also abhorred mysticism and "enthusiasm."[38] Their
interest in expanding religious toleration was lukewarm at best. After the
Glorious Revolution, the Latitudinarians' arguments for the reasonableness
of nature and the transparency of natural law justified their acceptance of
William's monarchy and permitted them to dismiss the importance of a strict
adherence to Church doctrine or scriptural fundamentalism.[39]

The biggest medical impact of the Clarendon Code came from the Act of
Uniformity, which required all students matriculating at Oxford and graduating at Cambridge to subscribe to the Thirty-Nine Articles of the Church
of England.[40] Because an English university degree was required in order to
become a Fellow of the London College of Physicians, this act (in theory)
prevented Nonconformists from rising to the top of the medical profession.[41]
Dissenters with foreign medical degrees could only become Licentiates,
allowed to practice in London at the discretion of the College. Anglicans
with foreign degrees could either become Licentiates or "incorporate" their
foreign degrees in England and become eligible for Fellowship. Most Anglicans received at least their premedical education at home and obtained an
English baccalaureate degree.

Anglican and Nonconformist physicians thus differed in their premedical education and, to some extent, in their medical training in addition to religion and professional status.[42] Foreign medical training exposed English Dissenters to new ideas and perspectives. This may have affected their approach to medicine more than any denominational tenets. Even Dissenters who became Anglicans retained the community, culture, education, and experience that had shaped them.

The College of Physicians was nevertheless much more heterogeneous in this period than it would be later in the century.[43] There were four reasons for this diversity. First, many Anglicans still pursued medical education abroad. Second, the College itself had decided to create honorary Fellows, although they had no vote in the College councils.[44] Seventy-three were created in 1664 and were exempted from the required oaths; after a hiatus, more were named in 1680.[45] The practice continued sporadically until 1726.[46] Third, the Crown awarded degrees by mandate, and recipients became eligible for regular Fellowship.[47]

Finally, and most importantly, James II issued a new charter of 1685 that added younger doctors such as Hans Sloane and his close friends Martin Lister and Tancred Robinson, all Fellows of the Royal Society.[48] As Sir George Clark noted, the King "was pursuing his unpopular policy of admitting religious dissidents of all kinds to offices, providing they would side with him." These new Fellows did not necessarily possess regular medical degrees, and they were not required to take the oaths or to pass the College examination.

Though some of these new Fellows, such as Richard Blackmore and Sloane, were Whigs, others were Catholics or future Jacobites.[49] They included the Tory John Radcliffe, physician to Princess Anne, who would be a troublesome member for years to come.[50] The bitterness of the medical politics of this period lingers in Gilbert Burnet's description of Radcliffe, whom he unjustly blamed for the death of Queen Mary: "an impious and vicious man who hated the Queen much, but virtue and religion more. He was a professed Jacobite, and by many thought a very bad physician."[51] Radcliffe was not, however, a hide-bound conservative. The son of the governor of the Wakefield House of Correction, a "man of strong republican principles," he was an admirer of Sydenham and would launch the career of the Whig Newtonian Richard Mead.[52]

James's list omitted several honorary Fellows and four existing Fellows, including Richard Morton, Edward Tyson's brother-in-law.[53] Others who were struck off included Nehemiah Grew (MD Leyden), son of the ejected and imprisoned Nonconformist minister Obadiah Grew; Thomas Gibson, the son-in-law of Richard Cromwell; and William Burnet, brother of Thomas and Gilbert Burnet.[54] Gilbert Burnet was then in exile because of his political views. Evidently these men were too closely associated with the Whig opposition to James.[55]

Following the Glorious Revolution, William and Mary proclaimed that the corporation charters James had issued were void. Instead of settling the issue for the College, this created a legal quagmire, dividing the College between the new Fellows who wanted to retain their membership and the smaller group of older members who wanted to return the College to its earlier status as an elite body of traditionally educated physicians. This internal feuding impeded College efforts to fend off encroachments from apothecaries and unlicensed empirics. Public perception of the College as a bastion of conservatism and resistance by Dissenters and fringe practitioners continued, although the College itself now encompassed a broader range of practitioners with different approaches to medicine.[56] Their numbers would gradually shrink again as the Fellows appointed by James retired or died.

Many Restoration Fellows had unconventional educations. As Dr. John Badger complained, in 1695 the College had 130 members (including Licentiates) "most of whom are no doctors; and scarce forty of them regular doctors in either of our universities."[57] Though fractious, these members gave the College a more intellectually vibrant character than it would enjoy later in the century. Naturalists such as Robinson, Lister, Sloane, Grew, and Tyson would become leading figures in the Royal Society. Morton wrote a classic book on consumption; Blackmore, a successful physician, would also become a very popular poet and playwright whose works sprawled all over the early-eighteenth-century landscape; we will consider some of his medical works below.[58]

Dissenters displaced by the Act of Uniformity needed new occupations, but even outside London medical careers presented new obstacles. Bishop's licenses were required for physicians practicing medicine outside London. The Archbishop of Canterbury, Gilbert Sheldon, vigorously enforced this power and demanded that his bishops report on the loyalty of medical practitioners toward both Church and Crown.[59] Men with a foreign MD were exempted from taking licenses, but most outcasts lacked the means to seek a second education abroad.[60] It seems likely that some ejected ministers practiced medicine clandestinely or reached an understanding with their bishops.[61]

Although all Nonconformists experienced civil disabilities after the Restoration, common adversity did not always produce amity. The Dissenters did not trust each other.[62] Protestants thought that Catholics were more loyal to the Pope than to the English constitution. Traditional Congregationalists, Independents, and Presbyterians thought that Arians, Socinians, and (later) Unitarians were infidels and that Ranters, Mennonites, and Quakers were dangerous fanatics or "enthusiasts." The Presbyterians themselves were divided between the moderate "dons" and the militant "ducklings."[63] Even denominations that made common cause in England engaged in fierce rivalry in the colonies where the dominance of the Church of England was less secure.[64]

In addition to English Nonconformists, many aliens and refugees from other countries had settled in England, especially London. Emigrants from France, Holland, and the Palatinate played an important role in shaping medical discourse. Like English Dissenters, they faced legal or social discrimination, but they continued to arrive, building a polyglot, multiconfessional London and bringing with them new European scientific and medical ideas.[65] There were also immigrants from other British nations, especially Ireland. Colonial immigrants were relatively uncommon in the seventeenth century.[66]

The flood of Scottish doctors would really begin in the mid-eighteenth century, but a handful of Scottish physicians and surgeons were influential in England before 1730.[67] The tolerant Scottish universities of the later eighteenth century would be the almost accidental product of bitter political infighting and patronage dispensed by a few Scottish magnates with their own motives for reducing the power of covenanting congregations.[68] At the turn of the century, Scottish culture was still firmly controlled by local presbyteries that supported a traditional, "scholastic," approach to theology, philosophy, and education, leading some Episcopalians to emigrate in search of greater intellectual freedom.[69] They would be joined by an increasing number of Presbyterians after the Act of Union. The English blamed the Scots indiscriminately for initiating the Civil War and saw them as grasping interlopers.

Jews had been expelled in 1290 from England (but not from Scotland or Ireland) by an act that was never repealed. A few had returned to England, often posing as Protestant refugees. The small community was expelled again in 1609 but began to trickle back by 1630.[70] Cromwell failed to get formal approval for Jewish resettlement, but he encouraged them to immigrate anyway, and his successors continued to extend their personal protection. A legal ruling gave the tiny Sephardic community the confidence to establish a synagogue and cemetery in 1656. All immigrants born overseas were barred from owning land, holding shares in merchant vessels, or conducting certain business, but in most ways Jews born in England were comparable in rights and disabilities to English Nonconformists. Their inability to subscribe to the Thirty-Nine Articles kept them from holding public office or graduating from the English universities unless they converted to Anglicanism.[71]

After 1689, Jewish synagogues were licensed in the same way as other Dissenting meetinghouses. The fabulously wealthy financier Samson Gideon obtained a private Act of Parliament to enable him to become a landowner in about 1753, but even he was unable to acquire a title.[72] Gideon's niece married Dr. Philip De la Cour, a Licentiate of the College of Physicians who would become a leader in the Licentiates' struggle to become Fellows. Henri Morelli or Morales, a friend of both Spinoza and Hans Sloane, became a Licentiate in 1684 under questionable circumstances.[73] Jews may also have experienced barriers to setting up as apothecaries or surgeons in London and other incorporated cities because of difficulties in becoming Freemen of the cities.[74]

The Huguenots who immigrated to England (especially in London) in significant numbers at the end of the seventeenth century also suffered the triple handicaps of nationality, religious nonconformity, and language. The Huguenots were supposedly French Calvinists, but in fact they were dissenters from French Catholicism who held a wide range of religious views. The Revocation of the *Édit de Nantes* in 1685 forbade Huguenots to practice their religion, required their infants to be baptized as Catholics, and ordered all Huguenots abroad to return or face confiscation of all their property. Despite savage penalties, it set off a mass exodus to the Protestant areas of Europe, particularly to Geneva and other Swiss cities, Holland, and England. With little property left to tie them to one place, Huguenot professional men were very mobile, and in the early eighteenth century a younger group who had been raised in Europe moved to England seeking opportunities or greater religious freedom. After arriving, some conformed to the Anglican Church, others continued to worship in separate French Huguenot churches, and still others adopted Socinianism, Deism, or other doctrines.[75]

Fluent in French and often in several other languages, Huguenots connected scientists, physicians, and authors across Europe and England. Like Jews, Huguenots born abroad were legally aliens and found full naturalization almost impossible to attain. They were also just as subject to popular prejudice: Londoners were apt to view them as hated Frenchmen instead of as persecuted fellow Protestants. During the turmoil over the Naturalization Bill in 1754, the motto of the opposition was "No Jews, no wooden shoes," referring to the stereotype of the French peasant in his *sabots*.[76] Visiting England in 1765, Pierre Jean Grosley was grateful that his poor English prevented him from understanding the epithets hurled at him by Londoners who resented his French appearance.[77]

Huguenots were especially attracted to chemistry in the early seventeenth century.[78] Hugh Trevor-Roper claimed that "chemistry in France was a Huguenot industry" not because the Huguenots were more "modern" than the Catholics but because they had been forced to be more cosmopolitan. Troubles at home led them to seek training in German universities, to serve German princes, and to work in German institutions. Because chemistry had originated as a German science, this inevitably exposed Huguenot doctors to Paracelsian ideas.[79]

This, however, oversimplifies the chain of causation. It was rebellion against traditional ideas that led French families to convert to Protestantism in the first place. They would at least have been sympathetic to the chemists' challenge to traditional learning before ever setting foot in Germany or other Protestant lands. This was also the case with many other "outsider" physicians whose inability to conform at home forced them to travel abroad where their nonconformity was further reinforced by a different education and experience.

ENGLAND AND HOLLAND

During the second half of the seventeenth century, England and Holland began to share closer cultural ties. Despite political differences, growing numbers of British students attended Leyden and other Dutch universities, where they imbibed the latest views about atomism and learned natural history and anatomy.[80] Each country served as a refuge for those escaping political or religious difficulties at home. Rotterdam reluctantly hosted some especially radical English Dissenters, including a group of Quakers.[81] At mid-century, and after the Exclusion crisis, English political refugees sought safe harbor in Holland; the stream flowed the other way at the end of the Dutch War in 1674 and became a flood with William III in 1688. Dutch naturalists, including Antoni von Leeuwenhoek, visited the Royal Society. A handful of Dutch immigrants (some with English ancestry) established themselves both as medical practitioners and as quacks in England, and refugees from elsewhere in Europe arrived in England after a stay in Holland.[82] For physicians, the traffic most often consisted of young Englishmen visiting Leyden and other Dutch centers of learning before returning to England to practice.

Because so many Britons studied in Holland, Dutch thought permeated British medicine.[83] Among the most influential Dutch Helmontians was Franciscus de le Boë Sylvius, who lectured privately in Leyden in 1639–1640, settled in Amsterdam, and returned to Leyden in 1658 to become a professor of practical medicine, adding chemistry instruction in 1666.[84] Sylvius, the first person on the Continent to demonstrate Harvey's theory of the circulation of the blood, adopted many of Van Helmont's chemical ideas but rejected his theories about *Archei*.[85] He taught that physiology could be reduced to chemical interactions, in particular the effervescence that resulted from mixing acids and alkalis, which he assumed was the same as fermentation.[86] He did, however, claim that phthisis (tuberculosis) could spread through contagion in 1679.[87]

Sylvius's less speculative version of van Helmont's theory was extremely influential in England: by 1676 the College of Physicians admitted examinees who grounded their arguments on his work or on the similar theories of Thomas Willis.[88] Robert Sibbald, a naturalist who would become the leader of the Edinburgh "Sydenhamians," was a pupil of Sylvius as was a Dutch Mennonite, Burchard de Volder, who brought Boyle's ideas to Holland, where he established a laboratory to demonstrate them.[89] De Volder would later teach both Bernard Mandeville and Hermann Boerhaave.[90] Another student of Sylvius was Johannes Groenevelt, who later settled in London, joined the "Oracle" practice at the Angel and Crown, and set off a medical *cause célèbre* at the end of the century.[91] Later we will discuss the work on *contagium vivum* published by the London physician Benjamin Marten, who was almost certainly trained by Groenevelt.

There is some evidence that ideas about *contagium vivum* were circulating in Holland during this period, and they may have supplied one source for British contagionism. Kircher's biographer, John Edward Fletcher, noted a group of contagionists in Holland associated with Gerard Blasius, Professor of Medicine in Amsterdam, and the Cartesian physician Cornelius Bontekoe.[92] Two doctors earlier active in Amsterdam, Zacutus Lusitanus (1575–1642) and Nicholas Tulp (1593–1674), had thought that cancer was contagious, a view that they shared with Sennert.[93] Both had blamed a poisonous emanation from their patients for transmitting the disease from person to person.

Groenevelt's near contemporary, the Amsterdam chemist and physician Steven Blankaart (1650–1704), edited the first Dutch medical journal, *Collectanea Medico-Physica*, a short-lived publication that operated like a medical *Philosophical Transactions*, publishing observations from a wide range of correspondents without imposing an editorial point of view.[94] One of Blankaart's publishers, Jacob van de Velde, visited London in 1684 and stayed with Groenevelt.[95]

Willem ten Rhijne, a schoolmate and close friend of Groenevelt's, supplied Blankaart with a description of the Asian practice of moxibustion for his *Treatise on Podagra and Gout* in 1684: a topic of great interest to members of Sloane's circle in London.[96] Blankaart supported Redi's arguments against spontaneous generation, even repeating his experiments, and he also published a work of his own on entomology describing observations he made with the aid of a lens.[97]

Blankaart, like Sylvius, ascribed diseases to a process that caused an imbalance between acidity and alkalinity in the body. His treatise on syphilis blamed a sharp acid for causing the disease, but he added that "we could advance another cause, of which no one . . . has thought . . . in the seed of men and that moist matter which women carry in their wombs and in their sheaths are found *small animals*, which being venomous, corrupt not only our genital parts, but even, growing in time to large quantities, thrust themselves everywhere *in our blood*, which they corrupt."[98]

Blankaart repeated the suggestion that a living entity was related to disease in his extremely popular *Physical Dictionary*. The entry on "*Febris*, a Fever" states that "fevers in general are divided into Intermittent, Continued, Continent, and Symptomatical. Scotus in his Magick assures us, *That the Blood in a Fever has Worms in it*."[99] Benjamin Marten would quote this comment in 1720.[100]

MEDICINE AND RELIGION IN ENGLAND IN THE LATE SEVENTEENTH CENTURY

Many Restoration medical reformers were hostile toward the Fellows of the College of Physicians who treated the rich under a royal charter guaranteeing their monopoly. A merely populist attack, however, would have demanded better access for the poor to the existing system: a tactic favored by the Society of Apothecaries. Instead, Helmontians championed an alternative "iatrochemical" medicine that rested on a chemistry still close to alchemy, opposing both Anglicans and more orthodox Calvinist Puritans.[101]

Helmontian medical writers continued to oppose bloodletting and to favor an ontological disease theory in opposition to the physiological theory of the Galenists. Allying themselves with a group of empirics, virtuosi, and courtiers (especially those associated with James, Duke of York), they tried to found a rival "Society of Chymical Physitians" with its own royal charter.[102] This Society suffered from internal divisions and collapsed after the Great Plague of London killed several of its members. Moreover, the College dismayed its external critics by establishing a chemical laboratory and competing on their own turf. Although the College co-opted the chemists' business, however, it did not adopt their ideology. By the end of the century, the College of Physicians had survived a barrage of criticism and remained relatively intact despite efforts by some doctors to break its hold over medical care in London and by others to reform it further from within.[103]

Whig politicians were discredited by the Rye House plot of 1683; many others were demoralized by the failure of Monmouth's Rebellion against the succession of James in 1685. The ejected Independent minister Matthew Meade was implicated in the planning of the Rebellion and fled to Holland. His young son, Richard Mead, would become one of the most successful physicians of the eighteenth century. Among those who narrowly escaped capture following the Rebellion was a young man named William Oliver, a medical student at Leyden who had been surgeon to Monmouth's troops.[104] He would survive to publish *A Practical Essay on Fevers* in 1704.

After a generation of persecution, the Toleration Act of 1689 exempted Dissenters from certain provisions of the Clarendon Code, but it did not relax the rules for the universities or Dissenting schoolmasters.[105] It allowed licensed Nonconformist meetinghouses but required everyone to have an Anglican marriage and burial. This provision was increasingly evaded by Dissenters during the eighteenth century with unfortunate results for students of vital statistics.[106] The Toleration Act offered toleration only to Protestant Trinitarians; it still barred atheism, Socinianism (Unitarianism), and Catholicism. This was reemphasized by the Blasphemy Act of 1697. The last person to be executed for heresy in Britain was Thomas Aikenhead, a hapless student at the University of Edinburgh.[107]

Some Dissenters chose to conform partially or occasionally; the morality of this practice was controversial, but many Dissenters and immigrants saw no ethical or doctrinal bar to taking Anglican sacraments while remaining members of the Nonconformist community. The rise of Anglican Latitudinarianism and the arguments raised during the Bangorian controversy (1717–1721) further undermined belief that religious oaths should be interpreted literally or narrowly.[108] Nevertheless, deprivations under the Clarendon Code created a long-lived political and cultural community out of dissenting groups that otherwise had little common ground.[109]

Radical authors continued to challenge the claim that only physicians with years of training should have medical authority or access to medical information.[110] During the Interregnum, the College had lost control over medical publishing. According to historian Elizabeth Furdell, by 1660 London printers were disproportionately Dissenters.[111] Under the Licensing Act of 1662, the College of Physicians regained the power to censor and license all medical works printed in London. The Act gradually became less effective and finally lapsed for good in 1694 without stemming the very profitable tide of books, pamphlets, and ephemera written by authors of all persuasions and intended for all sorts of readers, which increased steadily after 1660.[112]

Many of these works bore the name of the pharmaceutical reformer, astrologer, and republican Nicholas Culpeper.[113] During his life, Culpeper wrote and translated books for the publisher Peter Cole. When he died in 1654, Culpeper allegedly left many other translations of standard Continental medical works. With the cooperation of Culpeper's widow, Cole published most of these, adding the names of Abdiah Cole and sometimes William Rowland. In a decade, Cole produced nearly eighty titles, although many of these were retitled, abridged, or reshuffled versions of other works. These included translations of works by Thomas Bartholin, Jean Fernel, Felix Plater, and Daniel Sennert among others.[114] Following Cole's death, Culpeper's widow republished Sennert's *Practical Physick* in 1674.[115] Two years later, this was edited by Henry Care (1646–1688), "student in physick and astrology," author of many subversive works, including an anticlerical newsletter, *The Weekly Pacquet of Advice from Rome*.[116] Thus this small group had reproduced a library of works by authors who had modified or criticized Galenic medicine and made them available in English to ordinary English readers.

Translating such works forced writers to confront the disparate ways that different cultures interpreted their experiences. Not only did the classical languages differ from each other, but both the names and the concepts of diseases had changed in the transition from classical to modern languages. Fracastoro struggled with this problem throughout his work on contagion. For example, in his chapter on poxes and measles, he wrote of the fevers, "which the translators of Arab books call 'variolae,' [poxes] and 'morbilli' [measles and scarlet fever]. By the term 'variolae' they mean fevers which people commonly call 'varolae' . . . and they mean by 'morbilli' what people call 'fersae,' perhaps because of its fervid heat. The Greeks, however, in discussing these fevers used only the term exanthemata."[117]

Writing in the vernacular inevitably inclined authors toward vernacular conceptualizations of illnesses—but these conceptualizations also differed from place to place. Terminology circumscribed the ways that one could think about or write about diseases. The use of the same technical term—or common term—could also mask changes in the underlying conceptualization of its meaning or lead to serious misunderstandings. The Latin word "*semen*," for example, could mean "seed," "semen," "organizing principle," or "germ."[118] It was centuries before European medical authors jointly created a stable, widely accepted and mutually intelligible nomenclature for many diseases, making it possible for them literally to share notes on the behavior of the underlying illnesses, but the first steps toward creating one would be taken during the Restoration.

Authors hoped to broaden the market for their work by writing in English, ensuring a larger audience and a greater return.[119] Together with the reforming translators of medical works, they defended their publications by pointing to the parallel with Biblical translation, likening Latinate physicians to Catholic obscurantists. "Papists and the Colledge of Physitians will not suffer Divinity and Physick to be printed in our mother tongue," Nicholas

Culpeper wrote.[120] Richard Blackmore wrote in English to uphold "the true Dignity and Worth of Physick . . . as much as can be [against] Philosophical Notions and Scholastick Darkness."[121] They campaigned both for diffusion of medical knowledge among laymen (challenging the claim of the Fellows of the College of Physicians to exclusive authority) and for a more comprehensive and less classical approach to medical education (challenging the monopoly of Oxford and Cambridge). More traditional university-educated physicians fought a losing battle against English medical publication, which they feared might put dangerous information in the hands of unskilled laymen: as one Leyden-educated Scottish physician protested (in English) in 1694, access to medical works allowed "gardeners, old wives &c. [to] acquire as much knowledge as to kill, but seldom as to heal. In a word . . . it is the putting a weapon in a mad-man's hands."[122]

The Royalist newsletter *Mercurius Pragmaticus* criticized Nicholas Culpeper's English translation of the *Pharmacopoeia Londinensis* by describing him as a man who had

> commenced the several degrees of Independency, Brownisme, Anabaptisme . . . after that he turned Seeker, Manifestarian, and now he is arrived at the battlement of an absolute Atheist, and by two yeeres drunken labour hath Gallimawfred the apothecaries book into nonsense, mixing every receipt therein with some scruples, at least, of rebellion or atheisme, besides the danger of poisoning men's bodies. And (to supply his drunkenness and lechery with a thirty shillings reward) endeavoured to bring into obloquy the famous societies of apothecaries and chyurgeons.[123]

To the *Mercurius*, making medical knowledge available in English to the general public combined religious heterodoxy, rebellion, ignorance, greed, and the subversion of professional traditions in a single scandalous bundle. For their part, the sectarians insisted that Parliament must unseat Galenic medicine and replace it with "Chymistrie the handmaid of Nature, that hath outstript the other Sects of Philosophy, by her multiplied real experiences."[124]

DISSENSION WITHIN THE PROFESSION

The bitter feuds among the Fellows of the College of Physicians continued after the accession of William III. One group of Fellows tried to extend the authority of the College over all medical practice in London and throughout England. Despite opposition from another group, headed by Edward Tyson, the College tightened its control over the licensing of printing, prosecuted unlicensed practitioners more vigorously, and initiated an all-out war against the apothecaries.[125] The College was attempting to assert its commercial hegemony over medical practice. By sanctioning practitioners who were not members of the Church of England and/or who had not passed the College's classically grounded examinations, it was also attempting to establish its intellectual hegemony.

The apothecaries had obtained a charter in 1617 that permitted them to form a company separate from that of the druggists and forbade druggists to compound, prepare, or administer any medicines.[126] After rebuilding their hall in 1673, the apothecaries opened a chemical laboratory to facilitate both the manufacture and the sale of preparations, including Helmontian chemical remedies.[127] Apothecaries were more approachable and less expensive than the haughty university-educated physicians, whom a family might see only a handful of times in a generation (if that).[128] Customers often visited apothecaries to buy herbs, home remedies, patent medicines, confections, cosmetics, pamphlets, and books.

Apothecaries often acted as intermediaries between patients and physicians. The apothecaries would come to a coffeehouse (such as Batson's) where a physician attended at fixed times, describe their patients' symptoms, and leave with prescriptions. This gradually became less common, but patients still asked their apothecaries to recommend a physician, and physicians told patients where they could get prescriptions filled. This mutual dependence, however, did not prevent the College as an organization from trying to reassert its primacy in London medicine by setting up a controversial "dispensary" for the poor in 1696 (see below) or occasionally acting on complaints from patients.

The surgeons were still uncomfortably yoked to the Company of Barber-Surgeons. In theory, physicians restricted their work to "internal" problems that required medications or changes in modes of living, and surgeons limited their work to "external" problems and problems that required manual intervention, such as setting bones or lancing abscesses. In practice, these boundaries were not always clear. One area of contention was the management of venereal disease: an "external" problem (one with a visible lesion) that might also be treated with prescriptions. Nevertheless, the surgeons had a less contentious relationship with the College of Physicians than did the apothecaries.[129]

The College of Physicians had little control of medical practice outside London, although it could award an "extra-license" to provincial physicians. In Scotland, both Edinburgh and Glasgow had their own medical organizations: Edinburgh had a Corporation of Surgeons and a College of Physicians; in Glasgow, however, the physicians and surgeons were united in a single faculty.[130] Some provincial cities also had companies of barber-surgeons and/or apothecaries among their other professional guilds.

A proposal that the College establish a free pubic dispensary for the poor in 1696 also aroused controversy. According to historian Frank Ellis, this idea first appeared in an anonymous work entitled *The Accomplisht Physician* in 1670 that he attributed to Gideon Harvey, an "unprincipled quack."[131] It suggested that poor patients could be prevented from consulting quacks by appointing one or two junior physicians to serve every London ward for fixed fees. It also recommended a "*Pharmacopoeia Pauperum*" consisting of "cheap, few, and effectual Medicines" prepared by apothecaries chosen by the College. At that time, the College was unable to come to any agreement with the apothecaries. A rule of 1687 requiring all its members to give free medical advice to the London poor was never implemented.[132] After the City of London revived the idea in a letter of 1696, a narrow majority of the College agreed to create a store of medications for the poor to be provided at cost.[133] The College advertised this as a selfless attempt to bring medical advice to the poor, but the apothecaries saw it as an effort to drive them out of business. The dispensary proved very popular, and two more were opened.[134]

Although it ultimately defeated its enemies within, the College lost the larger war against the apothecaries when a 1704 ruling by the House of Lords for an apothecary, William Rose, effectively allowed apothecaries to practice physic.[135] The dispensaries gradually faded away; the last one, in Warwick Lane, closed in 1726.[136] The London poor did not have another charitable dispensary until John Wesley opened one in 1746.[137]

THE ORACLE GROUP

The innovative joint medical practice established at the Golden Angel and Crown in Cheapside in about 1687 epitomized many of the themes discussed above. The five Licentiates who opened the practice were from different countries, and they appealed to a polyglot clientele through an aggressive publicity campaign that included both advertising and publishing. The group comprised Richard Browne; Christopher Crell Jr., originally Christopher Crell Spinowski (MD Leyden); Philip Guide, a Huguenot educated at Montpellier; John Pechey (1655–1715, MA Oxford), who would become Sydenham's first translator; and Johannes Groenevelt (John Greenfield). Groenevelt was a Dutch immigrant and member of the Dutch Reformed Church who obtained his MD from Utrecht but had studied with Sylvius in Leyden.[138] Browne, who had matriculated but not graduated from Oxford, had also studied medicine at Leyden, though evidently he did not obtain an MD.[139] The clinic contained a stock of drugs (the "repository"), and they treated poor patients on the premises, combining the roles of physician and apothecary. By contemporary standards, this was ethically questionable because it created a conflict of interest: it gave physicians a financial stake in dosing their patients. On the other hand, some physicians, especially those who favored Helmontian medicine, argued that if they prepared their own medicines they could ensure their quality and purity.[140]

The group jointly published a small book called *The Oracle for the Sick* in about 1685, which explained to patients how to describe their own symptoms in order to receive treatment by mail. It included a set of dialogues relating to various illnesses and instructed patients to underline the phrases relating to their ailment. It also provided illustrations of a male and female body so that patients could simply mark or point to the place that was troubling them.[141] This must have been helpful to immigrants who did not speak English, but it was a far cry from the medicine favored by traditional physicians, which held that all treatment required an intimate knowledge of a patient's temperament.

Although all five were Licentiates, the College attacked them for publishing unapproved materials, consulting with unlicensed practitioners, advertising, malpractice, and failure to pay dues, arrears, and fines.[142] A separate malpractice case against Groenevelt continued until 1697 and became a *cause célèbre*.

In addition to testing the limits of acceptable medical practice, the members of this group lived on the political and religious edge. The radical physician Bernard Mandeville was evidently a close friend of his older compatriot, Groenevelt.[143] Born in Rotterdam, Mandeville had associated with Dutch anti-Orangist republicans. His first English publication, a poem criticizing the College of Physicians, prefaced the second edition (1703) of a book by Groenevelt.[144]

Christopher Crell, or Christopher Crell Spinowski, was the son and grandson of Polish Socinians. He and his sister had been adopted by a member of John Bidle's Socinian congregation in London, and it seems the philanthropist Thomas Firmin paid for Crell's education.[145] Crell contributed prefatory verses to a work by Sydenham published posthumously in 1695, and he also knew John Locke. His brother Samuel Crell, who carried on his grandfather's work, visited London frequently. In 1699, he stayed with Locke and evidently discussed Newton's secret anti-Trinitarian views with him. In 1711, Newton helped Samuel return to Germany, and in 1726 the two men met again to discuss Crell's anti-Trinitarian treatise on the Gospel of John. Newton gave him ten pounds, and Crell promised not to disclose this support: Unitarianism was a capital crime.[146] A third brother, Paul (or Pavel or Pawel) Crell, became the ward of Locke's pupil, the third Earl of Shaftesbury, who paid for his education at Cambridge and "treated him as a member of his family."[147] Thus all three brothers had ties to Locke.

John Pechey was a graduate of Oxford, where he obtained a BA and an MA (1678), but there is no evidence he ever obtained a medical degree, although he became a Licentiate of the College of Physicians in 1684.[148] He advertised very energetically, offering pills to cure scurvy and syphilis that could be taken without interrupting the patient's activities. He championed empirical medicine and opposed any mystification by physicians. He inserted a puff for his own pills in most of his publications.[149] Harold Cook suggests that he became acquainted with Sydenham no later than 1684 and published several translations of Sydenham's work. He also published his own work.[150]

Philip Guide, the final member of the partnership, was also a well-educated outsider. He had practiced very successfully in Paris before he fled to England with his family in 1681.[151] He had published two works in Paris; the first was inspired by Boyle's air pump.[152] The second was a treatise on venereal disease that argued that it was caused by an acid and could be cured only by mercury; it includes a mercurial preparation apparently of Guide's own devising.[153] Guide became a Licentiate in 1683 and published two more works. The first, on nutrition, appeared in 1699.[154] A work on fevers appeared in 1710. It dismissed both elemental and chemical theories, tracing fevers to internal friction of the body's particles, compression from a swelling of the walls of arteries, and putrefaction of the "nourishing juice" in the body.[155]

Conclusion

By the middle of the seventeenth century, British practitioners of Galenic medicine were under attack both from scholars who questioned its underlying elemental theory of matter and from populists and sectaries who challenged its traditional sources of authority. The Restoration saw a vigorous effort to restore political and religious order by proscribing all non-Anglican religious preaching and teaching, closing access to the universities, controlling entry into the medical profession, and prosecuting many individual Dissenters.

The Restoration religious settlement continued to influence medical education and the medical profession for two centuries. It deprived Oxford and Cambridge of talented men who could have enriched university education, thus undermining the training of the Anglicans who remained. By denying university degrees to Dissenters and barring them from some professions, these laws forced them to obtain educations elsewhere, impelled them to become more entrepreneurial, and, unintentionally, led them to become more cosmopolitan. The Code frustrated and handicapped Dissenters without crippling them completely.

Although this effort succeeded in inducing many sectaries to conform, silenced others, and consigned many to penury or prison, the effort to impose uniformity was fitful and was undermined by friction between the Crown and Parliament, political instability, and difficulty in controlling local administrators. Efforts by lawmakers and the universities to establish a more traditionally educated and doctrinally uniform College of Physicians were thwarted by royal interference and by the College's own poverty, which tempted it to raise funds by creating new sorts of members without examining their credentials. Efforts by the College to establish its hegemony over medical practice were thwarted not only by a lack of resources but by feuding among its members.

Further complicating the search for medical uniformity was the multicultural, multiconfessional, and polyglot nature of London and the interpenetration of English medical culture with medical training and thought from overseas. As a refuge for the victims of political and religious persecution elsewhere, England attracted scholars interested in unorthodox medical theories. They included many Paracelsians and Helmontians, who invigorated its native medical learning. As British influence grew during the eighteenth century, Britons would become cosmopolitan by traveling or serving abroad, but in this period the exchange was often in the other direction. Even the monarchs during this period brought foreign views and foreign attendants with them from France, Holland, and Hanover.

In addition to the admixture of overseas practitioners, English medical thought became more diverse because a broader swath of society participated in creating and consuming it. Not only did more people become literate and interested in obtaining medical information, more practitioners gained access to the means of producing it. As Furdell has noted, the popularization of medicine both reflected and affected political change.[156] Writing books and pamphlets on medicine became an alternative path to prosperity for many struggling practitioners.

CHAPTER 3

POPULIST WRITING ON DISEASES IN THE LATE SEVENTEENTH CENTURY

INTRODUCTION

One of the most opportunistic journalists of the Interregnum, Marcham-ont Nedham (or Needham) discussed the idea that microscopic animalcules might cause disease in an idiosyncratic work entitled *Medela Medicinae: A Plea for the Free Profession, and a Renovation of the Art of Physick Out of the Noblest and most Authentick Writers* (1665). This combined Nedham's own medical theories with Helmontian chemistry and a theory of living disease agents.

Weighing in at 516 pages, this was no mere pamphlet but an extended attack on "gentleman-physicians," the Galenic and Scholastic medicine they practiced, and the system of authority, laws, and traditions that conferred upon them the exclusive right to practice physic. The long title aptly describes the author's goals for physic: "shewing The Public Advantage of its Liberty, The Disadvantage . . . to the Publick by . . . Physicians,' imposing upon the Studies and Practice of others, the Alteration of Diseases from their old State . . . the Causes of that Alteration, [and] the Insufficiency and Uselessness of meer *Scholastic Methods* and Medicines, with a necessity of New." Nedham called for social, professional, ideological, scientific, and medical reformation, "tending to the Rescue of Mankind from the Tyranny of Diseases; and of Physicians themselves, from the Pedantism of old Authors and present Dictators."[1] Despite its heft, the book was intended for an audience of lay readers, not professional colleagues. Nedham explains and discusses medical terms such as "contagion," includes a thorough history of the concept, translates his Latin quotations and terms, and sometimes addresses the reader directly.[2]

Nedham's work attracted several rebuttals that further illuminate the relationships among Helmontian disease theory, contagionism, and ideas of disease specificity. These themes also appeared in other populist Helmontian works published during the Restoration, with other disease theories. It is possible that some of them reflect pre-existing lay theories, but our dependence on printed medical works makes it difficult to reconstruct a detailed and comprehensive picture of what ordinary Englishmen thought about the cause and nature of disease during this period. Even if these works did not reflect lay medical ideas, they surely contributed to shaping those ideas.

A THEORY OF LIVING PATHOGENS: MARCHAMONT NEDHAM (1620–1678)

A graduate of All Souls' College, Oxford, Nedham became the editor of a Parliamentary newspaper in 1643 but soon sided with the Independents. Forbidden by Parliament to publish, he turned to medicine in 1646. After supporting Charles I from 1647 until the King's execution in 1649, he worked for Cromwell and brought out a newspaper, the *Mercurius Politicus*, under the supervision of John Milton. It was during this period of close collaboration with Milton in 1650–1651 that Nedham wrote *The Excellencie of a Free State*: a work described as the "first sustained example of republican democracy in classical and Machiavellian terms . . . "[3] There are echoes of *Areopagitica* (1644), Milton's denunciation of press licensing, in the first chapter of the *Medela*, which argues that many great physicians have been unjustly censured by authorities.

Nedham's frequent and brazen changes of allegiance defy efforts to discern consistent political principles or loyalties.[4] For four years during the Interregnum, he spied for the government on the Fifth Monarchists and other radical groups while practicing chemical medicine. At the Restoration, he fled to Holland to avoid Royalists who maintained that the "restoration would be incomplete unless he were hanged."[5] After buying a pardon, he returned to London, where he resumed both medical practice and journalism for hire. He was paid by Charles II to write against Shaftesbury and even supported the ejection of Dissenting schoolmasters. His medical views remained unorthodox: he joined the effort to establish a "Society of Chymical Physicians" in opposition to the monopoly of the London College of Physicians.[6]

The *Medela* asserted that the nature of diseases had changed since classical times, making Galenic remedies outdated. Syphilis in particular had gradually changed its character, becoming hereditary and transmissible by asexual contagion. Scurvy had also changed, growing more dangerous, becoming hereditary, and causing many deaths that were attributed to other diseases. Together, syphilis and scurvy were altering the "whole Frame of Nature in Mankind, and all the diseases thereto belonging."[7]

Nedham pointed out that deaths from scurvy listed in London Bills of Mortality had increased from 5 in 1630 to 103 in 1656.[8] Even this understated the true mortality rate because physicians misdiagnosed many scorbutic diseases. Rickets had shown a similar transformation, increasing from 12 deaths in 1630 to 521 in 1660.[9]

Pox and scurvy now threatened everyone, demanding new methods of treatment and new means of investigation. There were several sorts of scurvy, each with its own means of transmission. One was due to bad diet, a second was congenital and incurable, and a third was due to contagion. Both pox and scurvy could be transmitted by direct contact, through lactation, or by the passage of imperceptibly small contagious particles through the air.[10]

In discussing the nature of these particles and the possibility of airborne transmission, Nedham cited Sennert, who had likened them to a ferment,

> which being received in a Body of the same Nature, induceth to it the like Disposition. It, being a small portion of particle of Contagious Matter . . . lights upon a sound Body . . . [Contagion] is a Body, of a fine invisible Nature, flowing out . . . after the manner of Atoms . . . Corpuscula [or] little Bodies, and Contagion is multiplied by these little Bodies. [*sic*] which like Seed, comprehend within themselves the whole Essence of the Disease.[11]

Qualities once thought to be incorporeal are now seen as "little indivisible invisible particles of Bodies."[12]

Nedham believed that these contagious particles, although they originated in the human body, were omnipresent in the atmosphere and infected virtually everyone. He therefore had no concept of contagion as a chain of case-to-case transmissions and lacked a doctrine of disease specificity. Different diseases could merge into one another to create a third disease.

> [Pox and scurvy] combining together to complicate themselves with all other Diseases, have now . . . so insinuated themselves, that they are become universal, a part of our Humane Nature, and consequently inseparable from us, as well as from our Diseases . . . a Third Monster is started out of them . . . being Tinctured in the Blood and Humors of Bodies [they] do pass into the Forms of such Diseases, as those Bodies which receive them are most inclinable to.[13]

Partway through, Nedham suddenly veered from contagious atoms to microscopic animalcules. He had read Athanasius Kircher's *Scrutinium Pestis* (1658) in which Kircher attributed the plague to a living contagion and claimed that he had seen the microscopic pathogens with his own eyes. Nedham declared that it presented "a Notion concerning Worms more finely improved, than ever I thought of before."[14] He appended a note explaining microscopes and included six of Kircher's microscopic "demonstrations" in his text.[15]

Microscopy was still in an embryonic state. Fracastoro's *Homocentrica* of 1535 mentioned the use of two superimposed lenses for magnification of celestial bodies, but compound lens microscopes did not develop until the 1590s, when they were employed by a small circle of Italian scientists.[16] The Paracelsian physician Pierre Borel published the first medical treatise to include observations made with a compound microscope in 1653, though his comments drew little attention. He mentioned "whales" or "dolphins" swimming in the blood, which probably referred to red blood corpuscles or *rouleaux* of blood cells. In 1655, Borel included a treatise on the microscope in a book on telescopes. The addendum, first published separately in 1656, became the first book devoted solely to microscopy.[17]

Borel claimed that "small worms" could be found in "*variolis*," ulcers, and other cutaneous diseases.[18] He remarked that a friend of his had observed a small "snail-like infection" on the penis of a soldier suffering from gonorrhea, which produced thirty or forty eggs from which small, hairy worms emerged.[19] He noted that "Alstedius" had asserted that during the plague the air was full of invisible animals born from the corruption of the air, which we breathe in.[20] He also commented that worms were said to appear in the blood of those with fevers and that it was probable that worms could be found in all decomposing matters.[21]

The first English work on microscopy, Henry Power's *Experimental Philosophy in Three Books*, appeared just one year before the *Medela*. Power observed that pond mites bred in stagnant water and thought similar animals might breed in putrefied air during plagues. He also referred to Moffet's itch mites, commenting that Moffet would have been even more impressed if he had seen them through a microscope, but otherwise he had little to say about illnesses.[22] It is not clear that Nedham himself ever used a microscope, but he probably knew about Boyle, who was making microscopic observations in 1663, and his assistant, Robert Hooke, who began microscopic demonstrations at the Royal Society in the same year and published *Micrographia* in 1665. Neither Boyle nor Hooke blamed animalcules for disease.

Nedham's support for the theory of contagious particles fell short of full contagionism both before and after he adopted a theory of *contagium animatum*. Nedham followed Kircher (and Hauptmann) in believing that the pathogens were spontaneously generated from putrefaction, and this inhibited his adoption of a concept of disease specificity or of person-to-person transmission. Nedham's "wormatic matter" was everywhere, lurking in everyone and exacerbating many diseases. Nedham's diseases were more the product of a teeming environment than of human contact. Although idiosyncratic in its conclusions and unoriginal in its arguments, Nedham's book represented the first published discussion in English of microscopic disease-causing animalcules, and it came from a Helmontian who, whatever his current political sympathies might be, had certainly worked with radicals in the Civil War period.[23]

Nedham's belief in spontaneous generation was not unusual. In 1668, three years after the *Medela*, Francesco Redi (1626–1697) published *Experiments on the Generation of Insects*, attacking this idea and demonstrating that maggots did not emerge spontaneously from meat.[24] Nevertheless, even Redi accepted spontaneous generation in certain cases, such as grubs in oak-galls and certain other plant parasites. Moreover, unless one adopted a strict theory of *contagium vivum*, the spontaneous generation of *diseases* did not pose the same problems as the spontaneous generation of *organisms*. A given disease could be caused by a poison, like mercury or arsenic, rather than by a living entity such as a worm or insect, and such poisons could conceivably be generated by spontaneous chemical reactions. For the next two centuries, most physicians adopted a compromise position.

Nedham's Critics: George Castle, Robert Sprackling, and John Twysden

The attacks on Nedham by authors including George Castle, Robert Sprackling, and John Twysden shed additional light on the way Helmontianism was understood in Restoration England.[25] Castle was a recent medical graduate of Oxford, where he had been associated with the circle of "physiologists," including Boyle and Locke. He argued that medicine could become a harmonious blend of Ancients and Moderns by combining Galenic remedies with recent discoveries in science and physiology, but only if unlearned practitioners were brought to heel: "ignorance (amongst some men) is become as necessary a qualification for the practise of Physick, as it us'd to be for Preaching."[26] He ridiculed Nedham for adopting Paracelsus' view "that Diseases, like Pompions and Turnips do grow from their peculiar Seeds; and that the distemperatures which we find in our Bodies, are but the blossoms, fruits, and products of them; and that they, like Animals, ingender and propagate their kinds."[27] Castle also objected that Nedham's theory came from Helmont, "who . . . makes a Disease, a real substantial thing, inherent in that which he calls the Archaeus, or vital Spirit."

Castle thus explicitly associated the ontological theory of disease and the doctrine of disease specificity with the works of Paracelsus and Helmont and with the medical "enthusiasts." He himself believed that the body was like a watch, and diseases were disorders of the springs and engines of the body.[28] Castle acknowledged that some diseases spread through contagion, but only when poisons were directly inserted into the body, as in stings or the bite of a rabid dog. The plague was due to a general but unspecified contamination of the air that curdled the blood in the same way that rennet curdled cheese.[29] The worms that appeared in the bodies of fever victims were spontaneously generated by the fermenting blood.[30]

Another critic, Robert Sprackling, a Candidate of the College of Physicians, published *Medela Ignorantiae* in 1665. Harold Cook suggested that more senior College physicians encouraged him to write.[31] Sprackling began with apparent moderation by claiming that Galenism, Paracelsianism, and atomism all had part of the truth. They simply divided the appearances of objects into successively smaller parts but did not significantly affect the current practice of medicine: "when new curious explications are found . . . they will not alter the Disease or Remedy."[32] Even if "tenuious bodies" emanated from the diseased, Sprackling argued that did not mean that they infected the sound. Many people tended patients with pestilential diseases and escaped infection because their "Pores and receptacles" were not adapted to "the magnitudes and figures of [these] Bodikins."[33]

Moreover, Sprackling added, many physicians believed in "vermination."[34] Writers who had little contact with real physicians often made much of ordinary observations that did not occur in every book but were well known to working physicians. Sprackling did not deny Nedham's account directly. Instead, he diminished Nedham's emphasis on the etiological importance of animate pathogens, first by describing "worms" as a consequence, not a cause, of decay, and second by denying that the "effluvia" emanating from the sick were alive. The fact that Sprackling did not simply dismiss the idea shows that even orthodox doctors felt that they had to grapple seriously with Kircher's arguments. Indeed, he conceded as much by saying that these ideas were commonly held by physicians.

The strategy Castle and Sprackling employed was typical of moderate physicians. They emphasized their own willingness to accept new ideas, but at the same time they attacked the education, training, knowledge, and competence of would-be reformers. They argued that Galenism, however deficient in theory, had served both doctors and patients well in practice for hundreds of years and then called for a compromise that would incorporate the best of ancient and modern medicine. Finally, they hinted that the reformers threatened the peace not merely of medicine, but Church and state as well.

A third participant in this controversy, John Twysden (1607–1688, MD Angers, 1646), on the other hand, maintained that Nedham was not contagionist enough.[35] Twysden stressed the stability of disease species. The pox and scurvy were diseases known in the ancient world, and they had not changed. They were common diseases, but not so common that they had infected the entire world through invisible emanations, atoms, or "bodikins." No one denied that emanations streamed from those with plague, spotted fevers, smallpox, leprosy, the itch, and like diseases. Nevertheless, not all diseases passed from person to person without contact in the same way as the contagion of the plague and malignant fevers.[36] Twysden showed that Sennert had actually written that the contagion of scurvy passed from person to person through the exchange of bodily fluids and had denied that it was communicable at a distance.[37] Twysden, who claimed to be a close friend of Robert Boyle, had himself tried to repeat Kircher's microscopic observations but had failed to do so.[38]

The physician Nathaniel Hodges, another member of the Oxford club, also commented on Kircher's theory of "animated Matter in the Air" in an ephemeral work written in about 1665. Hodges had no doubt that the plague was contagious and included a vigorous defense of that view, but he added that despite careful and industrious efforts, he himself had never been able to discern such insects. Perhaps the plague that afflicted Rome was of a different nature than the one that attacked London. Even if animalcules were found, it would not be very surprising as everyone knew that worms and insects were often generated by the putrefaction that accompanied malignant diseases.[39]

GIDEON HARVEY: SYPHILIS AND CONSUMPTION

The populist author Gideon Harvey extended the discussion of contagion to consumption: another undifferentiated disease in the seventeenth century.

Harvey was apparently born in Holland, but one or both of his parents may have been English, as he matriculated at Oxford in 1655. He returned to Holland and studied medicine at Leyden and elsewhere in Europe, including Paris.[40] In the Netherlands, he became a Fellow of the College of Physicians at The Hague and evidently served as physician to the exiled Charles II.[41] He claimed to have an MD, but the university is unknown. After he was appointed to the army in Flanders, he again traveled in Europe before settling in London. He never became a Fellow or Licentiate of the College of Physicians, and his relationship with the College deteriorated after 1678.

Although Harvey's therapies were often traditional, his approach was not. He published many works in English intended for the ordinary English reader. He translated what he considered technical words into more common words, encouraged patients to take more responsibility for their own health, criticized physicians and apothecaries, listed the prices for common drugs to promote comparison shopping, recommended many home remedies while hinting that he possessed powerful secret remedies, and provided detailed advice on preventing venereal diseases.[42]

In *Little Venus Unmask'd*, Harvey attributed the origin of the French pox (syphilis) to hybridization between the particles of two other diseases with similar symptoms. This had occurred during a coupling between two people: one possibly a Frenchman "troubled with an extreme firey itching manginess," and the other a "fretted Neapolitan Whore" suffering from "a deep fiery Scurvy." Steams from her blood mingled with the salt blood from the Frenchman, creating "little steemy bodies, or atoms." These remained in the whore to infect the next man who slept with her. Once within the body, they bred and multiplied, creating further little bodies of their own kind.[43]

In *Great Venus Unmasked*, Harvey explained that the particles that caused venereal contagion must be animated because only animated substances that possessed faculties of nutrition, accretion, and generation could be contagious.[44] Without these faculties they would be unable to persist. These particles were generated in a manner similar to the nits that grow on wool, which he had observed through a microscope. The combination of a theory of spontaneous generation with a theory of *contagium vivum* enabled Harvey to offer a comprehensive explanation of the origin and behavior of venereal diseases, but as in the case of Nedham, this came at the expense of a theory of the fixity of disease species.

In the *Morbus Anglicus: or, The Anatomy of Consumptions*, Harvey discussed wasting diseases and consumptions caused by grief and love, by cancer, excessive study, worms, magic, scurvy, smoky air, and too much or too little sexual intercourse. He argued that consumption was among the most ancient of diseases, adding that "next to the Plague, Pox, and Leprosy, it yields to none in point of Contagion (*catching*)."[45] When a healthy young man married a consumptive wife, he often followed her to the grave. Many people contracted the disease "by smelling the breath or spittle of Consumptives, others by drinking after them; and . . . wearing the Cloaths of Consumptives, though two years after they were left off."[46] Nevertheless, he did not suggest isolating or avoiding consumptive patients or their possessions, and he soon wandered off the topic of contagion into a welter of competing varieties of consumption, anecdotes relevant and irrelevant, and remedies. He included the three visible worms that infested the digestive tract among possible causes of consumptions but did not consider them a major cause.[47]

In other works, Harvey attributed plague to "Pestilential Miasms" and "Arsenical fumes" rising into the air from the earth, scurvy to "Miasms, or perfect Scorbutic seminaries," and smallpox and measles to saline and sulphurous particles of different sizes. These were emitted by the animate and inanimate bodies on the surface of the earth and were transformed by sunshine into poisons.[48] He consistently blamed venomous particles in the air for most contagious diseases, and he believed that each sort of particle was distinctive, but only in the case of syphilis did he insist that the particles themselves were animate.

EVERARD MAYNWARINGE (1628–1699?) AND SYPHILIS

Everard Maynwaringe, son of a Kentish minister, attended St. John's College during the Civil War, graduating MB in 1655, but he gained his MD from the University of Dublin the same year and seems to have spent some time in Ireland.[49] During the Interregnum, he also visited America. Settling in London as a "doctor in physic and hermetick philosophy" in about 1663, he became a founding member of the Society of Chymical Physitians and served as physician to the Middlesex workhouse during the plague. Unlike many of the other "Chymical Physitians," he survived to publish medical treatises throughout the second half of the century. He was prolific, publishing works on scurvy, tobacco, health, pain, consumptions, chemical remedies, and professional disputes.[50]

Maynwaringe's work also shows the connection between the rejection of Galenism, the adoption of contagionism, the belief in specific disease entities, and the construction of new disease taxonomies in the late seventeenth century. In a book on consumption, he explained the Helmontian theory that the life of men depended on a soul, a vital spirit, and ferments in the different parts.[51] An innate or vital spirit was contained in a seed or *Archaeus seminalis*. This seminal agent carried on all the vital functions of the body, and its decline or corruption caused diseases.[52] Fermentation was a vital process that both produced and cured diseases. The temperaments and humors were merely effects of the vital principles, not things in themselves.[53] Phthisis came from breathing unwholesome air, including the breath from a phthisis patient. Infected bedfellows were especially dangerous.[54] Too frequent copulation could deplete the seminal principle and result in a "Spermatick Consumption."

Maynwaringe's book on venereal disease, *The History and Mystery of the Venereal Lues*, appeared in 1673.[55] The idea that syphilis was contagious was not new in 1673; Fracastoro had discussed the contagion of syphilis in 1546.[56] Many others had agreed; Maynwaringe listed a dozen, including

Fernel and Sennert.[57] However, Maynwaringe offered an exceptionally clear and concise account. He divided contagious diseases into three kinds: those that were communicated over a length of time, such as phthisis, scurvy, and cachexia; those that infected rapidly, such as venereal disease and the itch; and finally those that were especially severe and could infect at a distance, such as the plague.[58]

Contagion had four different meanings: a preternatural effect impressed on the body, an entity or venomous substance, an action, or a medium of communication and contact. The act of infection involved the infecting person, the substance being communicated, and the person being infected.[59] The person who transmitted a disease did not lose it, but emanated a virulent exhalation. Because the virulence of these emanations varied, disease was not always contagious, and some people were more resistant than others.[60] Copulation with strangers was more dangerous than between constant bedfellows because it was "fiercer." Infected beds and clothing could also transmit the disease, since the virulent miasmas were attracted to the pores of warm bodies. Miasmas consisted of corpuscles or "subtile finer invisible bodies, which are minute small portions *exhaling* from infected bodies."[61] After entering a victim's body, they fermented and became virulent.

Maynwaringe's *Inquiries into the General Catalogue of Diseases* complained that Galenic physicians mistook symptoms for diseases, creating almost as many "Fevers" as there were diseases.[62] He provided a brisk summary of the elaborate and confusing welter of diseases based on supposedly preternatural (altered) states of the primary qualities (hot, cold, wet, and dry); and secondary qualities (such as taste and odor) accepted by Galenic physicians. Condemning listings of diseases by the affected parts of the body or by the similarity of some diseases to others, he stated that

> setting forth Diseases in their due Classes, is the Basis of Practice; if the first general divisions be false; the subdivisions contained under them must needs be wrong . . . There are [only] three things preternatural . . . the Morbific Cause, the Disease, and the Symptom. The Morbific Cause generates a Disease; the Disease begets Symptoms, all Preternaturals are comprised under these three Heads . . . Qualities cannot be Diseases.[63]

Diseases, he argued, were either spiritual or corporal, either a defect of the vital principle or of the passive body it governed. His categorization of spiritual diseases was relatively straightforward, including such ailments as debility, perturbation, syncope, apoplexy, and hysteria, but his division of corporal diseases was vague and chaotic, interrupted by a rambling postscript about the provision of medicines for the military.

Nedham, Harvey, and Maynwaringe all discussed venereal disease. Harvey's biographer, H. A. Colwell, thought he was "feathering his nest financially no matter how much he may have been damning himself professionally."[64] In fact, it is not clear that publishing on this topic condemned early modern authors to ignominy, although, as we noted earlier, some

physicians disapproved of any medical publications that encouraged readers to choose their own course of treatment.[65] Because of the embarrass-
ment it caused, publications on venereal disease were especially attractive to
consumers who did not want to consult a doctor in person, giving rise to a
very robust vernacular literature. Because most seventeenth-century authors
viewed syphilis as a contagious disease, these works may have contributed to
the dissemination of ideas about contagionism.

THE PERSISTENCE OF HELMONTIAN IDEAS

During the Interregnum and early Restoration, Helmontianism had been
associated with political subversion and religious heterodoxy.[66] In the midst
of a heated exchange of pamphlets between two York physicians, for example, the older Presbyterian author defended traditional Galenism and accused
his younger rival of sectarianism: "His Philosophy is . . . altogether novel and
precarious . . . I wish (for his own sake) his Divinity may be better; for I have
seldom seen anyone so Sceptical in Reason, but the same had been Heterodox, if not Heretical in Religion."[67] Indeed, perhaps he was "inclinable to
be a Quaker."[68]

Even historians such as Christopher Hill, who were sympathetic to the
radical Helmontians, once thought their ideas died out in the late Restoration, killed by a combination of political repression and their own inadequacies.[69] Further research, however, has shown that Helmontian ideas and
laboratory practices were far more influential in the development of modern
chemistry than previously believed.[70] Helmontianism was evolving rather
than decaying.[71] J. Andrew Mendelsohn found that Restoration Paracelsians
successfully retooled their metaphors and language to appeal to the Crown
and Court.[72] Lawrence Principe pointed out that both eighteenth-century
élogists and later historians exaggerated the break between the chemistry
of the early Enlightenment and the alchemy that preceded it.[73] Anna Maria
Eleanor Roos has traced the gradual transformation of Helmont's "vital principle" during the long eighteenth century beginning with its depiction as a
volatile salt, through its characterization as a saline spirit and an "acidifying
principle," and ultimately into an understanding of the role of oxygen in
combustion and respiration.[74]

Studies of the Helmontian George Starkey, a leader of the ill-fated Society
of Chymical Physitians, have revealed Boyle's unacknowledged debt to Starkey's careful laboratory work.[75] Historian William Newman has shown that
Starkey, who used several pseudonyms, was the author of popular alchemical
works published under the name of Eirenaeus Philalethes, which left a lasting impact on such important figures as Isaac Newton, Gottfried Wilhelm
Leibnitz, and Georg Ernst Stahl, in addition to many radical Dissenters and,
later, the Swedenborgians.[76]

Helmontian chemistry and chemical medicine inspired many investigators at Montpellier and at *the Jardin du Roi* in Paris where many Englishmen studied, and it also accompanied European immigrants into England.[77]

An investigation of the chemists and pharmacists who were operating shops or manufactories in Restoration England might also reveal more about the Helmontian legacy. When Hans Sloane first arrived in England, he studied at the Society of Apothecaries and lodged for four years with Nicholas Staphorst, the Society's "chemical operator" and author of a work on "spagyric" (chemical) medicine.[78] We can assume that Sloane also studied with Staphorst.

It was in Staphorst's shop in about 1679 that Sloane first became acquainted with Frederick Slare, FRS, who would become a lifelong friend and associate.[79] Slare's cousin, housemate, and fellow countryman, Theodore Haak, a founder of the Royal Society, had been a close associate of Samuel Hartlib, one of Starkey's patrons.[80] It was no accident that Slare happened to be in Staphorst's shop, as he was also deeply interested in chemistry.[81] William Yworth, an immigrant from Holland, opened a Spagyric Academy in London in 1691, and received funding from Newton at the turn of the century. He practiced as a chemical physician in Suffolk until his death in 1715. Newton also worked very closely with a young Swiss alchemist, Fatio de Duillier.[82]

Chemical remedies, especially mercury and antimony, continued to find a buoyant market, as did the proprietary medicines that often contained them.[83]

CONCLUSION

Although the political radicals were defeated, the Dissenters suffered religious discrimination, and the "alchemists" or "chymists" were discredited by the scathing criticisms of Boyle and his colleagues, Helmontian medicine survived during the Restoration. Its central concepts continued to attract support by authors during the second half of the seventeenth century and to demand serious consideration even by physicians who did not adopt them.

Among these concepts were the beliefs that disease was a "thing," that disease came from outside the body, that fermentation offered useful analogues for disease processes, that the body's own vital force must be respected, that some diseases came from a challenge to the vital force by an inimical "idea" or entity, that disease symptoms arose from the body's effort to expel this pathogen, that new diseases could appear, and above all, that disease must be approached as a vital process and not merely as a mechanical derangement or an imbalance.

Because the Helmontians did believe in "germs" in the sense that they viewed disease as the invasion of a body by a hostile vital force, it might be said that they adopted a germ theory of disease, although they did not always conceptualize these germs as physical particles. However, because they also emphasized the omnipresence of these hostile forces, they were not necessarily contagionists. Diseases might be specific, but they were also everywhere: for Van Helmont himself, hostile *Archei* were present in every bite of food; for Nedham, "wormatick matter" pervaded the world, different sorts blending into new compound diseases.

Proponents of animate contagion faced three major hurdles: the first was that they were challenging established practices in a conservative profession, the second was that there simply was not enough clinical evidence to clinch their case, and the third, highest hurdle, was that correlation is not causation. The claim that animalcules had been observed in the bodies of disease victims did not mean that they were responsible for diseases. The Helmontians emphasized the permeability of the divide between inorganic and organic: not only did they believe in spontaneous generation, they often interpreted it as broadly as possible.[84] Thus new life forms and also new diseases were constantly arising spontaneously in patients and in the outside world.

Nevertheless, when discussing certain specific diseases, some Helmontian authors asserted that they were contagious. It is not clear how much they drew from a well-established tradition of "lay" or "popular" contagionism and how much from the published work of medical philosophers such as Sennert and Van Helmont, but the two strands of thought were compatible and both furthered the sort of populist medicine the reformers had championed. If they themselves did not see the connection, their enemies certainly did: opponents of medical reformers alleged that early modern contagionism and religious sectarianism shared a similar approach to the nature of creation and man's place within it.

After the Great Plague of London and a series of attacks from their critics, the chemists' hopes of a newly constituted medical profession faded as the sectaries and Nonconformists painfully negotiated a transition into a much less receptive world. Just as the sectaries survived on the fringes of British life, however, so did the new emphasis on individual diseases. Like Golems, once diseases acquired identities, they became hard to dispel.

CHAPTER 4

THE SEARCH FOR MIDDLE GROUND: DISEASE THEORY AS NATURAL HISTORY

INTRODUCTION: MECHANISM AND HELMONTIANISM IN ENGLAND

Although many Catholics and Anglicans saw atomism as tantamount to materialism and atheism, in the mid-seventeenth century, René Descartes and Pierre Gassendi had tried to rescue atomism by stressing the passivity of matter and arguing that God made the atoms and gave them their motion.[1] Thus, the cost of their effort to make atomism respectable was a reemphasis on dualism.[2] Although Descartes's works were placed on the Catholic Index in 1663, his corpuscularism was tolerated, if considered very controversial, in Protestant Holland, and it filtered into Britain. The Royalist British physician Walter Charleton (1620–1707) published a treatise in 1654 introducing Gassendi's work to English readers: this may have induced the London College of Physicians to deny him a Fellowship in the following year.[3] Charleton emphasized the immortality of the soul and the contingency and passivity of matter.[4]

Following the war, several English authors, including the chemist Robert Boyle, also argued that atomism could fortify religion rather than undermine it.[5] Isaac Newton never fully committed himself to an atomic theory of matter, but between the publication of the *Principia* in 1687 and the 1730s, it was deployed by many "Newtonians" who hoped to defeat the "radicals, the materialists and the epicureans" by stressing the idea that the forces that controlled atoms were spiritual and external to the atoms themselves.[6]

The atomists insisted on the real existence of immaterial entities: the soul, God, and, in the realm of physics, forces, although they had difficulty explaining how these entities affected the lifeless mechanical world of matter and motion.[7] They emphasized the creationist and providential elements of corpuscularism and cosmology, the importance of natural theology, and the argument from design.

Dualistic atomism and Newtonian mechanism formed the basis of the "holy alliance" between Anglican religion and physical science in eighteenth-century England. An emphasis on the orderly design of the universe meshed with a Latitudinarian belief in reason rather than revelation as a basis for Christian faith and with the Whig doctrine of government based on natural law.[8] Like the universe, the nation should be orderly, harmonious, and law-abiding, directed by a benevolent but distant ruler.[9] Because it rested on a divide between God and nature, this "holy alliance" between Church and state was always a troubled marriage. If God, the Great Clockmaker, were permitted to become too remote, religion could trail off into Deism or atheism, losing its mysteries, its scriptural foundations, and its connection with the immaterial. If, on the other hand, the emphasis was placed on the need for constant tinkering, God could become indistinguishable from his creation, leading to pantheism, mysticism, or "enthusiasm."

Assiduously promoted by a succession of Whig governments who needed a bulwark against a disaffected Tory lower clergy, Newtonian Latitudinarianism became the surest road to success for young men: dons, clergymen, politicians, and even physicians aspiring to a career based on intellectual abilities and Court favor. Its chief scientific promoters were the mathematicians of the Royal Society, and, as a basis for scientific research, it served them well despite criticism.[10] Traditionalists pointed to the dangers inherent in stripping religion of its basis in authority and unbroken tradition: they did not agree that reason alone offered an adequate basis for faith. Nor did they agree that arguments based on "natural law" or on contract theories of government offered a sufficient excuse for deposing an anointed king.

In the hands of unskilled practitioners, "Newtonianism" became a dogma rather than a science. In theory, Newtonianism rested on the scientific method elucidated by Bacon: carefully designed experiments, inductive reasoning, empiricism, and the eschewing of *a priori* arguments. In fact, many "Newtonians" reasoned by logic, not by experiment, and did not question their original assumptions. What they took from Newton's work was its apparent mechanism, its support for natural religion, and the argument from design; they neglected its sophisticated mathematical analysis of data derived from decades of astronomical observation and well-considered experiments.[11]

In place of either tradition or immediate clinical experience, many "Newtonian" physicians emphasized the role of logic and reason in medical practice and therapy, but they also incorporated Galenism's stress on a moderate life and its goal of restoring physiological equilibrium. Nevertheless, mechanism was especially inadequate as a foundation for medicine because it relied on mathematical analysis; even if physicians mastered calculus, it offered few answers to everyday medical dilemmas. "Newtonians" explained physiological processes by the flow of particles through specially shaped pores, attributing illness to a disjunction between the shape of these pores and the size, shape, cohesion, or number of particles, but their theories remained wholly speculative. "I am sorry to see so useful a *Science* as the Mathematicks prostituted thus, and applied to Matters that are not capable by any of its *Rules* of being made evident and demonstrable," William Oliver lamented in 1704.[12]

Although Newtonianism supplied a new theory, the Newtonians' medical practice employed traditional Galenic remedies, especially bloodletting, purging, and treatment by opposites.[13] They still assumed that illness resulted from a humoral imbalance that was either internal or due to climate and lifestyle (the "non-naturals" of sleeping, exercise, eating, etc.). From the Glorious Revolution through the early eighteenth century, this was the approach favored by many members of the London College of Physicians, but Helmontian ideas did not entirely disappear.[14]

The mechanists and the chemists both questioned classical medicine, but the two movements differed in emphasis. For Paracelsus, the pursuit of knowledge had required philosophers to be attentive to the object itself, to "'learn of,' to 'listen to,' to be 'taught by' the object . . . the knowledge, for example, that enables a pear tree to grow pears and not apples."[15] Van Helmont had cautioned against efforts to apply to living men the lessons learned from *in vitro* experiments such as distillation, chemical decomposition, fermentation, or even anatomy. He viewed efforts to compare events in the outside world with those inside the body as "poetical, heathenish, and metaphorical, but not natural, or true," because man was created in the image of God, not the outside world.[16] Disease represented a vital process that was not analogous to "cooking" or to condensation. Helmontians would seek to convey the experience of each disease: to uncover its individual nature.

Whereas the Newtonians sought to discover by experiments the mathematical laws established by a distant mover that governed the relationship between things, the sectaries had sought to obtain through direct experience insight into the individual essence of things themselves and thus to commune with an immanent God. Because they considered experiments inadequate and prone to error, they promoted what has been described as a "vulgar" or "low" Baconianism, "a promiscuously eclectic process of collecting, labeling and classifying" observations and objects.[17]

Some doctors continued to insist on Helmont's ontological disease theory and to emphasize disease specificity; others sought a compromise. Boyle and his followers rejected the vitalist perspective inherent in Helmontianism, criticizing the alchemists for personifying nature and asserting that the experimental investigation of both natural and artificial "contrivances" offered important insights into physiological processes. To the experimental community, the radical emphasis on the primacy of personal experience was destabilizing and threatening. It made replication impossible and devalued both authority and communal assent: two elements of natural philosophy that were highly prized in the postwar climate. Learned physicians associated the Helmontian idea of individual disease entities with popular ignorance and superstition. For example, the anonymous author of a tract entitled *The Censor Censured or the Antidote Examin'd* warned against those who "would perswade us to believe that Diseases are substantial forms which enter in and take possession of our Bodies, as the evil Spirits did heretofore . . ."[18] Diseases, like atoms, were best seen as inactive.

Many physicians sought an intellectual and professional balance by making a thoughtful selection from jumbled heaps of Baroque medical systems, incorporating the best of both the old and the new without becoming too blindly attached to any single version. Whereas the mathematical approach led to physiological experimentation, the experiential approach led to the detailed case history. In the eighteenth century these approaches led to methods of investigation that combined elements of both and proved more suited to medicine, such as the clinical therapeutic trial and the grouping of systematic case histories for quantitative analysis.

THE OXFORD CIRCLE AND THE SEARCH FOR COMPROMISE

In 1654, Boyle moved from London to Oxford, leaving the radical "invisible college" for the world of the professional academic scientist. Behind him were aggrieved critics of the academic establishment and the College of Physicians. Ahead were the tolerant "Latitudinarians" who defended the universities and believed that reason, not emotion, should be the foundation of religion.[19] Boyle joined a group of skilled mathematicians who believed that the key to the advancement of knowledge lay in mathematics rather than experimentation.[20] At first they were suspicious of Boyle because of his lack of mathematics and his interest in chemistry. "The Oxford scientists were very critical of the practical aims of mere 'sooty Empiricks.' They ridiculed the 'company of meer and irrational Operators, whose Experiments may indeed be serviceable to Apothecaries, and perhaps to Physicians, but are Useless to a Philosopher, that aimes at curing no disease but that of Ignorance.'"[21] Boyle, however, believed experimentation was essential to natural philosophy because it promised to curtail the endless disputation that had stemmed from a reliance on logic alone.[22]

Although some Oxonians considered chemistry a dubious enterprise, Oxford was not lacking in experimental scientists. Indeed, Interregnum Oxford hosted one of the richest scientific cultures that ever existed: more than one hundred virtuosi lived within a few minutes' walk of each other.[23] As Robert Frank has shown, much of this was due to the influence of William Harvey, who had lived at Oxford from 1642 to 1646 while serving as physician to King Charles I.[24]

Harvey's contemporaries were cautious about his work, but it was enthusiastically adopted by members of the next generation who realized that his discoveries would require an entirely new approach to physiology. Several of them received university appointments from Cromwell's Parliamentary commissioners: although it replaced some Cavalier academics, the Cromwellian government defended the universities against radical and Anabaptist critics, sometimes literally at sword's point. At the same time, it encouraged scientific innovation and made several distinguished appointments. For example,

William Petty studied medicine and taught anatomy there from 1646 until 1652.[25]

The members of this group were seeking a middle way: to incorporate the new science and promote institutional reform without destroying the existing institutions of learning. Moderate Oxford Royalists and the university-educated scientists introduced by Parliament found common interests and common ground. Although some of the Parliamentary scientists were forced to leave at the Restoration, many of them remained, accepted the settlement, and hoped to achieve reform from within.

Boyle struggled to develop a new "corpuscular" science free from both the atheism of classical atomism and the "enthusiasm" of radical religion and to make it more appropriate for university-educated gentlemen. His initial solution was to depict all matter as composed of lifeless particles whose only attributes were motion and extension. Despite his intense interest in medicine, Boyle became increasingly disenchanted with alchemy. He rejected both the "elements" of the Galenists and the "principles" of the Paracelsians and was especially opposed to Paracelsian monism.[26] The Helmontian alchemists, however, had developed their own corpuscular theory that held that the smallest particles were generated from the primal element, water, by a formative "seed" implanted by God. They did not consider their matter theory incompatible with Boyle's.[27]

Among the virtuosi at Oxford during these years was the Puritan Parliamentarian Thomas Sydenham (1623–1689) a Fellow of All Souls College who would finally obtain a Cambridge MD in 1676.[28] Sydenham, who would work closely with Boyle, adopted Boyle's theory of matter, though he opposed the idea that experimental science should have a role in clinical medicine. He was equally hostile to the idea that physicians should rely on traditional learning for their medical knowledge.

The Oxford circle contained several other founding members of the Royal Society, which would be established by a charter of Charles II in 1660. Before 1700, at least nine distinguished members of the Society—John Wilkins, Walter Charleton, George Ent, John Locke, Edward Tyson, Frederick Slare, Robert Hooke, William Petty, and John Ray—expressed or allegedly expressed some interest in contagion. Boyle himself used the phrase "contagious diseases," but he used the word as equivalent to "infectious," and his theory of disease was entirely miasmatic.[29]

Other members of this circle held divergent views. Thomas Willis attributed disease to ferments arising in the body. As part of a universal taxonomy of knowledge, John Wilkins created a classification of diseases that included a list of contagious disease species. Others speculated about the role of "worms" or "animalcules" that pervaded the environment. John Ray would develop a new theory of classification for all living entities that rested on a biological conception of species. The rejection of spontaneous generation that was embedded in his work also entailed a reevaluation of the relationship between insects and disease. Edward Tyson, John Locke, and several other natural philosophers were also speculating that certain diseases were the result of living entities.

THOMAS WILLIS AND "FERMENTS"

The Anglican Royalist Thomas Willis (1621–1675) was practicing as a physician in Interregnum Oxford.[30] He would receive both an Oxford MD and the Sedleian professorship of natural philosophy in 1660. A founding member of the Royal Society, he settled in London in 1666 and became an honorary Fellow of the College of Physicians.[31] Willis believed that particulate contagion played a role in the transmission of fevers, which could arise from substances either already in the body or introduced from outside.[32] He rejected the Galenic humors, adopting the Paracelsian elements and Van Helmont's "ferments," but he substituted their effects on the physical motion of the corpuscles of the body for Helmont's sentient *Archeus*.[33] Willis could not entirely shake the Galenic dichotomies between hot and cold qualities, so he depicted fevers as products of a cooking process that heated and roiled the blood and saw chronic diseases as the effect of coagulation, just as water boils when hot and forms ice when cool.[34]

By "ferments," Willis meant substances that created rapid motion in particles of the blood. We would describe it as effervescence. Any "humor" that contained a large proportion of particles of any of the three principles—salt, spirit, and sulfur—could act as a ferment. By "humor," Willis seems to have meant any "organic" fluid, since he gives as examples the dregs of beer, wine, and blood. Earth and water were inert substances that did not ferment.

A belief in the essential similarity between acute disease and fermentation persisted long after the other elements of Willis's etiology were discarded, and nineteenth-century research on fungi and yeasts would provide important insights into the transmission of diseases.[35] Willis's own mechanistic conceptualization of fermentation, however, discouraged an ontological conceptualization of diseases. A seed always grows into the same distinctive plant that generated it, but many seventeenth-century authors believed that when a "ferment" is added to grapes it makes wine, if it is added to hops and barley it makes beer, and if it is added to wheat it makes bread. It was the nature of the object being fermented, not the nature of the ferment, that determined the outcome. They did not view ferments as living entities that grew in a host but as potentially harmful chemicals that effervesced or heated up when added to a fluid.[36] Newton, for example, saw fermentation in 1706 as an "active principle" that moved particles in a manner akin to gravity.[37]

For Willis, there were only three vital principles (his three elements) and thus only three effective fermentations. Because the fermentation that caused disease could arise from either within or outside the body, he thought that differences between diseases arose from differences in the host (the nature or location of the humor that was invaded or differences in host resistance) or in the concentration and proportions of the three elements within the fermenting fluid. Lacking a complete concept of disease specificity, Willis did not infer the natural history of different disease entities from the many detailed case histories he included in his work.[38]

THOMAS SYDENHAM (1624–1689), ROBERT BOYLE (1627–1691), AND JOHN LOCKE (1632–1704): CHEMICAL AND MECHANICAL CAUSES

Thomas Sydenham saw long and hard service as a captain in the Parliamentary army before studying medicine at Oxford.[39] After the Restoration he settled in London. Sydenham never joined the Royal Society, and he did not become a Fellow of the College of Physicians, though he did obtain the College license to practice in London in 1663. His relationship with the College seems to have been prickly from the start, perhaps because of his unconventional therapies and his disdain for classical learning, but he gradually built a moderately successful practice. His publications gained him a growing circle of admirers throughout Europe, and after his death he became perhaps the best known of all British physicians after Harvey. In particular, Herman Boerhaave, the very influential Professor of Medicine at Leyden, greatly admired Sydenham and borrowed several critical components of his disease theory.

Sydenham's name became synonymous for a "Hippocratic" clinical or empirical approach to medicine, but just as the "Newtonians" did not always reflect the thought of Isaac Newton himself, so the "Sydenhamians" did not always represent Sydenham's own approach to disease, which was by no means free of theory. In particular, his conceptualization of disease categories was idiosyncratic and depended on a concept of biological species that was very different from the one that would develop soon after his death in 1689.[40] His use of a botanical analogy has misled many historians who have assumed that Sydenham's concept of species was similar to the one that would be developed by the botanist John Ray and his successors.

Sydenham objected to the view that disease was "but a confused and disordered effort of Nature thrown down from her proper state, and defending herself in vain."[41] The preface to his *Medical Observations* called instead for a natural history of diseases drawn "with the same care which we see exhibited by botanists in their phytologies," because identifiable diseases belonged to distinct species in the same way that plants do.[42] Referring to the regular symptoms of the quartan ague, it noted that "it cannot easily be comprehended how the disease in question can arise from a combination of either principles or evident qualities, whilst a plant is universally recognized as a substance, and as a distinct species in nature."[43]

Despite references to both "morbific causes" and the need to classify diseases as botanists classify plants, the metaphor stopped there.[44] The idea of morbific causes as "seeds," which combined the idea of specific and discrete entities with the idea of generation or vitality, seems never to have occurred to him. He viewed malign particles as inert, like minuscule birdshot. For example, although he suggested in passing that Jesuit's bark cured intermittent fevers because it prevented "the germ of the disease" from reproducing in the body, he did not draw the further conclusion that this implied some form of living organism.[45]

Sydenham postulated three general causes of diseases: the temperament of the individual patient, the influence of perceptible external factors such as climate and seasons, and a third imperceptible external cause: the so-called "epidemic constitution" of the atmosphere, which contained invisible morbific particles.[46] The first cause produced chronic ailments that were more or less the same every year; the second resulted in seasonal outbreaks of acute diseases, such as colds, dysenteries, and agues; the third created very widespread and unpredictable epidemics, such as the plague. All three factors interacted to determine the characteristics exhibited by all diseases occurring at the same place and time.[47]

For the origin of these morbific particles, Sydenham turned to the work of his close friend Boyle, whose *Treatise of Some Unheeded Causes of the Insalubrity and Salubrity of the Air* suggested that effluvia emanating from deep within the earth permeated the atmosphere and insinuated themselves into the body through the lungs and the pores of the skin.[48] Boyle thought that these subterranean bodies might have "a kind of propagative or self-multiplying power," but he added, "I will not here examine whether this proceed from some seminal principle which many chemists ascribe to metals and even stones, or, which is perhaps more likely, to something analogous to a ferment . . ."[49]

Sydenham did adopt an ontological theory of disease.[50] Moreover, in his therapy he intended to treat the disease, not the person. That is, he ignored the individual constitution of the patient and prescribed specific regimens according to what he had found successful in a previous patient with the same symptoms. Orthodox Galenists despised this "empirical" method, viewing it as a way of throwing treatments at diseases instead of taking the time to develop a thorough understanding of the individual patient. It was also much better suited to the practice of a physician who treated large numbers of the poor, ill with epidemic fevers, than one who treated smaller numbers of the wealthy, afflicted with chronic diseases.[51]

Sydenham's understanding of the nature of disease species and the relationship among diseases was entirely different from the taxonomy that would emerge from the work of John Ray. Sydenham's theory stemmed from Aristotelian logic, which assigned every object to a category that was determined both by its membership in a larger group, its *genus* or kind, and by the "specific" characteristics, the "*differentia*" that separated the members of that group into discrete subgroupings. For example, in the definition of man as a rational animal the species, "man" cannot be understood except as a combination of *both* the genus and the *differentia*: the adjective "rational" is a specific term, one that divides men from other animals, but it is not a "species" in itself. There might be other "rational" entities that are not animals, for example, angels or computers.[52]

Sydenham recognized several epidemic diseases, including "Continued Fever," "Intermittent Fever," "Pestilential Fever," smallpox, "Cholera morbus," dysentery, measles, bilious colic, and epidemic coughs. He initially saw the plague as a more serious form of pestilential fever. The genus, or overriding nature, of these fevers was not determined by the course of symptoms in different individual patients but by the epidemic atmosphere that prevailed in any given set of years. All patients who fell ill during a given time had the same genus of disease. The species of each of these diseases was determined by the alteration of a particular "specific" fluid, or humor, in the body.[53] "Whereas all species, both of plants of animals . . . subsist by themselves, the species of disease depend on the humours that engender them."[54]

Thus, the smallpox of one epidemic period was "generically" (i.e., in genus or kind) different from the smallpox of another, being the product of the reigning epidemic constitution. On the other hand, all outbreaks of acute disease that occurred during this same period were attributable to the same "common and general producing cause" outside the body.[55] The prevailing epidemic constitution altered all humors to a certain degree; when someone living during that constitution was exposed to a disease that attacked a particular humor, he or she became ill with a particular species of disease, but one that also manifested the tendency of the prevailing epidemic. One disease could turn into another as soon as the atmosphere changed.

For Sydenham, diseases came in vintages. "Smallpox" was not a separate disease but a distinctive appearance of disease. The *actual* disease was formed by a combination of that appearance with a particular epidemic constitution, "the Smallpox of the years 1667, 1668 and part of 1669." The smallpoxes of two different constitutions resembled each other just as cases of frostbite in two cold winters resembled each other. In each case the epidemic constitution happened to attack the same humor.

Sydenham did identify what he called "intercurrent" and chronic diseases.[56] Intercurrent diseases were continued fevers that usually occurred at a more or less constant rate, independently of the epidemic constitution. Although such individual errors as a premature change of dress or exposure to cold after exercise caused most of these, they could also come from an extreme change of temperature. As this affected many people at once, these diseases sometimes appeared as epidemics. Among them he distinguished "scarlet fever, pleurisy, bastard peripneumony, rheumatism [rheumatic fever], erysipelas, and quinsy" as separate diseases. He noted that these diseases were occasionally contagious, but he saw them as processes, not independently existing entities; there was no need for a physical connection among victims to spread these diseases.

He also described groups of "combinations of fevers" placed in categories according to their ability to occur together. This is best seen as a sort of timetable or seasonal chart of epidemic prevalence, not as a system of classification. A botanical analogue would be a handbook listing plants that emerge in the spring. For example, epidemic plague might appear with sporadic fevers and also with other "pestilential" fevers. Sporadic plague might appear with autumnal agues and lingering intercurrents. Although he claimed to be discussing "specific" diseases, this language is easily misread: in most cases

Sydenham did not believe that there were different kinds (species) of diseases or disease entities that remained constant over time and space.

It is possible that the introduction of cinchona, or "Peruvian bark," led both Sydenham and Boyle to a new interest in disease specificity. Sydenham sometimes used cinchona as a diagnostic aid: fevers that responded to cinchona differed from fevers that did not. This may have encouraged him to group the various periodic agues together as a single category of fever: the "intermittents" that were distinct from all the "continuous fevers" that did not respond. However, this difference in response could also be due to a difference in the affected humors, not a difference in the morbific particles themselves.[57]

Several scholars have concluded that Sydenham was changing his approach to disease near the end of his life, becoming less influenced by Boyle and more by Locke; others have reversed the influence, claiming that Sydenham's views and Locke's own medical experience left the younger man dissatisfied by both mechanical and Helmontian theories.[58] Their ideas became so similar that scholars are still unable to decide which of them wrote two short works, *Anatomie* (1668) and *De Arte Medica* (1669).[59]

Kenneth Dewhurst wrote that Sydenham's own clinical experience with smallpox and cinchona "caused Sydenham, towards the end of his life, to consider jettisoning humoral pathology as it gradually began to collapse under the weight of various *ad hoc* hypotheses thrust upon it by empirical facts."[60] A third possible factor was his experience of the plague in 1665. Sydenham admitted that he had begun to suspect that the atmospheric constitution alone was not sufficient to cause plague:

> Either the disease itself must continue to survive in some secret quarter or else, either from some *fomes* or from the introduction from pestilential localities of an infected person, it must have become extended. And even in these cases it cannot become epidemic except with the conditions of a favorable atmospherical diathesis.[61]

Plague epidemics now required a conjunction of the atmosphere and a specific disease that had its own individual nature and continued existence.

Locke, however, may have moved in the other direction, although his published work expunged some of his earlier thoughts. Historian Patrick Romanell has published a long, unfinished manuscript note entitled "*Morbus*" that Locke wrote in the mid-1660s, weighing the differences between Helmontian and Cartesian medicine.[62] Locke commented that some things were produced by seminal principles, others by a bare mixing of the parts. The seminal principles caused contagious diseases including the itch, plague, and ulcers, whereas other diseases were produced merely by the mixture of chemicals, in a manner similar to the effervescence of acids and volatile salts.

Locke thought that the seminal principles were

> some small and subtile parcelles of matter which are apt to transmute far greater portions of matter into a new nature and new qualitys, which change could not

be brought about by any other means, soe severall seeds set in the same plot of earth, change the moisture of the earth which is the common nourishment of them all into far different plants which, differ both in their qualitys and effects, which I thinke is not donne by bare streining the nourishment through their pores.[63]

From a very small and almost imperceptible beginning, these seminal principles or potent *Archei* gradually alter the substances they "work upon" to their own nature, "make them obey their motion, and quite alter the natur of those bodys which are fit to receive their impression."[64] Unlike Sydenham and Boyle, therefore, Locke thought that the pathogens that caused some diseases were self-replicating and that these seeds or germs caused a disease process that was distinct from simple effervescence.[65] However, he still limited their effect to very few diseases. Locke never published this fragment.[66]

Locke became increasingly pessimistic about whether it would be possible to visualize germs or ferments even with the "assistance of glasses or any other invention," or to determine by dissections what kind of ferment separated any part of the fluids in the viscera.[67] He also grew more skeptical about Helmontian chemistry.[68] By the time he published the *Essay Concerning Human Understanding* in 1690, Locke appeared to be committed to a mechanistic view of nature.[69]

Sydenham has been described as both a Galenist and a Helmontian, but neither label is helpful. He is best understood as an eclectic. His minimalist and anti-Galenic approach to therapy aligns him with the Helmontians, populists, and naturalists, but his corpuscular theory of matter was inspired by Boyle, and his miasmatic theory of disease was similar to the view expressed by other moderates associated with Boyle, such as Castle or Hodges, whom we met as critics of the Helmontian author Marchamont Nedham and defenders of the College of Physicians.

CONTAGION AND THE ROYAL SOCIETY: ROBERT HOOKE (1635-1703), GEORGE ENT (1604-1689), AND WALTER CHARLETON (1619-1707)

At the same time that Locke was ruminating about seminal principles, other Fellows of the Royal Society were also thinking about contagion. Robert Hooke, a Royalist who had assisted Boyle at Oxford, became curator of experiments for the Society in 1662. In 1665, after the plague forced the Society to suspend its meetings, Hooke, who was about to leave London with John Wilkins, wrote a letter to Boyle showing that he had adopted a contagionist, anti-miasmic view of plague very similar to that of Twysden. He commented,

I cannot, from any information I can learn of it [the plague], judge what its cause should be, but it seems to proceed only from infection or contagion, and that not *catched*, but by some *near approach* to some infected person or stuff;

nor can I at all imagine it to be in the air, though there is one thing which is
very different from what is usual *in other hot summers*, and that is a very great
scarcity of flies and insects."[70]

There is no evidence anyone (including Boyle) paid much attention to
this comment.[71] Hooke's *Micrographia* would appear in September 1665,
only a couple of months after he wrote this letter, but it does not mention
this topic.[72]

When the Society reconvened in the spring of 1666, Dr. Walter Charleton,
a Royalist physician who had been a student of John Wilkins, raised the question of the cause of the epidemic.[73] Perhaps because he worked with the
Paracelsian physician Theodore de Mayerne, Charleton had been an enthusiastic Helmontian and was Helmont's first English translator.[74] He was a
founding member of the Society, but despite his MD from Oxford, he was
denied a regular Fellowship by the College of Physicians. He was admitted as
an honorary Fellow in 1664, when the College was raising funds, and finally
became a full Fellow in 1676, rising to become President.

Charleton told the Society that his friend George Ent had been the first person in England to suggest that plague was caused by a "vermination of the air"
before Kircher published the idea in Italy.[75] If this is correct, Ent must have
shared this view before 1658, but Ent does not seem to have left any written
record of his own.[76] Recently, Charleton continued, Kircher's English friend
"Dr. Bacon, who had long practiced physic at Rome" had said that "there was
a kind of insect in the air, which being put upon a man's hand, would lay eggs
hardly discernible without a microscope." When these eggs were fed to a dog,
it had developed all the symptoms of the plague. Charleton offered to bring
"Dr. Bacon" to confirm this event for the Society and suggested interviewing
the managers of the city pesthouse about their observations. This suggestion
was approved by the Fellows, but the observations were never recorded.

Despite the interest expressed by several Fellows of the Royal Society in
the subjects of contagion, the causes of epidemics, the presence of animalcules, and similar topics, these ideas did not coalesce into what historian
Robert Frank has called "a research tradition" that developed momentum
based on a continuity of interest and direction.[77] The Fellows were distracted
by more dramatic research projects, including experiments with blood trans-
fusion, and the community of investigators that had coalesced during the
Interregnum was either disintegrating or evolving, depending on one's perspective. Perhaps they simply did not see this as a promising avenue for investigation. The lack of a journal that addressed strictly medical topics may also
have discouraged further activity.

At least some of the philosophers held beliefs about contagion that were
very similar to those expressed by more populist Helmontian authors, such
as Maynwaringe, but they did not publicly advocate for these ideas. The first
work to offer a full exposition of Kircher's ideas was thus published by the
mercurial and suspect Marchamont Nedham, not by the more reputable and
cautious philosophers in the Royal Society.

CONTAGION AND CLASSIFICATION: JOHN WILKINS
(1614–1672) AND WILLIAM PETTY (1623–1687)

At about the same time as Sydenham and Locke were writing *Anatomia* and *De Arte Medica*, John Wilkins, an Oxford virtuoso and a founding member of the Royal Society, was creating a "universal language" that he hoped would be a better reflection of the natural world than European languages.[78] Wilkins thought that replacing Latin with a new language would reduce sectarianism by removing ambiguity in the understanding of Biblical texts and might also enable scientists who spoke different languages to communicate with each other about natural phenomena.[79] He published his first attempt as *An Essay towards a Real Character, and a Philosophical Language* in 1668.[80] This work also unintentionally showed how contagionism shaped the understanding of acute diseases.

In order to create a "philosophical" language, that is, one where words represented an underlying relationship between things, Wilkins first needed a taxonomy of conceptualizations, like a filing system for ideas: a "regular enumeration and description of all those things and notions to which things are to be assigned." He employed the young Cambridge naturalist John Ray to produce a botanic classification according to his predetermined schema.[81] Ray chafed at trying to fit a catalogue of plants into this system, but the experience led him to refine his own ideas about classification and the characteristics of species.

Wilkins discusses diseases as categories in chapter 8 "Concerning the Predicament of *Quality;* the several Genus's belonging to it, namely I. *Natural Power.* II *Habit.* III. *Manners.* IV. *Sensible Quality.* V. *Disease; with the various Differences and Species under each of these.*"[82] He distinguishes between notions of the "causes of disease" and notions of "the diseases themselves," whether they were "peculiar to some parts" or "common to the whole body," recalling Fernel's diseases of the whole substance.[83] Wilkins divided the general causes of disease into intrinsic or extrinsic. Intrinsic causes included errors of the humors, plethora, ill humors, distemper, and inflammation, and the obstruction of parts and vessels. Extrinsic causes arose either from violence or from "other bodies of a malignant dangerous quality," including poisons and "insensible Effluvia."[84] Diseases of the whole body could be either noninfectious or infectious. Noninfectious diseases included diseases of parts of the body, usually accompanied by wasting, and those seated in the humors from "some putrid matter causing preternatural heat." These included both intermittent and continued fevers, hectic fevers, and consumption.

Wilkins then listed the infectious diseases that were due to extrinsic putrid effluvia. The French pox was an infectious disease caused by sexual contact. Other effluvia caused infectious diseases that were usually accompanied by marks on the skin. Wilkins divided them into "Spots in the skin," "Breakings out in the skin," and "roughness in the skin." Diseases that caused "Spots in the skin" included malignant fevers (synonyms: spotted fever or purples) and the plague (synonyms: pestilence, pest, pestiferous, pestilential, the sickness, murrain). Diseases that caused "Breakings out in the skin" were subdivided into those that were more dangerous (smallpox and measles) or less dangerous (itch and tetters, including ringworm and shingles). Diseases that caused roughness in the skin were leprosy and "scurf."[85]

Wilkins was thus classifying syphilis, the plague, malignant fevers (but not continued fever), smallpox, measles, the itch, tetters, leprosy, and scurf as infectious diseases caused by a contagion that entered the body from outside. Among noninfectious diseases were intermittent and continued fevers, gout, erysipelas, and tumors (including scrophula), cancer, carbuncles (including plague sores), wens, warts and chilblains, catarrh (in the category of diseases of the head), consumption (a disease of the middle region of the body), scurvy (a disease of the spleen), and diarrhea and dysentery (diseases of the guts).[86]

Wilkins considered "contagious" and "infectious" to be synonyms. Under "Contagion" he listed concepts such as "Infection, taint, catching, run, spread, diffuse."[87] His classification includes some redundancies. For example, plague is classified as an infectious disease due to external effluvia, but the carbuncle or plague sore that signified the plague appears in an unrelated category of diseases characterized by a swelling of the parts. This problem of nonexclusive categories would dog many later efforts to classify diseases. Wilkins uses the words "genus" and "species" to refer to ideas of distinct kinds of disease, but he doesn't clearly apply these words to diseases themselves. Each "genus" had its own kinds, or divisions into species, which then had its further kinds or divisions into further species.

Wilkins's list of contagious diseases is similar to other early modern lists, such as Sennert's. He groups these diseases etiologically, by their kind of cause, and he views infection as one of those causes. Unlike Sydenham, he does not group diseases by their date of occurrence or the place or sort of place where they occurred.[88] He does not discuss where the putrid effluvia originate or whether they are animate, particulate, chemical, or aerial. The absence of any reference to contemporary theories of disease may be due to the concision of his classification or the fact that Wilkins was not medically trained, but it may also reflect Wilkins's goal of achieving a "universal" account of nature that would command universal agreement. In other words, Wilkins sought a noncontroversial taxonomy.

Contagionism also influenced the work of William Petty, a founder of the Society, who had studied medicine at Oxford and Holland and served as physician general to Cromwell's army in Ireland.[89] In July of 1665, Petty, with Hooke and Wilkins, had escaped the plague by staying in Durdans, near Epsom.[90] In a letter to his cousin in 1677, Petty, an avid reader of Hooke's *Micrographia*, commented that animals became stronger as they became smaller. The most powerful armies on earth were insignificant compared to the millions of invisible animals that traveled from country to country, even from Africa to England, and caused the plague, advancing and retreating in a day.[91]

Petty hoped his "Political Arithmetic" would cast new light on health issues. In his *Observations upon the Dublin Bills of Mortality* (1681), he recommended improving the information included in the bills and included sample forms for collecting data.[92] The weekly forms he designed included not only the plague but other contagious diseases in every parish with separate columns for "Spotted Fever, Measles, Small Pox and Plague."[93] In the monthly bill, he combined these into a single column (contagious diseases) and added two columns for "Sudden Death, Quinsey, Plurisie, Fever" (acute diseases), and "Stone, Gout, Dropsie, Consumption" (chronic diseases). His forms for quarterly reports reduced diseases to three categories: contagious, acute, and chronic, "in order to know how the different Situation, Soil, and Way of living in each Parish, doth dispose Men to each of the said three Species." The proposed form for an annual bill included a list of "Casualties and Diseases" under 24 heads, such as "Palsey, Consumption and French Pox, Fever and Ague, and Scowring, Vomiting, Bleeding."[94]

Thus, Petty, like his friend Wilkins, considered spotted fever, measles, smallpox, and plague to be contagious diseases, and this belief dictated his new classification, though it is not clear whether he also viewed diseases in the other columns of the bills as potentially contagious. Unlike Wilkins, he applied these ideas directly to a plan to collect and use medical information. His colleague and fellow arithmetician William Graunt, on the other hand, believed that epidemics were caused by bad air and the season or climate. Graunt discussed the difference between acute and chronic diseases but made no effort to separate contagious diseases as Petty had. His list of 18 "notorious diseases" is set out in the traditional alphabetical order.[95]

CONCLUSION

Although several members of the Royal Society mentioned the topic of contagion, and even indicated avenues for additional investigation, they never coalesced into a research community that focused on it. They probably found the subject interesting but did not give it a high priority. What they did find important was the effort to develop a new way to conceptualize and order all natural phenomena, including illness. This effort consumed the energies of some gifted Fellows over a long period, and they believed that the models that they developed superseded the traditional system, which was based on Galenic elements.

Despite this, many medical authors, including Sydenham, Boyle, and Willis, continued to understand illness as an event that depended on the interaction between a particular individual temperament and the qualities of the atmosphere. This picture of disease as a varying interaction between extremely particular factors (the patient's own unique physiology) and extremely general factors (the atmosphere, which affected everyone) left little space for an intermediate level in which differences between acute illnesses might depend on differences among the agents that caused them. Although they were translated from four Galenic qualities first into three chemical "principles" or "ferments" and then into "corpuscles," their presumed agents of diseases did not have unique, individual natures, and so the diseases that resulted from these undifferentiated agents had to be matters of quantity, not kind.

A belief that diseases were contagious implied that the matter being transmitted from person to person in a continuous chain could multiply itself or generate additional matter through a vital process. Disease theories based on transmission by inanimate particles alone could easily accommodate miasmatism, but contagion was more resistant. Because miasmatism assumed that diseases that co-occurred in the same time and place shared the same cause and that diseases that occurred at different times and places did not share a common cause, miasmatism discouraged the development of an ontological theory of disease.

Helmontian medicine, on the other hand, saw diseases as alien invaders, each of which had its own nature. It did not have to reject corpuscular matter theory to do so because it was based on a monist view in which matter was integrally combined with a vital agent in all living entities. Although contagion was not necessary to this idea of disease, it was not inexplicable. Some Helmontians wrote about contagion; others wrote about morbific agents that were ubiquitous.

Boyle, as we have seen, wrote that disease agents might have some sort of "propagative or self-multiplying power" similar to that of a ferment. His work and the comments by Sydenham on the plague suggest that both men were searching for explanations that might bridge the gap between mechanistic miasmatism and Helmontianism, but neither seems to have felt he had achieved a definite solution. Sydenham's taxonomy reveals that his approach was fundamentally miasmatic.

Wilkins and Petty do not seem to have struggled with the problem of harmonizing disease theory with matter theory but merely to have borrowed their disease concepts from elsewhere. Historian Barbara Shapiro remarked that Wilkins was unconcerned about the religious implications of mechanism and did not attempt to resolve the inherent conflict between providence and scientific regularity.[96] His approach to disease seems to reflect a similar unconcern with tracing illnesses to fundamental causes; it may simply capture a lay taxonomy and a lay vocabulary, or it may anticipate what historian Stephen Gaukroger has termed a "horizontal" or phenomenological approach to natural philosophy that uncoupled empirical findings from underlying mechanist theories.[97] Gaukroger traces the development of this approach to their colleague and collaborator John Locke. Whatever the cause of this uncoupling, it enabled Wilkins and Petty simply to assume that some diseases were contagious and thus to classify them as individual entities.

Table 4.1 Excerpts from Wilkins's List of Diseases

Causes of Disease

  Intrinsic:
  Errors of the humors
  Plethora
  Ill humors
  Distemper
  Inflammation
  Obstruction of parts
  Obstruction of vessels

 Extrinsic:
  Wounds
  Bruises
  Poison
  Contagion

Kinds of Diseases

 Diseases of the Parts
  Gout
  Erysipelas
  Tumors
   Scrophula
   Cancer
   Carbuncles
   Wens
   Warts
   Chilblains
  Catarrh
  Consumption
  Scurvy
  Diarrhea
  Dysentery

 Diseases of the Whole Body

  Noninfectious
   Diseases of the habit of the body; wasting
   Diseases of the humors from putrid matter causing
    Preternatural heat [fevers]
     Intermittent fevers
     Continued fevers
     Hectic fevers
     Consumption
  Infectious, caused by extrinsic putrid matter

  Spotted diseases
   Spots in the skin
    Malignant fevers (purples) [typhus]
    Plague (pestilential fevers, murrain)
   Breakings out in the skin
    Smallpox
    Measles
    Itch
    Tetters
     Ringworm
     Shingles
   Roughness in the skin
    Leprosy
    Scurf
  French pox

Source: John Wilkins, *An Essay towards a Real Character, and a Philosophical Language* (London: 1668)

CHAPTER 5

ANIMALCULES AND ANIMALS

INTRODUCTION: THE ROYAL SOCIETY AND THE ROYAL COLLEGE OF PHYSICIANS AT THE TURN OF THE CENTURY

In the early eighteenth century, the College of Physicians tightened its membership rules and procedures. As the Fellows admitted by James's charter began to die out, the College slowly became more homogeneous and less fractious. Hans Sloane, one of the intruded Fellows, proved to have remarkable diplomatic skills and became President of the College in 1719. He was re-elected annually until 1735, the longest presidency that the College had ever seen. From 1727 to 1741, he also served as President of the Royal Society.[1] Despite Sloane's prominence, in the 1720s and 1730s, the College was dominated by physicians trained at Cambridge and Leyden who shared a common approach to medicine.[2] After 1740, the rise of the Scottish medical schools would create a chasm between the increasingly homogeneous Fellows and the graduates of Scottish universities, most of whom were limited to practicing as Licentiates.

Unlike the College, the Royal Society was open to any serious male scientist of good family and reputation regardless of religious allegiance. Doctors constituted a significant proportion of the Society's members who worked for a living, with an increasing number of surgeons joining the physicians and a handful of apothecaries as the century progressed.[3] The Society also contained a significant body of aristocrats and wealthy gentlemen who were not working scientists. This does not mean that they contributed nothing to the Society. Many had patronage at their disposal; this made membership of the Society more attractive to those hoping for opportunities, clients, patients, or positions. In addition, many had completed their education with a "grand tour" on the continent, gathering an international roster of friends and correspondents.

One advantage the Society held over the College was readier access to a large network of potential correspondents throughout the world.[4] This proved to be critical to the development of Enlightenment science and played an important role in shaping eighteenth-century contagionism. The desire to communicate about diseases with distant correspondents who often did not speak English forced British medical researchers to address issues of classification, the identity of diseases, and the meaning of similar manifestations of illness in patients separated by time and space.

Some members of the Society believed that the Baconian natural history pursued by Fellows such as Tyson and Sloane was degrading to a man of learning. The naturalist Martin Lister wrote Tyson, praising his work "though some of late of our people have bespatted this kind of studie; as despicable and unbecoming a Physician."[5] By 1689, when Sloane returned from Jamaica, disagreements within the Society and the inactivity of its officers had left the Royal Society in a state of collapse, inducing him to found his own club of botanists at the Temple Coffee House..[6] That seems to have been the nadir of the Society; Sloane himself became Secretary in 1693. His work and the active presidency of Isaac Newton (1703–1707) helped restore the Society's finances and reputation, although continued friction between the mathematicians and the naturalists was reflected in hostility between Newton and Sloane.[7]

If it had been possible to poll the medical members of the Royal Society in the early eighteenth century, it is likely that a majority of them would have favored the neo-Galenism taught by Hermann Boerhaave in Leyden, but its members held a wide range of ideas about disease. Some members of the Royal College of Physicians were also open to contagionism; after all, Sloane himself, who would play an essential role in the introduction of inoculation, was the President of the College before he became President of the Royal Society.[8] The Newtonian James Jurin, who would support the campaign for inoculation by gathering and publishing statistical information about its result, was also active member of both organizations.[9] Nevertheless, as a body, the College would be far less hospitable than the Society to inoculation.[10] The College was also hobbled by the fact that its governance rested mostly in the hands of eight Elects who held office for life. This ensured that the College was controlled by its older members, who were often the most conservative.[11]

The structure of communication within the profession also affected the nature of knowledge production. College Fellows were busy men who had little to gain from writing for publication. Their ability or desire to support medical research was limited, and much of their communication took place within an oral, face-to-face culture, training their own students, consulting with colleagues, and meeting for discussions in the College itself. Before 1768, the College held endowed lectures, but there was no regular journal or publication program to disseminate medical papers. At the end of the seventeenth century, even the College lectures nearly had lapsed.[12]

The Royal Society did have a journal, the *Philosophical Transactions*, but it also limited its coverage of medical topics. In 1723, Jurin wrote to Anton Deidier, a Montpellier physician who had published an article on the plague in the *Transactions*. Jurin wrote that Sloane had "strongly approved" of Deidier's works on tumors and on venereal disease, but "it would not be in accordance with our custom to publish tracts of quite this sort in the *Philosophical Transactions*, which are intended rather for scattered pages reporting experiments and observations which otherwise would easily be lost."[13]

It is clear that Jurin was not being disingenuous because he and Sloane then jointly referred Deidier's works to the Royal Society's own bookseller and publisher, Samuel Palmer, who printed both tracts in a single volume in 1724.[14] In addition, Sloane himself nominated Deidier for Fellowship. When Deidier sent the Society his complete *Treatise on the Plague* in 1725, however, Jurin could not arrange for publication, although he had read the entire synopsis to the Society. He confessed that he was unable to publish medical works in the *Philosophical Transactions* regardless of their merit if they discussed medical ideas instead of experiments or observations:

> Dissertations on medical subjects are not very acceptable for the most part to our Noblemen, who comprise the greater part of the Royal Society, because they are regarded as hardly relevant to the purposes of the Society—if one may except singular Observations and Experiments such as your highly praiseworthy recent ones on Plague . . . Theses which relate to the Theory of Medicine, particularly those which more or less depend on Hypotheses, are regarded, as I have said, as less appropriate to the purposes of our Society.[15]

Thus, the *Philosophical Transactions* would publish medical "experiments" or data if papers were short and descriptive rather than theoretical. Indeed, between 1720 and 1779, the *Philosophical Transactions* published some 452 articles classified by one historian as "medical."[16] Inoculation met this criterion because its outcomes could be expressed in a few sets of figures, but longer articles about the nature of smallpox itself or about medical philosophy did not, though the Society may have discussed such topics at its meetings.[17]

The Royal Society, as a research institution with no responsibility for the welfare of patients, could afford to be adventurous, whereas the Royal College, as a licensing body conducting regular examinations, was forced to establish standards of ethical medical practice and ensure that young doctors did not stray too far from traditional norms. Those norms were still based on classical ideas that left little room for contagionism.

BIOLOGY AND CONTAGIONISM IN THE LATE SEVENTEENTH CENTURY: LEEUWENHOEK, SLARE, REDI, BONOMO, AND CESTONI

Although several Fellows of the Royal Society had considered animate contagion, particulate contagion, and microscopic "animalcules" that might be contagious, they published very little on such topics. The Society did, however, publish the letters of Antoni van Leeuwenhoek, a Delft tradesman who began writing to the Royal Society in 1673, and its participation gave his work both attention and credibility. In his eighteenth letter to the Royal Society, dated 1676, Leeuwenhoek wrote of seeing minuscule *animalcula* in rainwater and peppercorn infusions.[18] Hooke was so intrigued by Leeuwenhoek's findings that he learned Dutch in order to read his letters.[19] In 1677, Hooke told the Society that he had reassembled his microscope in order to confirm Leeuwenhoek's observation of "incredibly small" creatures in pepper water. These may have been bacteria. Leeuwenhoek's thirty-ninth letter, dated 1683, gave an unambiguous account of bacteria.[20]

Leeuwenhoek never suggested that the *animalcula* that he saw with his lenses were harmful to humans, considering them to be commensals. In 1677, however, a medical student named Johan Ham van Arnhem came to him with a sample of semen from a man suffering from venereal disease.[21] Hamm had observed the sample with a lens and saw animalcules in it that he thought were due to putrefaction in the fluid. Leeuwenhoek, however, correctly identified them as spermatozoa. This spectacular discovery and the ensuing controversy over generation diverted attention from the discussion of the relationship between animalcules and diseases that had brought Hamm to Leeuwenhoek in the first place.[22]

Leeuwenhoek corresponded with members of the Royal Society for fifty years, (1673–1723), and several Fellows visited him, including the Irish epidemiologist Thomas Molyneux, the physician Alexander Stuart, the apothecary James Petiver, and Hans Sloane, who traveled to Delft in 1700. Leeuwenhoek became a Fellow in 1680. Some officers of the Society, including Henry Oldenburg, Nehemiah Grew, and Robert Hooke, welcomed his correspondence, but during the periods when Edmond Halley controlled the *Philosophical Transactions* (1686–1693 and 1714–1721), his letters were not published.[23] Perhaps Halley, an astronomer and ally of Newton's, shared the latter's enmity toward Robert Hooke and this hostility spilled over to Leeuwenhoek. Once Sloane became Secretary in 1693, Leeuwenhoek began writing to him; from 1702 to 1712, he wrote to Sloane exclusively, and Sloane published every letter.[24]

We have a hint that members of the Royal Society were considering a theory of *contagium animatum* in light of Leeuwenhoek's work from a postscript that Frederick Slare added to a letter that he brought to the Society in 1683. The letter, from Dr. Wincler, chief physician of the Prince Palatine, concerned an epizootic of "murrain" among local cattle. Wincler himself believed that the outbreak was due to poisonous vapors from the bowels of the earth, but Slare, noting Wincler's observation that "cattle secured at rack and manger were equally infected with those in the field," speculated that the murrain might be carried by a volatile insect that could only make short flights.[25] Slare expressed the wish that Leeuwenhoek had been able to attend the dissection of the infected cattle, hinting that he might have found microscopic organisms. His suggestion would later be cited in *A New Theory of Consumptions* by Benjamin Marten, who was evidently an avid reader of the *Transactions*.[26]

Slare personified both the cosmopolitan world of evangelical Protestant-ism and the transition from Helmontian "chymistry" or alchemy to skepti-cal chemistry. His father, Frederick Schloer (DD, Geneva), an immigrant from the Palatinate, was the minister of a Calvinist parish in Dorset.[27] Slare had an MD from Utrecht but was trained at Leyden. His lifelong friend-ship with Hans Sloane dated from the years when Sloane was lodging with the apothecary and chemist Nicholas Staphorst.[28] He was also one of the most important English patrons of the Halle Pietists, sending many boys to Halle for education, supporting English charity schools inspired by Halle's example, and providing lodgings in his own home for the Pietist preacher William Boehm.[29]

Slare became an assistant to Boyle in the 1670s. Hooke brought him to the Royal Society in 1679 to demonstrate Leeuwenhoek's experiments on spermatozoa. He became an FRS himself in 1680.[30] In February of 1682/3, Slare and Edward Tyson had been jointly named curators at the Royal Soci-ety: Tyson to prepare anatomical demonstrations and Slare to provide chemi-cal experiments.[31] By 1692, after a long series of chemical experiments, Slare concluded that Nedham had been wrong to believe that diseases entered the body when particles of air were ingested with food and drink. Instead, he wrote, "Contagious Diseases are Communicated to the Blood by Inspira-tion into the Lungs."[32] This is as notable for its tacit acceptance of the rest of Nedham's argument as for its correction of one portion. His last publi-cation was a defense of inoculation against the criticisms of William Wag-staffe in 1725.[33] Two years later, Slare wrote a letter implying that physicians opposed inoculation because they lost income when they had fewer gravely ill patients, suggesting that he shared Mary Wortley Montagu's jaundiced view of the profession.[34]

In 1668, Francesco Redi (1626–1697) had cast serious doubt on the the-ory of spontaneous generation (at least of maggots and worms), initiating a debate that would be temporarily resolved in favor of antispontaneism by Louis Joblot in 1718.[35] Redi had observed the life cycle of the housefly and shown that it laid eggs on meat, which developed into maggots and flies. He also showed that maggots did not appear in meat that was covered with a cloth too fine to admit flies. Two of Redi's associates: the physician Giovanni Cosimo Bonomo and an apothecary and microscopist, Giacinto Cestoni, col-laborated to argue that the skin disease scabies was due to a mite.[36]

Thomas Moffet, a Paracelsian, had first seen the itch mite under a magni-fying glass in 1590 and described it in his *Insectorum sive Minimorum Ani-malium Theatrum* of 1634, which was translated into English as the *Theatre of Insects* in 1658.[37] This work, compiled by Moffet from the work of other naturalists and his own observations, had been retrieved, edited, and pub-lished after Moffet's death by the Paracelsian court physician Theodore de Mayerne, who added a preface.[38] In 1657, August Hauptmann, who had attributed all human and animal diseases to microscopic "insects," "worms," and "acari," published a very rough sketch of the acarus (itch mite) he had observed with a microscope.[39] Michael Ettmüller would publish a more defined picture and a description of itch mites, which he called "sirones" or "cirones" in the *Acta Eruditorum* in 1682. His article received little attention in England, but Andry would reprint the plate.[40]

In a letter to Redi, Bonomo summarized earlier theories of the itch:

> This contagious Disease owes its origin neither to the Melancholy Humour of Galen, nor the corrosive acid of Sylvius, nor the particular Ferment of Van Helmont, nor the Irritating Salts in the Serum or Lympha of the Moderns, but is no other than the continual biting of these Animalcules in the Skin . . . from here we come to understand how the Itch proves to be a Distemper so very catching; since these Creatures by simple contact can easily pass from one body to another . . . and a very few of them being once lodged, they multiply apace by the Eggs which they lay. Neither is it any wonder if this infection be propagated by the means of Sheets, Towels, Handkerchiefs, Gloves, etc. used by Itchy persons.[41]

Although it was known that the victims of the itch had small "worms" on their skin, it had been assumed that these were the result, not the cause, of the ailment. Spontaneous generation enabled the acari to "breed" in the putrefying humors caused by the ailment. Bonomo inverted this argument, showing that a living parasite caused the symptoms of the disease. He also explained how it was transferred from person to person, causing fresh cases.[42] The letter appeared in Italian in 1687, but, like Ettmüller's work, it was not widely known in Britain until Richard Mead published an abstract and translation in the *Philosophical Transactions* in 1702/3.

Bonomo's letter did not completely settle the issue. Mites cause scabies by burrowing into tunnels under the skin, irritating it and causing it to react by creating fluid-filled vesicles that eventually fill with pus. Mites are only present in these vesicles very early in their formation. Although Moffet had clearly described the presence of the mites in their tunnels, Bonomo and Cestoni only spoke of finding them in the vesicles. Later microscopists searched the fluid from these vesicles without finding any mites, raising doubts about the whole chain of argument among scientists hostile to theories of *contagium vivum*.[43] By the early nineteenth century, French physicians were so dubious about the idea of specific contagious diseases that they actually "lost" the itch mite. A student who had described and demonstrated it in a thesis of 1812 was thought to have perpetrated a scientific fraud.[44]

DISPUTING SPONTANEOUS GENERATION IN ENGLAND: JOHN RAY AND HIS ASSOCIATES

The argument against spontaneous generation was popularized in England by John Ray in his extremely influential and popular work, *The Wisdom of God Manifested in the Works of Creation* (1691), which confronted atheism by showing that the great extent and complexity of the universe required a divine designer and creator.[45] Although pious, Ray was no fundamentalist: he contended that the Bible had been designed to accommodate the limited understanding of its early readers.[46] Ray devoted a large section of Part 2 of this work to refuting the belief in spontaneous generation. He viewed this section as essential to his argument from design because if lower animals could be generated spontaneously there would be no theoretical bar to the claim that men also arose spontaneously and did not require divine creation.[47] Although some authors had opposed Redi by pointing to the existence of worms that appeared to be bred in the intestines of man, Ray believed that they originated in "seed" and added, "moreover, I am inclinable to believe, that all Plants too, that themselves produce Seed . . . come of Seeds themselves," citing the Italian physician and naturalist Marcello Malpighi.[48]

Some people, he argued, believed that lice bred on children's heads and that mites and maggots bred in cheese, but in fact all such creatures came from eggs laid by insects that were attracted to these places.[49] The intestinal worms probably entered the body when their eggs were swallowed with food.[50] Finally, Ray noted, if spontaneous generation were possible, new monsters would be constantly created. This problem led even Lucretius to argue that all species must come from seeds of the same species. "Some," he commented, "have been offended at my consistent Denial of all Spontaneous Generation, accounting it too bold and groundless," but his position was supported by Malpighi, Redi, Leeuwenhoek, the Dutch microscopist Jan Swammerdam, and many others, including his friends Martin Lister and Tancred Robinson.[51]

Ray's argument shows that he thought that the theory of spontaneous generation was still widely held and required detailed refutation.[52] His references to other members of his circle suggest that they were all collaborating on this project. Ray also revealed his belief in the fixity of species and his adherence to vitalism, a viewpoint expressed also in Nehemiah Grew's *Cosmologia Sacra*.[53] All living things, he believed, shared a vital force that was unique and could only have come from God. It was not inherent in matter because matter could not organize itself.

A rejection of spontaneous generation was also essential to Ray's system of classification because it meant that new individuals could only come from like parents. Rejecting the traditional classifications of plants by alphabetical order, use, or location, he developed a "biological" concept of a "species" as a group whose members were determined by common reproduction. A species of animals contained those animals that could generate fertile offspring together and a species of plants included the kinds of plants that grew from the seed of a single parent plant.[54] Although he used fructification and procreation as a guide for classification, he was prepared to modify this principle when it seemed to conflict with obvious natural groupings.[55]

EDWARD TYSON (1650–1708) AND PARASITIC INFECTIONS

Edward Tyson obtained his MD from Cambridge in 1673 and moved to London in 1677.[56] Once in London, Tyson, who had a consuming interest

in anatomy and zoology, became an intimate friend of Robert Hooke and
was soon submitting anatomical observations, often aided by a microscope,
to the Royal Society. Although he was a Fellow of the College of Physicians, Tyson was a reformer and frequently a troublesome member, "the
great solicitor, and Fomenter of the Differences and Divisions in the College" as the aggrieved College registrar later complained.[57] After the Glorious Revolution, he was among a minority of Fellows opposed to an effort
to salvage the charter granted to them by James II that had increased their
powers.[58] He was evidently sympathetic to the apothecaries in their struggle
with the College. After he had been elected Censor, he opposed efforts by
his colleagues to regulate misbehavior by medical practitioners in London
and stalled an effort to convict the Dutch immigrant physician and College
Licentiate, Johannes Groenevelt, of malpractice.[59]

A young Danish physician named Frands Reenberg visited Tyson and
wrote in his diary on January 10, 1679/80 that Tyson had told him that

> he had sometimes when letting blood found worms in the blood, as also in
> hairs of the body, and others in the intestines, and he was of the opinion that
> this was a cause of not a few diseases as in the plague, the itch, in the resulting
> pustules, bubonic swellings, ulcers of various kinds, gonorrhea (in the semen of
> this he had noticed that worms were found), veneral [sic] diseases, rottenness
> of bones, and that this threw light on the fact that [mercury] is the best remedy
> for worms in the intestines, and by virtue of the same was beneficial in venereal
> disease, and in gonorrhea.
>
> He had observed that each kind of tree or plant usually had a particular kind
> of worm in it, especially in those species where worms are found; thus the oak
> has its kind, the willow its kind . . . He thought that oak-galls and oak-apples
> develop in the same way in which swellings are caused by the sting of a nettle,
> by the injection of a drop of poisoned juice between the skin and the flesh—in
> the same way, the semen of insects . . . penetrates the surface . . . of the leaves,
> and remaining there between the layers separates them, and gradually rises into
> a swelling . . .
>
> In the dried blood of a seal which he dissected he observed several fibres,
> which he thought to be worms; from the muscles of the nostrils or nose he had
> squeezed out with his nails an oblong object (which is commonly believed to
> be worms), and by the aid of the microscope he had seen it by its movement
> to be a worm.[60]

Tyson never published these ruminations, but he opposed spontaneous
generation. His article on hydatids, published in the winter of 1690/1 after
a three-year delay, showed both experimentally and through microscopic
observation that these were a sort of "Worms or Insects *sui generis*" and were
plentiful in the bodies of "rotten sheep." He blamed this parasitic infection
for hydropsy in sheep.[61] In 1696, he commented at a meeting of the Royal
Society that hydatids were also frequently found in women suffering from
dropsy "in incredible quantities . . . but Contained in Severall Cists."[62]

Tyson never drew these ideas together, but he had certainly attributed diseases in plants, animals, and humans, including plague, the itch, ulcers, and

venereal diseases, to parasitic infections. Tyson's views were similar in some ways to Nedham's but far better grounded. They would have commanded more respect from his colleagues, but he was never inclined to theorize in print about anything beyond his anatomical observations.

NICHOLAS ANDRY DE BOISREGARD (1658–1742) AND HUMAN WORMS

Ray's belief that different species of minute harmful "worms" were ingested with food to cause disease was echoed by the extremely influential book of a French doctor, Nicholas Andry de Boisregard, published in 1700. In English it was entitled *An Account of the Breeding of Worms in the Human Bodies with Three Letters to the Author* (these were by Nicholas Hartsoeker and Georgio Baglivi).

Andry provided a detailed account of "worms," which can be understood as including animalcula, and described them as a cause of disease. In his "Maxims concerning worms in human bodies," he stated that

> Worms as all other Animals, come from Seed, which contains them in little. The Eggs of Worms enter our Bodies with the Air, and our Food, and ofttimes into our Flesh by the outside . . . When the Eggs of Worms are entred [*sic*] in our Bodys, the Worms shut up in those Eggs, breed, provided they find in us a proper Matter for making them breed . . . Since Worms are engendred of Seed, it's impossible there should be any new Species of them. The Air is full of the Seed of Worms, as are also Rain, Water, Vinegar, sour Wine, Stale Bear [*sic*], Cider and sour Milk . . . All the parts of the Body are subject to Worms . . . The Urin & Blood of those who have the small Pox, has Worms in it. The Pustules of the Small Pox are full of Worms . . . In the Venereal Distempers there's no part almost of the Body which is not gnaw'd with little imperceptible Worms; and it is those Worms that occasion most of the Ravages . . . A Physician ought carefully to examine with a Microscope, the Blood that he orders to be taken from his Patient, to see if there be any Worms in it . . . The Doctrine of Acids and Alkalis, ill understood, often hinders the timely giving of purging Medicines which would banish Worms.[63]

Andry saw the "worms" as constituting various species, and he attributed scabies, cancer, some cases of dropsy, tumors, deformities, and jaundice to their activities in addition to smallpox and venereal disease. He claimed that most illnesses that were attributed to witchcraft were actually caused by worms, and he also suggested examining the milk of wet-nurses with a microscope to make sure it was free from worms, but he did not associate specific sorts of worms with specific diseases. Nor did he discuss the implications this theory might have on ideas about different mechanisms for disease transmission: as with Nedham, he saw the worms as omnipresent. Moreover, Andry did not distinguish between the large worms, such as tapeworm and roundworm, which were visible with the naked eye, and microscopic "worms." Within a generation, Andry's book, which found a receptive audience in England, was to become a source for the contagionist work of Benjamin Marten.

THE RINDERPEST EPIZOOTIC OF 1711 IN ITALY: RAMAZZINI, COGROSSI, VALLISNERI, LANCISI

At the turn of the century, most physicians attributed illness to an internal mechanical dysfunction. The very influential Leyden Professor, Hermann Boerhaave (1668–1738), who taught nearly two thousand medical students after 1700, believed that the body was composed of a series of microscopic and submicroscopic vessels, through which increasingly rarefied fluids circulated. Disease resulted from the lodging of improperly assimilated particles in these vessels, where they obstructed the circulation, causing friction, heat, and swelling: the symptoms of the many sorts of "fevers," which all resulted from the same physiological process.[64] Despite the popularity of this theory, the belief that some diseases were caused by specific entities survived the turn of the century.

European interest in contagionism revived when an outbreak of rinderpest, a viral disease of cattle, spread to Italy from Dalmatia in 1711.[65] The history of this epizootic has been recounted in detail by Lise Wilkinson and C. A. Spinage.[66] Outbreaks of cattle diseases caused panic in communities that depended on local markets for fresh meat, milk, and cheese. Furthermore, many Europeans relied on oxen to draw carts and plow fields, so rinderpest also threatened transportation and grain supplies. Cattle epizootics portended widespread hunger, the ruin of farmers, and the devastation of both rural economies and urban markets.

The nature of the disease immediately at hand often favors the development of particular sorts of explanations for disease transmission, which are then applied to other ailments. Rinderpest favored the development of contagionism. First of all, the disease was in fact contagious; like measles, to which it is closely related, it spreads primarily through droplet infection from the breath of animals near one another.[67] Moreover, it has a comparatively short incubation period and latency, facilitating the identification of contacts between sick animals. The spread of the epidemic could actually be traced to certain fairs and markets, or in some cases, to particular animals.

Second, it was clearly specific, at least insofar as it only appeared to affect a particular kind of animal.[68] Authors commented on the relative immunity of farmers and herders: this discouraged explanations that relied only on a general atmospheric or waterborne poison. Third, it was much easier to trace and control the movements of cattle than of humans. Herds of cattle are relatively discrete entities, and the moments when they mingle with other herds or with cattle coming from a distance are usually limited and well remembered. A given practitioner was called in to view the entire herd, whereas when a human disease began to spread through a city, recognition of an epidemic might be delayed by recourse to a large number of different practitioners. Fourth, there were fewer compunctions about isolating diseased animals than quarantining ailing people; dissection was common and aided in more definitive diagnosis, and the slaughter of the affected animals offered a practical (if appallingly expensive) prospect for control.

Finally, cows lead relatively routine lives, and when they fall ill the treatments applied are somewhat limited (at least by expense), although they included many remedies routinely used on humans, including bleeding, setons, cinchona bark, snakeroot, purging, tar-water, and fumigations, to name just a few of the more common ones.[69] Many of the explanations invoked for human diseases—excess indulgence, late nights, too much good/rich food, anxiety, evildoing (by the cattle), and divine wrath—could not be introduced to confuse the situation, although some did claim that the suffering of the cattle was divine retribution for human sins. Because the practices of animal husbandry did not change significantly from year to year, it was less likely that practitioners would look to dramatic changes in the way the cattle lived to account for the problem, though some blamed changes in the weather. In human society there was always some new activity, luxury, or vice to blame. To Europeans, rinderpest was manifestly a newly introduced ailment. It did not stem from some "constitutional" weakness or lack of self-control in the cattle, but it did seem to follow armies and their huge cattle trains.

As the disease spread through Northern Italy, the Venetian Senate turned to the physicians at the University of Padua. The Professor of Medicine, Bernardino Ramazzini (1633–1714), took up the challenge, attacking those who thought it was undignified to discuss animal diseases.[70] He made this disaster the subject of his annual lecture to the faculty and students and published it as a Latin treatise in 1712. It was republished several times, not only in Ramazzini's collected works but as a separate volume in Leipzig in 1713 and in an Italian translation in 1748.[71]

Ramazzini noted that the illness did not affect other animals or man, and so could not be caused by some general poisoning of the air or pastureland. He rejected astrological arguments. He thought oxen fell ill by inhaling an infectious particle because the seeds of infectious diseases often multiplied and propagated if they found a suitable subject.[72] Ramazzini did not view these seeds as animate.[73] Comparing the illness to human eruptive diseases such as smallpox, Ramazzini decided syphilis, brought to Italy by a few sailors, offered the best analogy for rinderpest, as both diseases had spread quickly throughout Europe from a small beginning. He thought it could be controlled by cleanliness, the isolation of infected animals, and the fumigation of their stables.[74] He commented,

When we consider to what extent this present contagion has been spread through the secretions and the excreta of ailing and dead oxen, how stables and pastures have been contaminated to the injury of other oxen subsequently using them and how hides of dead oxen are polluted (for the infection can long persist in hair) I say that we most certainly must not be astonished that the infection has spread far and wide.[75]

Ramazzini's comments were echoed by the papal physician, Giovanni Maria Lancisi, best known for his writings on malaria, who wrote a short essay on the original outbreak in Padua. In 1713, as the disease continued to spread through Italy and reached the city of Rome, Pope Clement consulted him and adopted his draconian recommendations that all infected or suspect cattle be slaughtered. Despite the furious controversy this measure provoked, it succeeded in containing the epizootic. Two years later, in 1715, Lancisi published his "historical dissertations" on the cattle plague.[76]

The only part of Italy that avoided the epidemic was Tuscany, where the Duke of Tuscany ordered a military *cordon sanitaire* to protect his borders from the importation of cattle. At this date, governments seeking to avert human plague epidemics were also beginning to deploy lines of soldiers along their frontiers.[77] In 1712, the Elector of Hanover borrowed the idea to halt the spread of an apparent epidemic of plague that had been introduced into his territories from Hamburg—an experience that may have prompted the Hanoverian government in Britain to support Mead's recommendations to prevent plague in 1720.[78]

Whereas Ramazzini and Lancisi saw rinderpest as a contagious disease that was transmitted by some sort of poisonous substance, another Italian author unambiguously attributed it to a living entity. This was Carlo Francesco Cogrossi (1682–1769), who defended this "new idea" in an exchange of letters with his former professor, Antonio Vallisneri (FRS, 1662–1730).[79] Cogrossi, who had also been a student of Ramazzini, had recently graduated from Padua and was working as a doctor in Crema. Vallisneri was a Venetian liberal (historian Jonathan Israel calls him a "crypto-radical") and a founder of the Venetian *Giornale de'Litterati* in 1710.[80] In the same year, he published a work on the generation of worms in human bodies.[81] Cogrossi's correspondence with Vallisneri appeared in Milan in 1714 as *Nuova Idea del Male Contagioso de'Buoi* (New Theory of the Contagious Disease among Oxen).[82]

In his portion, Cogrossi referred to Leeuwenhoek, to Kircher, and to the work of Redi and Cestoni on the transmission of scabies. He wrote that although many believed that the disease was contagious, and was due to "a poison or fermenting substance," he himself believed that it was caused by tiny poisonous insects that could "penetrate into the most hidden recesses of animals," passing from one animal to another of a similar kind and entering the blood after passing through their throats, noses, or even their skin.[83] Cogrossi argued that the epidemic could be traced to a single Hungarian ox, whose "malignant ferment" had "infected with its effluvia" the entire region of Lombardy.

Admitting that if he spoke with some of his academic colleagues ("long-robed peripatetics") they would laugh at the idea that "animals so tenuous" could exist, Cogrossi pointed out that if these effluvia were merely "atoms," that is, particles, it was difficult to see how they could have been so numerous as to have infected such a large number of animals. But if the effluvia of the first infected ox had acted as a catalyst for the generation of more effluvia, it was easier to explain the proliferation of the disease. Although this might be explained chemically, Cogrossi himself was convinced that the pathogen was a living entity:

> There can be invisible insects from which the epidemic of oxen derives in the very manner in which the itch in man derives from the grubs which have been discovered on the skin . . . If these exhalations . . . were nothing but a heap of very agile and very tiny poisonous atoms, one must realize how numerous they were to be divided among so many infected and afflicted animals . . . The manifest and visible spread of the insects . . . suggests a . . . ready example for understanding the spread of the contagious disease. Only two of these insects carried into Italy . . . could in successive generations have prepared an innumerable host.[84]

The introduction of these "imagined mites" and their multiplication explained the spread of an epidemic without the need to invoke traditional astrological or environmental theories without "blaming poor and innocent Saturn, without accusing the extravagance of the seasons, the corruption of waters or of the pasture, without supposing that in the first ox to be afflicted this malignant ferment arose from an accidental heaping of fluids."[85]

Vallisneri replied that this theory was in accord with his own ideas.[86] He had just been waiting to put his views in writing until he finished the observations he had been making with Dr. Bernardino Bono. Using "very sharp microscopes," Vallisneri and Bono had seen that the blood of oxen was "filled with tiny little worms" that were a little larger than the worms found in semen.[87] Vallisneri agreed that living contagion offered the best explanation for the way epidemics spread and multiplied.[88] The time required for these worms to reproduce also explained why there was a period of incubation for diseases. He thought that they might remain latent in places where the circumstances did not favor their growth, but when food or air became damp or dirty they could multiply to such numbers that they became fatal.[89] Dr. Bono had also examined the excrement of a man with dysentery and found it to be "brimful of an infinity of tiny little worms which are different from the ordinary ones."[90] Cogrossi thought these little worms could be carried by the wind on little bits of straw, hay, hair, or grass.[91] They could also deposit their eggs in the throat and nostrils of oxen. These eggs explained why infected items could also spread contagion.[92]

The *Giornale de'Letterati d'Italia* reviewed the *Nuova Idea* together with a history of earlier works about *contagium vivum*.[93] Lancisi, the Pope's physician, also discussed Cogrossi's comments at considerable length in a Latin history of the cattle plague. Lancisi considered the theory "very probable but not certain," noting that he himself had only found these insects on the surfaces of bodies.[94] Redi had merely shown how quickly they could penetrate an unprotected sample.[95]

As Wilkinson has noted, Lancisi did not really consider the possibility that the organisms Cogrossi was postulating "philosophically" could be small enough to elude the resolution of existing microscopes, although it is clear from Cogrossi's comments that he believed they were very small.[96] In fact, because rinderpest was a viral disease, the pathogens were too small to see with any optical microscope.[97]

Lancisi himself, like earlier Italian physicians, attributed the epidemic to "a pestiferous ferment," consisting of tenuous corpuscles like the ones that caused bread to rise and wine to ferment, or like a viper's bite, which caused its victim's body to swell.[98] Whereas Cogrossi viewed rinderpest as a separate and distinctive illness uniquely caused by successive generations of a particular sort of insect, Lancisi did not clearly see it as a specific disease but rather as an ailment that might have several causes.

When Lancisi later wrote a book on swamp fever (malaria), he concluded that there was a need for further microscopic observations of the blood of victims but added that, even if "worms" were found in the drawn blood, it would be doubtful whether they had caused the disease or had been produced by the process of the disease, which "broke down the fluids," setting free the "minute ovules" that had been trapped in the blood.[99] In other words, he doubted that even the best microscopic observations could ever settle the question of cause and effect. Like Lancisi, most medical authors of the eighteenth century remained agnostic about the nature of pathogenic substances.[100]

Few English authors would have read the exchange between Cogrossi and Vallisneri directly, as it was published in Italian, and there was no English translation. However, readers interested in the cattle plague or in agricultural "improvement" could have consulted Lancisi's Latin history of 1715, which included Cogrossi's speculations as well as Lancisi's more moderate views. The English medical community had a keen interest in Italian microbiology.[101] We know that Richard Bradley read Ramazzini's work blaming rinderpest on infectious particles, because he referred to it in his own book on the plague. It is also likely that Richard Mead, who graduated from Padua in 1695, knew these works. Mead was fluent in Italian and was very interested in Italian research. The outbreak of the plague at Marseilles led to a revival of the discussion and a lively debate among Italian medical authors, who published many discussions of the theory between 1720 and 1733.[102]

RINDERPEST IN ENGLAND IN 1714: BATES

Although cattle disease may not have been an urgent professional issue to most English physicians, it was a major concern to landowners and the government, especially when it began to spread in Britain. When it broke out in London in 1714, George I asked his surgeon, Thomas Bates, to investigate.[103] In his account, written that year and published in the *Philosophical Transactions* in 1718 (the year he became an FRS), Bates recommended the separation of cattle into small "parcels" as soon as a single case was found, in addition to the kind of hygienic precautions also recommended by the Italian authors.[104] Bates believed that the disease had initially been caused by a drought, which had led to poor and less digestible grass in the spring, but he had no doubt that it was contagious. He commented,

> Contagious Diseases [are] very seldom infectious to different Species; but . . . Contagions may be communicated to the same Species, by touching the Woolen, Linnen, etc. to which the Infectious *Effluvia* of the Diseases had adhered, tho' the two Bodies should be at a very great distance; and I verily believe that more Hundreds died from Infection, which was carried by the Intercourse that the Cow keepers had with each other, than single ones from the original Putrifaction. The Nature of Contagious Diseases are but little understood.[105]

Bates recommended avoiding sick animals, destroying sick cows, and burying them at least six feet deep. He also recommended keeping new cows off the pastures of infected cows for at least two months and out of their stalls for three months and thorough cleansing of their barns. He suggested that the government offer compensation of forty shillings for each destroyed animal to encourage the slaughter of infected animals. The government followed his recommendations, and this epidemic was contained. The cost included fifty-eight thousand pounds sterling and six thousand cattle, but this was very small compared to losses elsewhere.[106] Europe as a whole lost an estimated two hundred million head of cattle to rinderpest during the eighteenth century.[107] At least half a million cattle died in the next British epidemic of 1745–1757, which revived interest in earlier works; contributors to such popular periodicals as *The Gentleman's Magazine* would cite Bates, Lancisi, and Ramazzini.[108]

As Wilkinson notes, efforts to control rinderpest "taught useful lessons in epidemiological control, especially with regard to isolation and quarantine."[109] The symptoms of rinderpest in cattle resemble the symptoms of measles in humans.[110] Rinderpest offered an obvious analogy with severe human diseases; if it could be controlled, so might human diseases such as the plague.

CONCLUSION

The publications of Leeuwenhoek, Redi, Bonomo, and Cestoni and their colleagues created a new context for eighteenth-century British contagionism. Natural philosophers had observed and described minute organisms and had shown that living agents might transmit a cutaneous disease. An increasing number of naturalists agreed that humans suffered from worms, mites, lice, and perhaps other ailments that came from the eggs of microscopic insects; that insects might cause deformities among plants through a "contagious vapour" emitted from their eggs; and that virtually all these organisms arose only from other organisms of the same kind and could not be spontaneously generated in some alien substance.

The rejection of spontaneous generation, however, complicated the issue of the nature of disease instead of resolving it. Jesuit authors such as Kircher had accepted the Aristotelian idea that some living animals were superior and propagated themselves through reproduction, whereas others, especially loathsome animals such as worms and maggots, were inferior and were spontaneously generated by filth.[111] The early modern scale of being from insects to humans was also a moral and spiritual scale. Disease could thus be viewed as the invasion or pollution of victims by a lower (vicious) life form or as the result of an internal corruption. Historian William Eamon has pointed to parallels between the belief that disease resulted from physical pollution or corruption and the post-Tridentine Catholic emphasis on the value of exorcism to purge demonic possession: "where it was often impossible to tell whether diseases were of natural or demonic origin, the figure of the exorcist and the charlatan merged."[112]

Many Calvinist authors, including Leeuwenhoek, on the other hand, resisted the idea that even tiny insects and microscopic animals were agents of corruption. They argued that God's providence could be found in the perfection of even the smallest beings and argued against the idea that any living being could be generated from filth or corruption.[113] If, however, animalcules were not the spawn of filth, it was harder to reconcile the view that animalcules or insects acted as agents of disease and death with the idea that all their species had existed since their creation by a provident deity.[114] The association of the idea of *contagium vivum* with Kircher and the Jesuits and the philosophical problems posed for disease theory by a rejection of the belief in spontaneous generation meant that many mainstream Protestant authors hesitated to blame internal insects for disease. They preferred to attribute contagion to inanimate forces, to leave the nature of the agent unspecified, or at least to depict insects as mechanical agents, biting and gnawing the body, not as inimical vital entities growing and multiplying within their host.

Moreover, the theory that small organisms caused a given disease did not necessarily imply that these organisms migrated from host to host. It was equally plausible to assume that they were omnipresent in the environment. As microscopic insects did seem to be omnipresent, the idea that they caused disease may actually have discouraged contagionism.

Contagionists, on the other hand, had no compelling reason to think that the agents of disease were animate. Animation offered the simplest explanation for the explosive growth of victims during an epidemic of an acute disease, but the boundaries between living and nonliving were still fluid.[115] Ferments and molds might spread and multiply, but their nature remained to be investigated, as did catalysts, poisons, and putrefaction.[116] The building blocks of disease theory could be assembled in many different ways, although the idea of disease specificity narrowed the range of evidence that a medical investigator had to consider in choosing between explanations for a particular acute ailment. As there are, in fact, a host of different diseases with many different sorts of causes, the problem was always to find the best match between the available evidence and the most plausible explanation.

Nevertheless, throughout the late seventeenth century and into the early eighteenth century, some naturalists did propose an animate origin for some diseases. The idea that more than a small number of diseases were contagious was probably still a minority view among university-educated physicians, but by 1714 the view that some diseases might be contagious was not disallowed. Sifting through their options in 1714, a few Italian authors believed that a theory of *contagium vivum* could explain the behavior of rinderpest in animals. English leaders agreed that the disease was transmitted by contagion and imposed an energetic quarantine to circumscribe it, although there is no evidence that they blamed the contagion on an animate agent. Just six years later, Benjamin Marten would publish a complete theory of *contagium vivum* for human diseases.

More important, acute disease entities would increasingly be understood as disease "species" in Ray's biological sense: experiences whose individual manifestations were related to each other by generation through time and space because a material "specific essence" was passing from person to person.[117] It would be reasonable to refer to this cluster of ideas as a "germ theory of disease"—especially as a "germ" is literally the inner, vital portion of a plant seed—but none of the naturalists discussed above published these ideas as a coherent, worked-out theory of disease. Instead, it was sometimes raised, only to be dropped, or it formed a background hypothesis for discussions of other topics, such as vital statistics, taxonomic theory, parasitology, anatomy, the chemistry of respiration, and generation.

During the long eighteenth century, the concept of "species" as a tool of formal classification would be articulated and then debated by many naturalists. Just as chemists developed tables of affinity and a new vocabulary that both reflected and transformed their conceptualization of the nature of elements, and natural philosophers created new classifications for plants and animals, so new medical taxonomies reflected and transformed ideas about the nature of diseases.[118] The discussion about how plants and animals should be classified into groups or kinds, and the uncertainty about how artificial classification was related to the "natural" order of the world, interacted with the idea that diseases might also be conceptualized as forming different species—especially when they were attributed to a living entity or at least to a certain kind of matter.[119] Medical authors and teachers struggled to delineate the essence of a disease by analogy with the essential vital force that many believed was transmitted during the generation of plants and animals and shaped them into certain predetermined forms.[120] Contagionism played a role in that effort, but one that bore fruit only gradually.

CHAPTER 6

ENGLISH CONTAGIONISM
AND HANS SLOANE'S CIRCLE

INTRODUCTION: SLOANE'S EARLY YEARS
AND ASSOCIATION WITH SYDENHAM

At the center of every medical web in the early eighteenth century was Hans Sloane (1660–1753): Fellow (1685), Secretary (1693–1713), Vice-President (1712–1727), and finally President (1727–1741) of the Royal Society; and Fellow (1687), Censor, Elect, and President (1719–1735) of the College of Physicians; the only person who has been President of both the College and the Society. Sloane also held many other offices, including Physician Extraordinary to Queen Anne, Physician-General to the army under George I, and First Physician to George II.

Botany, especially medical botany, was Sloane's greatest scientific interest. Sloane played an important role in the introduction of cinchona into widespread medical use and was largely responsible for the fourth edition of the *London Pharmacopoeia* (1724), the first to contain a list of "simples." His role in the introduction of inoculation for smallpox and his (unpublished) advice on the plague will be discussed below. Sloane's circle was nearly coterminous with the circle of contagionists; examining his own biography and the nature of his coterie enables us to place contemporary contagionism in a specific political, religious, and scientific context: one that embraced both religious and cultural outsiders, appreciated "vulgar Baconianism," and remained distinct from both Galenism and mechanism.

Historians have depicted Sloane as a pillar of the scientific and medical establishment: a "Court Whig" who had little personal involvement in medical reform beyond his advice on inoculation. Sloane's social tact and skill are unquestionable, as is his high status in medicine and science. He could not have been effective if he had not excelled at climbing the social ladder. However, he also quietly served as a patron and encourager of many Dissenters,

Nonconformists, medical reformers, and contagionists. It was Sloane whose extensive correspondence, tact, and determination swayed the balance over inoculation; Sloane, who together with Richard Mead and John Arbuthnot, advised the Privy Council to maintain a contagionist plague policy; and Sloane who asked the Royal Society to write to Leeuwenhoek about investigating itch mites and smallpox pustules to see if he could find animalcules. As we shall see, Sloane supported Richard Bradley's work over decades despite Bradley's personal shortcomings. In addition, there were ties between Sloane and Johannes Groenevelt, one of the Oracle doctors; this gains significance from the connection that has come to light between Groenevelt and Benjamin Marten. Looking behind the scenes, we find Sloane's influence behind many of the contagionist developments of the early eighteenth century.

Sloane's background was unusual for a man who attained such distinction in early-eighteenth-century England; he was born an Ulster Presbyterian of Scottish descent. His maternal grandfather had been chaplain to Archbishop Laud; his father Alexander (d. 1666) had been an agent for James Hamilton, first Viscount Clandeboye of County Down and later served as a commissioner of array for Charles II.[1] Sloane's paternal ancestors seem to have come from Ayrshire and to have migrated to Ireland with the Hamiltons, with whom their fortunes were entwined.[2] His stepfather, John Baillie, served the Hamiltons. A disaffected member of the Hamilton family described John Baillie as a perpetual turncoat, "first a royalist, then a Cromwellian; Presbyterian, Anabaptist, Episcopalian by turns."[3] Hans's eldest brother, James Sloane, who also served the Hamiltons, was an attorney and later became MP for Killyleagh. A second brother, William, married into the Hamilton family and become a fabulously wealthy merchant, leaving a fortune valued at one hundred thousand pounds at his death in 1728.[4] The connection with the Hamiltons may also have given Hans Sloane a link to Robert Boyle, son of the Earl of Cork.[5] Sloane's parents were well connected, and it is clear that he and his brothers had substantial support from some quarter or he could not have spent four years studying in London and abroad. It was surely helpful to him to have two brothers who were also prospering, and in later years he was able to help them as well.[6]

Hans followed his brothers to the Killyleagh parish school in Lisnagh and from his earliest days loved natural history. In about 1676, at the age of 16, he contracted phthisis and spent three years confined to his home. After recovering, he went to London to study physic, chemistry, and anatomy. He stayed in a house next to the laboratory of the Apothecaries' Hall and lodged with the surgeon Nicholas Staphorst, a relative and pupil of the Oxford chemist Peter Stahl.[7] Staphorst, who assisted Boyle in his chemical research, taught Sloane how to prepare and use chemical medicines.[8] From the beginning, therefore, Sloane's medical education was unconventional for a physician and tilted toward iatrochemistry.

Sloane also studied botany at the Apothecaries' Garden in Chelsea and established a friendship with Robert Boyle and John Ray.[9] He was introduced to Ray by his friend Tancred Robinson. In 1683, with Robinson and another student, he traveled to Paris, where he studied medicine at the *Hôpi-tal de la Charité* and botany at the *Jardin du Roi* under Joseph Pitton de Tournefort, who then sent the three students on to Montpellier with a letter of introduction to Pierre Chirac.[10] At Montpellier, Sloane developed a close friendship with the botanist Pierre Magnol—so close that he remained in Montpellier for a year while his two friends went to Italy; they all returned in 1684, thus just missing the Revocation of the *Édit de Nantes* in 1685.[11] He obtained his MD by examination from the University of Orange in the South of France.[12]

At the time of Sloane's tour, botanists were debating the question of how to classify plants. Although Aristotle had discussed classification methods in general, until the sixteenth century herbals and other botanical works had grouped plants either alphabetically or by their medicinal uses. In 1583, Andrea Cesalpino had published a work that for the first time grouped plants by their fruits and flowers (fructification).[13] Magnol's most important work, *Prodromus Historiae Generalis Plantarum*, which would appear in 1689, was an attempt to classify plants into "natural" groups, which he referred to as "families," in itself suggesting that members of these groups were related by generation.[14] In 1694, Tournefort would publish *Éléments de Botanique*, which argued for using the flowers and fruit of plants as the basis for clas-sification; this generated a dispute with John Ray.[15]

Upon his return to London at the end of 1684, Sloane visited Thomas Sydenham, bearing a letter of introduction from Boyle that described him as "a ripe scholar, a good botanist, a skillful anatomist."[16] Sydenham report-edly replied, "That is all mighty fine, but it won't do. Anatomy, Botany—Nonsense! Sir, I know an old woman in Covent Garden who understands botany better, and as for anatomy, my butcher can dissect a joint full as well; no, young man, all this is Stuff: you must go to the bedside."[17] Sloane appar-ently took this advice, as he was soon living in Sydenham's own house, accom-panying him on calls to his patients, and acting as his substitute when the elderly Sydenham was ill. His biographer wrote that Sydenham "helped him . . . to a medical practice, and whole-heartedly promoted his interests."[18] This was shortly after Sydenham had altered his view of cinchona and modified his approach to disease. The introduction of cinchona as a medical "specific" made a deep impression on Sloane; many years later he was to compare the resistance to inoculation to the way cinchona had been received at first.[19] His intimacy with Sydenham, always adventurous in adopting medical ideas, perhaps predisposed Sloane to be open-minded.

One key to Sloane's epistemology lies in Sydenham's colorful comment about anatomy and botany, which has been repeated countless times as evidence of Sydenham's hostility to anatomy or medical research. In fact, it merely displayed his hostility to book learning. There is another impor-tant facet of this comment, however. Sydenham told Sloane to "go to the bedside" for his knowledge—in other words, to focus on clinical study, but he was also implying that butchers and herbalists, common people usually depicted as ignorant and superstitious, had their own knowledge to impart.

At the bedside, the patient is a more-or-less passive object, offering at most an account of symptoms to the practitioner, who also notes the patient's signs, observes the behavior of the illness, constructs an independent narrative, and determines the treatment. Consulting an herbalist about botany is a different sort of undertaking. Despite his or her humble status, the herbalist is the one with the unique knowledge, and researchers depend on the herbalist's own terminology and narrative about the uses of a given plant, although they may later supplement the herbalist's nomenclature with their own. This ethnographic or ethnobotanical research is respectful toward local, traditional, or "folk" knowledge in a way that is distinct from other early modern methods of knowledge production, such as rationalism or inference. Historian William Eamon described this approach as "medical primitivism" and pointed out that it justified resistance to "the political control that academics maintained over institutions of learning and the professions."[20] Whereas populist authors instructed ordinary people, medical primitivists hoped to learn from them.

We have little evidence that Sydenham himself was actually interested in this sort of inquiry. His advice was more a sneer than a research program. Sloane, however, carried it out. He was willing not only to consult with anyone and everyone about their own knowledge but also to report and publish both their observations and his own with a minimum of refashioning or editorializing. Whether this is viewed as self-effacing or merely as gullible, it challenged eighteenth-century ideas about the proper boundary between science and superstition and the role of medicine as a learned profession. It has also made it difficult for Sloane's biographers to come to grips with his own character. As Sloane's biographer commented, "When he gives an opinion . . . Sloane is often right. But it is not always easy to guess what he himself believes."[21]

In 1687, shortly after becoming a Fellow of the College of Physicians, Sloane embarked on a hazardous voyage to Jamaica as personal physician to the Duke of Albemarle, who had just been appointed Governor. When he heard of the project, Sydenham said Sloane would do better to drown himself locally.[22] Because of various obstacles, including the death of the bibulous duke, the trip lasted only 15 months. Sloane took full advantage of this opportunity, observing, preserving, drawing, and recording an extraordinary range of phenomena, including the culture and music of the African slaves.[23] He had negotiated a generous advance and a large salary, which he invested in cinchona for resale in London. He also experimented with the local chocolate.[24] During his stay, Sloane met Elizabeth Langley Rose, who was then married to Fulke Rose. She would marry Sloane in 1695, after her husband's death. Until her own death in 1724, he benefited from her share of her late husband's Jamaican sugar plantation, which was worked with slaves.[25]

Following the Duke's death, the dowager Duchess decided to return home. Sloane accompanied her, and they arrived on May 29, 1689. Thus, Sloane was fortunate enough to have been absent from England during the Glorious Revolution, and so was not forced to choose sides when animosities were at their height. He remained in the household of the Duchess for six more years, and her patronage helped him to establish an extremely lucrative and successful practice.

Sloane's *Catalogus Plantarum . . . in Insula Jamaica*, a description of every plant he had seen in Jamaica, appeared in 1696. Sloane believed that one accomplishment of this work was that it drew together other observations of the same plants from books in several languages. Ray, who helped him edit this work, noted his success in "contracting and reducing to one many plants distracted into many species by the unskilfulness of some and misapprehension of others."[26] He thought Sloane had proffered a great service to botanists by "distributing or reducing the confused heap of names, and contracting the number of species."[27] Sloane's catalogue of botanical species, therefore, anticipated the rationalization that Cullen would later bring to species of disease.

Sloane was very sensitive to criticism. The antiquary Thomas Hearne wrote in 1705 that Sloane was hesitating to publish his book on Jamaica because "he was well assur'd there were 2 or 3 ill natur'd men who would write ag[ains]t it."[28] He was correct. When his account of the diseases he saw in the West Indies finally appeared in the first volume of *Voyage to the Islands* in 1707, it quickly drew a parody from satirist William King.[29] Nevertheless, it was an important work that still attracts scholarly attention.[30]

Historian Wendy Churchill has argued that Sloane's experiences as a physician in the West Indies led him to abandon the Hippocratic approach to disease that blamed illness on the environment and its effect on humoral balances. The transformation occurred when he realized that the diseases he saw in Jamaica were very similar to those he had seen in England and that they affected men and women in the same way.[31] She comments that "Sloane argued that disease manifested identically in different locales and thus yielded to the same medical treatment. This applied not only to the white, male body, but to all types of bodies, regardless of race or sex. Sloane believed that illnesses presented the same symptoms in different bodies, and hence, they should all receive the same types of treatments."[32]

In other words, Sloane believed that diseases, like plants, could be gathered into species that remained fixed through time and space, each with its own characteristics and natural history. Sydenham had also claimed this, but, as we have seen, his medical theory of the "epidemic constitution" conflicted with this claim. Sloane did not adopt Sydenham's theory, but he did not substitute an alternative theory. Because he was not certain how symptoms should be grouped together to form species, his medical treatise consisted of a discrete set of case histories that appeared to be unconnected and unorganized.[33] They couldn't be used to illustrate something because there was as yet nothing for them to illustrate. They existed as separate experiences that he brought home in much the same way as he transported pressed plants, minerals, or feathers without attempting to set them in a sequence or framework. His notes were like a pail filled with stones, shells, corals, and fossils before there was a consensus on how to classify them.

In fact, historian John Thackray describes Sloane's own mineral collection in just these terms:

> "It is clear . . . from the way . . . he arranged his collection that Sloane believed that the fossil bones, shells and leaves he collected were the remains of once-living animals and plants. Equally, he recognized that crystals, earths and metals were natural and lifeless constituents of the Earth. However . . . he simply followed the opinions of others . . . In this, as in so many other areas, Sloane was not a controversialist . . . Sloane was clearly more attracted to the specimens . . . than to the theories . . . "[34]

This approach can be viewed simply as a naive delight in acquisition for its own sake or as a radical empiricism fueled by profound skepticism.

Despite a voluminous correspondence, Sloane has left little evidence about his own position on many subjects. For example, he was probably brought up as a Presbyterian, but he seems to have conformed to the Anglican Church after moving to England.[35] He was buried in the churchyard of his parish in Chelsea, but Anglican burial was required by law.[36] Because he had been named a Fellow of the College of Physicians by James II, he never had to subscribe to the Thirty-Nine Articles of the Church of England or pass the College examinations.[37] He may have been one of the Presbyterians who drifted toward Deism or Socinianism in the early eighteenth century.[38] There seems to be no record of his personal religious views, though his actions suggest that he did not forget his origins. He interceded with Queen Anne on behalf of Ulster Presbyterian ministers who had been ejected from their parishes for refusing to take the oaths required by the Test Act. It was also common for Presbyterian ministers to send him young men with the request that he advise them or assist them in beginning their careers.[39] In his will, Sloane named five members of the Moravian Church to the committee of Trustees responsible for the disposition of his collection; historian Marjorie Caygill describes this as "something of an old man's fancy," although she also views this will as "adroitly worded by an old and worldly man who knew . . . how to manipulate the power structures of the eighteenth century to his advantage."[40]

Sloane joined with the astronomer Edmond Halley in a plan to propose the notorious anti-Trinitarian William Whiston for Fellowship in the Royal Society in 1720; they failed because Newton, then President, had quarreled with Whiston.[41] Historians have suggested that Newton's opposition to Whiston was partly motivated by his desire to conceal his own heterodox views; but apparently none has noted that Sloane was not deterred by the same fear of guilt by association.[42]

Sloane served as a patron for the tiny Jewish scientific and medical community: when Meyer Low Schomberg passed his Licentiate's exam at the College of Physicians but lacked the means to pay for his license, it was Sloane who suggested that the college take his bond for the money.[43] Together with Cromwell Mortimer and Alexander Stuart, Sloane sponsored Jacob de Castro

Sarmento for an Aberdeen MD in 1739: the first doctor's degree awarded to a Jewish practitioner in Britain.[44] Emanuel Mendes da Costa recalled that he had become interested in natural history in about 1736 because of patronage by Sloane and Mead among others.[45] Sloane also sponsored the French chemist Étienne François Geoffroy for a Royal Society Fellowship in 1698, when Geoffroy was still a medical student.[46]

In his will, published in 1753, Sloane wrote that "from my youth I have been a great observer and admirer of the wonderful power, wisdom and contrivance of the Almighty God, appearing in the works of His creation," and he expressed the hope that his huge collections would contribute to "the confutation of atheism and its consequences, the use and improvement of physic and other arts and sciences, and [the] benefit of mankind."[47] This preface expresses the utilitarianism and optimism of the Enlightenment and Sloane's own commitment to a piety founded on an appreciation of the Book of Nature. It also echoes Ray's *Wisdom of God Manifested in the Works of Creation*. If Sloane had faith in anything, it was in the pursuit of natural history.

Members of Sloane's Circle

Sloane knew everyone. There were only about 67 Fellows of the College of Physicians at the turn of the century, so it is not surprising that he knew all the important London physicians.[48] By virtue of his position as corresponding secretary of the Royal Society and editor of the *Philosophical Transactions* (which he revived), he served as the natural contact point for scientists and physicians throughout Britain and in many parts of Europe. In addition, as a fashionable physician, he saw or consulted for thousands of well-born patients. His most intimate scientific friendships seem to have been with other botanists. His background, education abroad, travels, and friendships with many overseas scientists made him more cosmopolitan than many of his colleagues. During Sloane's tenure as Secretary of the Royal Society, the percentage of Fellows who were graduates of universities outside England rose to one-third.[49]

It is not surprising, therefore, to come across Sloane's name in the biography of virtually any early-eighteenth-century naturalist, medical writer, or physician, but many of those whom he assisted or befriended lived on the fringes of the social, medical, or religious establishment. Thomas Fuller might write in a dedication to Sloane that every author should "be in Physic (tho' not in Divinity) a Free-Thinker."[50] Sloane himself, however, seems to have had a soft spot for free-thinkers of many different descriptions, including physic, divinity, and politics.

In light of his close association with Sydenham, his preference for empirical rather than theoretical practice, and his love for botany and natural history, it is evident that Sloane was a "Sydenhamian" rather than a "Newtonian." In fact, his relations with Newton during the latter part of Newton's presidency of the Royal Society were stormy.[51] The Tory satirist William King complained in 1700 that Sloane's "trifling and shallow Management" as Secretary of the Royal Society showed he was unqualified for the responsibility. King lamented that Sloane's correspondents were equally lacking in learning and understanding.[52]

From the beginning of his career, Sloane revered John Ray, a Cambridge Platonist who had given up his College Fellowship in 1662 because of a quibble over the required oaths. He received many gifts from Sloane in his impoverished old age.[53] Above, we noted that John Wilkins had employed Ray to develop botanical categories for his universal language. Ray's own method of classifying plants departed from traditional Scholastic methods, which divided plants into groups from the top (largest groupings) down. Ray built his taxonomy up from the bottom: grouping plants (and, later, animals) "empirically" by common characteristics into species as the smallest discrete units for classification.[54] His conceptualization of species as the basic building block of all taxonomy became widely accepted, especially in England. In its strategy it resembled Geoffroy's table of chemical relationships.[55] At the end of the eighteenth century, William Cullen would use a similar method to construct species of diseases.

Early in his career, Sloane also made friends with the privateer Thomas Dover, who was apprenticed to Sydenham about 1687, at about the time Sloane left for Jamaica.[56] Sloane gave letters of recommendation to Dover when he began to practice medicine in Bristol in 1697.[57] Following Dover's return from the South Seas, Sloane proposed his admission as Licentiate to the College of Physicians in 1721.[58] A few months after Dover's admission, William Wagstaffe accused him of stealing one of his patients who subsequently died, and Dover was "admonished" by the College.[59] Wagstaffe would be one of the most determined opponents of smallpox inoculation.

Dover is best known as the creator of "Dover's Powder," a mixture that originally contained ipecacuanha, opium, licorice, and sulfate of potassium.[60] Ipecacuanha contains the alkaloid emetine, which is effective against amebic dysentery, a common and dangerous disease. The drug had been brought from Brazil to France in 1672 but proved difficult to harness because it induced vomiting so rapidly.[61] Dover's formula made it possible for patients to keep it down. The fourth edition of the *London Pharmacopoeia* included ipecacuanha in the list of "simples"—a new feature that was mostly Sloane's creation—so Sloane deserves some credit for bringing this drug into widespread use.[62] Kenneth Dewhurst, writing in 1957, commented that Dover's remedy, an "emblem of sleep and gentle perspiration" was still in use around the world.[63] The combination of ipecacuanha and opium was later appropriated by the quack doctor Joshua Ward.

Dover was also a believer in mercury, which he recommended for many different diseases. He himself took one ounce a day but sometimes recommended as much as a pound in cases of intestinal obstruction, believing that it could clear a passage through the body by sheer weight. This drew criticism from other physicians, who called him "an Empyric, Quack, and Nostrum-Monger."[64] Dover's work was notable, however, because he experimented with varying doses, trying to uncover the relationship between the dose and the patient's response.[65]

His final publication, the *Ancient Physician's Legacy to his Country*, which first appeared in 1732 and ran to eight editions during the century, was addressed directly to laymen, "that any Person may know the Nature of his own Disease."[66] It is filled with bitter criticism of the College of Physicians, which he described as "a Set of Gentlemen, who like Moles work under Ground, lest their Practices should be discover'd to the Populace."[67] He also accused the apothecaries of venality and the Tory Dr. John Radcliffe of incompetence. It is interesting that Sloane, President of the College of Physicians, maintained his friendship with this raffish doctor who was so hostile to the College.

Another friend of Sloane's was William Coward, (1657?–1725, MD Oxford, 1687), the mortalist Dissenter.[68] In 1704, the House of Commons had declared Coward's works offensive and ordered them burnt by a hangman. Sloane corrected the proofs of Coward's *Ophthalmoiatria* (1706), which ridiculed the Cartesian idea that the pineal gland was the seat of the soul. Sloane reportedly argued with him about the wisdom of expressing his views so openly, but the two remained friends.

Sloane was fluent in French, evidently an unusual trait among British scientists of his time, and he cultivated friendships abroad.[69] The French physician Antoine Deidier was introduced as a correspondent to Sloane by yet another liminal figure: John Thomas Woolhouse (1666–1734). The descendent of a long line of oculists, Woolhouse had attended Trinity College, Cambridge, as a scholarship student. After graduating in 1686/7, he became groom of the chamber to James II. Historian Elizabeth Furdell describes him as one of the "irregulars" who treated Queen Mary and proposed iridectomy for her sore eyes.[70] By 1711, he was living in Paris, where he worked as a surgeon, taught many students, served as physician to the hospital for the blind, and became oculist to the King of France.[71]

Some sources describe Woolhouse as a "charlatan."[72] This is probably because eye surgery was the preserve of practitioners who lacked formal qualifications. Some traveling oculists pitched tents at fairs and couched cataracts on the spot. Like venereology or the sale of patent medicines, eye surgery was a dubious and not very respectable paramedical occupation, but Woolhouse was still elected FRS in 1721 and became a member of the Royal Academy in Berlin in 1723. Sloane's receptiveness was rewarded by Woolhouse's service as an intermediary and agent for the Royal Society in Paris. In addition to encouraging Deidier, he carried on an active correspondence with Jurin and Sloane about such topics as inoculation, the purchase of books, and the Bills of Mortality.

Jacob de Castro Sarmento, whom Sloane recommended for an Aberdeen MD, was a Jewish refugee from the Portuguese Inquisition, which had imprisoned his parents. An enthusiastic botanist, he proposed the creation of a botanic garden in Portugal using seeds from the Apothecaries' Garden in Chelsea, which Sloane had recently restored, and he wrote the first pharmacopoeia for an English hospital.[73] He also published (under his initials) the first separate pamphlet on smallpox inoculation to appear in England.[74] He and Philip de la Cour, a Licentiate, co-founded the Jewish hospital or *Beth Holim*, which they briefly served as physicians. De Castro Sarmento appears to have become a Deist later in life.[75]

Among Sloane's most intimate friends later in his life was Cromwell Mortimer (d. 1752, MD Leyden): a Dissenter whose father's first wife had been Richard Cromwell's daughter.[76] Mortimer dedicated his Leyden dissertation to Sloane, who may have been a relative of his.[77] In 1728, he was elected to the Royal Society and in 1730 became its "second" or acting secretary. In 1729, Mortimer moved to Bloomsbury at Sloane's request and acted as his assistant, prescribing for Sloane's patients. He became unpopular with more traditional members of the medical profession after he published a proposal to allow individuals to pay a guinea per year and be assured of medical care for any symptom of illness that might appear.[78] After his death a correspondent of the *Gentlemen's Magazine*, presumably nettled by this plan, described him as "an impertinent, assuming empiric."

Mortimer, who had been an enthusiastic Newtonian physiologist in the 1720s and 1730s, had a change of heart and became the first medical author to criticize mechanist physiology for assuming that animal heat was caused by friction between particles. Instead, he concluded it was the product of chemical effervescence.[79] His repudiation of the mechanical philosophy has been described as marking a turning point in British science, but it could also be seen as the revival of an outlook that had been present just below the surface all along.[80]

Another younger man who became one of Sloane's closest friends was the Quaker merchant Peter Collinson (1694–1768, FRS, 1728), a Londoner who had been raised in Cumberland and shared Sloane's love of botany. Collinson, a successful woolen draper, traded with America and used his American contacts to promote the exchange of seeds and of scientific information. As a young man, he helped Sloane arrange his collections; later he added to the collections in Sloane's museum.[81] A Fellow of the Society of Antiquaries, Collinson also became one of the founders of the Foundling Hospital.[82] When Linnaeus visited London in 1736, he was brushed off by Sloane but found a warm welcome from Collinson, and the two continued to correspond after Linnaeus's return to Sweden.[83]

Collinson would himself establish very close friendships with the younger Quaker physicians John Fothergill (1712–1780) and John Coakley Lettsom (1744–1815). Sloane, Mead, and Fothergill joined the subscription that Collinson organized to fund the work of the American naturalist John Bartram and share the seeds of American plants that Bartram sent back to England.[84] Sloane also maintained friendships with other Quaker naturalists. He was a close friend of the reformer John Bellers, whose son, Fettiplace Bellers, introduced him to James Logan of Pennsylvania, one of Benjamin Franklin's earliest patrons. Logan also corresponded with Collinson and Fothergill.

Other younger men whom Sloane helped included Philip Miller (1691–1771) and William Watson (1715–1787, MD Halle, 1757). Miller was the son of a Scottish nurseryman who shared Sloane's reverence for Ray.[85] Sloane appointed him to the post of head gardener of the Apothecaries' Physic Garden at Chelsea despite objections to Miller's nationality.[86] Watson probably met Sloane in 1730 when, as an apothecary's apprentice, he won a prize from the Society of Apothecaries for a paper on botany.[87] Sloane and Philip Miller sponsored him as a Fellow of the Royal Society, where he won the Copley medal for his experiments on electricity in 1745.

THE TEMPLE COFFEE HOUSE BOTANY CLUB

Sloane was probably the founder of the informal "Temple Coffee House Botany Club," which flourished between about 1689 and 1706 during a period when the Royal Society was in disarray. The Club met on Friday evenings and combined social activities with botanical discussions, experiments, communications from abroad, and an exchange of seeds and specimens. On summer Sundays and holidays the members often joined together for botanizing expeditions into the countryside.[88] This was probably where Sloane first met Richard Bradley. In addition, its members included the apothecary James Petiver, William Sherard, Tancred Robinson, Samuel Dale, and Martin Lister, all good friends of Sloane's, though his relations with Sherard would later cool.[89] Alexander Stuart, later the first physician to the Westminster Hospital, collected specimens for Club members during his service as a ship's surgeon in Asia.[90]

The members of this club formed a network of researchers interested in natural philosophy, including pursuits that would become biology, botany, taxonomy, and microscopy. Sloane both contributed to this network and benefited from it, receiving information, accounts, and objects from a widely dispersed area. As we have seen, Sloane also cultivated relationships with foreign-born refugees, who often had connections with family and friends abroad. These contacts and his patronage relations with young outsiders fed Sloane's huge collections, which ultimately became the British Museum, another important refuge for outsiders.[91]

Sloane quietly made critical contributions to the boom in botany and horticulture that took place in the century that followed his return from Jamaica in 1689 with his notes and herbarium.. When William and Mary came to the throne of England, they were so frustrated by the shortage of local garden plants that they imported them from Dutch nurseries.[92] Sloane rescued the dilapidated and sparsely planted Physic Garden of the Society of Apothecaries, and Miller transformed it from a small working garden to a major scientific institution.[93] He revolutionized the art of ventilating and heating buildings in order to cultivate a wide range of tender plants and established a scientific exchange of seeds, bulbs, and plants that spanned the world.[94] Techniques for stoving, ducting, and ventilation that were developed to preserve tropical plants in greenhouses were later adapted for use in residences and institutions, including ships, prisons, and hospitals.

Botany was a humbler pastime than astronomy or physics; it did not necessarily require expensive equipment or a staff of assistants or years of advanced instruction in mathematics. The objects it studied were (literally) much less elevated. Instead of staring up at the heavens by night, botanists dug in the dirt like any laborer or housewife. Botany was primarily a practical trade, not an intellectual pursuit. It was closely associated with old wives, humble herbalists, and the tradesmen—apothecaries and druggists, gardeners, dyers, cooks, soap boilers, and perfumers—who needed to recognize the herbs they bought, used, and sold. Whereas astronomers courted noble patrons by naming heavenly bodies after them, few aristocrats of the seventeenth century were flattered to have gangly plants or even weeds named in their honor.[95] Many Tories considered this sort of science uncouth, but it reminded Dissenters that God's work could be found in even the humblest creatures.[96]

Botany furnished analogies for ideas about the classification, propagation, and spread of disease that were as fruitful and significant as early modern ideas about microscopic animals. As we have seen, ideas about "seeds," the disavowal of spontaneous generation, and the new "biological" conceptualization of species of plants all played an important role in the ongoing discussion about whether the rules that applied to botanical entities also applied to disease entities.

One of Sloane's strategies for elevating the status of botany as a learned pursuit was to lease land in Chelsea to the Society of Apothecaries for their Physic Garden in perpetuity on the condition that they present fifty new plants annually to the Royal Society; the list of plants was duly recorded in the *Philosophical Transactions*.[97] The beautiful illustrations Sloane commissioned for his *Natural History of Jamaica*, which appeared in two folio volumes in 1707 and 1725, also helped raise the prestige of botany.[98]

As we noted earlier, John Locke worked very closely with Thomas Sydenham, whom Sloane assisted during Locke's exile in Holland. Locke retained his interest in botany, distributing seeds to friends in England and pondering the interlinked issues of classification and generation. After returning to England, he corresponded with Sloane on medical and scientific topics.[99] In 1694, they discussed spontaneous generation (Locke doubted whether spontaneous generation could be completely ruled out), and in 1699 Locke sent Sloane a copy of the second edition of his *Essay Concerning Human Understanding* with a request for Sloane's comments on new chapters about enthusiasm and the association of ideas.[100] Locke must have believed that Sloane was sympathetic to his ideas to make this request.

Locke also requested medical advice from Sloane about diabetes and the diseases then current in London. Sloane's reply on diabetes reveals a characteristic skepticism, "I have been many times consulted in cases where people have thought they had a diabetes but I was of another opinion . . . I never saw but one woman that had a true diabetes."[101] Another physician treated her with *Theriac Andromache* despite Sloane's misgivings, and she soon died.[102] His conduct in this case was also characteristic: he stood aside but still observed the outcome with interest. Sloane's comments on London fevers revealed a more conventional view: he was seeing both mild intermittent fevers and continued fevers, the latter accompanied with headaches and spots. He treated both kinds with venesection, purges, and cinchona. Some fevers combined the symptoms of both the other kinds but also responded to cinchona.[103]

Among Sloane's more specialized botanical friends was Nehemiah Grew, a protégé of John Wilkins and the son of Obadiah Grew, an ejected Presbyterian minister. Nehemiah's older stepbrother, Henry Sampson (1629–1700, MD Leyden, 1668), also an ejected minister, helped him become established in the scientific world.[104] Jonathan Israel comments that "having graduated from Leyden in 1671, [Grew] . . . had a detailed knowledge of the Dutch intellectual scene and presumably knew Dutch."[105] He was named Curator of the Anatomy of Plants by the Royal Society in 1672 and later lectured for professors at Gresham College.[106] Grew also became a Secretary to the Society from 1677 to 1680 and served as an editor of the *Philosophical Transactions*. Hooke allowed him to use the Society's microscope. Together they tried to duplicate Leeuwenhoek's observations of animalcules in pepper water in 1677, and it was Grew who prodded Leeuwenhoek to pursue his research into spermatozoa and generation.[107] An atomist, Grew wrote a popular book on natural theology in addition to several important botanical works.[108] He was one of the popularizers of the theory that plants reproduced sexually and with Malpighi, he is often described as a founder of botanical anatomy and physiology.[109]

Sloane's friend Tancred Robinson was a Yorkshireman. He appears to have been an Anglican because he graduated from St. John's College, Cambridge. He was Sloane's companion during their studies abroad, and he was also very close to John Ray, whom he introduced to Sloane.[110] Robinson contributed a section on the chemistry of plants to Ray's greatest work, the *Historia Plantarum Species* (1693–1704), which laid the foundation for empirical botany and botanical taxonomy, and he also contributed to Ray's *Wisdom of God*. A Fellow of the Royal College of Physicians, he later became physician to George II.

One of Robinson's first contributions to the *Philosophical Transactions* was a letter addressed to another member of this group, Martin Lister (1639–1712), the recipient of the first letter describing Chinese methods of smallpox immunization. Like Robinson, Lister was a Yorkshireman and a graduate of St. John's College, Oxford, though he had studied medicine in Montpellier from 1663 to 1666 and traveled with John Ray in France.[111] Lister developed his own idiosyncratic theory of animate contagion, arguing that smallpox had originated in the bite of poisonous insects and that syphilis first developed from eating venomous iguanas, both subsequently spreading by contagion from person to person.[112]

Lister had been among the first botanists to speculate in 1673 that fluids might circulate in plants in a way that was analogous to the circulation of the blood described by Harvey—an idea inspired by observations of Ray's.[113] Lister and Robinson jointly helped organize the opposition to John

Woodward's theory of the origins of the earth; Woodward also became a bitter opponent of Sloane's.[114] Robinson, along with Lister and Ray, was bitterly attacked by one of Woodward's defenders, a clergyman named John Harris.[115] Like Sloane, Lister would be a target of William King's satire.[116]

It is evident that this group of men formed a tightly woven scientific network of natural historians. This is confirmed by a list of the correspondents of Sir Robert Sibbald, the leading "Sydenhamian" in Edinburgh, who was also a correspondent of Sloane's. According to historian Roger Emerson, Sibbald's network of foreign correspondents included Hermann Boerhaave, Nehemiah Grew, Martin Lister, James Petiver, John Ray, William Sherard, Hans Sloane, Edward Tyson, Joseph Pitton de Tournefort, and John Woodward among others, whereas "Scots like Professor James Gregory and Archibald Pitcairne had quite different contacts with Padua, Leyden and Oxford . . . and with Newtonian circles in London."[117]

NATURAL PHILOSOPHY, TORY ATTACKS, AND *THE TRANSACTIONEER*

The Tory writer William King ridiculed Sloane, Petiver, and the members of the Temple Coffee House Botany Club in his parody *The Transactioneer*, published anonymously in 1700. King had been one of the "Ancients" and "wits" who attacked Richard Bentley and the other "Moderns" during the Battle of the Books satirized by Jonathan Swift in *The Tale of a Tub*.[118] He became a member of the "Brothers Club" of Tories who produced the *Memoirs of Martinus Scriblerus*, an extended parody of modern learning.[119] *The Transactioneer* was one of several attacks King made on members of Sloane's circle, parodying their "myopic affection for inconsequential fact."[120] Claiming unconvincingly that he spoke more in sorrow than in anger, King confessed that he himself was not a member of the Royal Society, but his respect for natural philosophy was so great that it pained him when Sloane cast both science and the Society into disrepute.[121]

Historian Richard Olson has suggested that what offended the Tory "wits" about the Royal Society was the hubris of claiming powers, knowledge, or responsibilities that they should have entrusted to God's providence. They especially detested "Cartesian, Hermetic, and Epicurean natural philosophies, all of which shared unacceptable emphases on the discovery of final or ultimate causes."[122] King, however, did not attack Sloane for meddling too much in theories best left to philosophers and divines but for the "vulgar Baconianism" of the articles that appeared in the *Philosophical Transactions* during Sloane's editorship. King's method was to summarize actual articles in such a way as to emphasize the comedy of the events themselves or the circumstantiality of their reporting. For example, in one exchange the naive interlocutor exclaims:

Gent[leman]: O good Sir, Pardon me, be as Circumstantial as you please. It's a very Philosophical Transaction indeed: A Woman boiled Herbs and Bacon for

Supper; the Children Purged; the good Man Slept longer than ordinary; went to Work at Mr. Newports . . .

Transact[ioneer] [Sloane]: Truly Sir, we ought to be particular in the Circumstances of Things so Remarkable; for this Herb is described and figured in several Authors, and therefore we ought to take Notice of its Effects.[123]

King's point is that accounts that are "particular in the Circumstances of Things" are anything but "Philosophical." Such particularity should be left to vulgar empirics.

King implied that the compilation of details from far and wide in the *Philosophical Transactions* inappropriately gave ordinary people the same epistemological standing as learned scholars. What we might view as ethnographic reports, he depicted as complete credulity. He ridiculed as trivial pedantry the concern of Sloane's friends for vocabulary, taxonomy, and exact description. He pilloried their refusal to entertain hypotheses and their focus on the mere collection of heaps of facts that were not marshaled into theses: in other words, he attacked the natural history approach to knowledge production favored by members of Sloane's botanical network and suggested that what they saw as basic science was merely base. They proceeded (at best) by induction, unguided by any prior hypotheses or true learning. He complained that their work reduced the status of science and implied that Sloane had packed the Royal Society with his own cronies.[124] King did not attack Sloane's circle as the embodiment of "science" but as the sponsor of the wrong kind of science: a vulgar, uncourtly, and unphilosophical undertaking.

Historian Stephen Gaukroger has contrasted the "vertical" reasoning associated with mechanism with the "horizontal" logic that explored associations between phenomena without deriving them from more fundamental corpuscular causes. He claims that Boyle and Newton focused on their results instead of creating a rationale for their practice: "Boyle was committed to mechanism . . . yet his own experimental practice led him to reject . . . foundational considerations, mechanist or otherwise." Locke, on the other hand, developed a rationale for studying horizontal or phenomenological relationships, thus laying the foundations for the "collapse" of mechanism.[125]

Gaukroger argues that in the course of writing and revising his *Essay*, Locke demonstrated that natural philosophy does not need to be grounded in a more fundamental matter theory to make legitimate claims based on "natural experiments" that do not rely on seeking universal knowledge.[126] He writes that Locke believed that "Causation, in real cases . . . is typically a mixture of 'horizontal' and 'vertical' causation, and in the medical cases . . . it is the 'horizontal' ones, that is, the causal relations between phenomena, that are doing the real explanatory work."[127]

Sloane also believed that empiricism was the only road to real certainty:

The Knowledge of *Natural-History*, being Observations of Matters of Fact, is more certain than most Others, and in my slender Opinion, less subject to Mistakes than *Reasonings*, *Hypotheses*, and *Deductions* are . . . these are things we are sure of, so far as our Senses are not fallible; and which, in probability, have been ever since the Creation, and will remain to the End of the World, in the same Condition we now find them.[128]

His epistemology, however, seems to be parallel to Locke's, not derived from it.

Sloane's practice demonstrated that he believed scientific truth was also something that could be obtained from anyone—it did not require interpretation or translation by a learned natural philosopher or gentleman to be worth reporting.[129] T. C. Bond has shown that Sloane deliberately followed a *laissez-faire* editorial policy in the *Transactions*, printing many letters sent to him verbatim, allowing contradictory claims, and accepting observations he himself considered "unfounded."[130] Bond argues that this promoted the development of a dialogue that became self-regulating as correspondents wrote in to correct or dispute claims from earlier articles. It is indisputable that Sloane rescued the *Transactions* from an early death and that the volume of submissions surged along with the number of readers. The inclusion of many entertaining articles on "curiosities" and anomalies that so annoyed the literati must have helped to attract readers to the *Transactions*, as did the many controversies that began to erupt in the journal.

One of the authors whom King singled out for ridicule was Martin Martin, who published a detailed article about a visit to the remote Scottish island of St. Kilda in the *Transactions* in 1697.[131] In this case, Sloane was vindicated: Martin's work attracted so much interest that he published a more extended version as a book and then made further explorations for his *Description of the Western Islands of Scotland* in 1703. This work is still an important source for local historians because of its detailed depictions of a rapidly disappearing culture. It later served as a source for some of the vivid descriptions of Scottish life in James Thomson's extremely popular poem *The Seasons*.[132]

Martin's original article contains the first mention of the "Boat Cough" that only afflicted the residents of St. Kilda after a visitor arrived. This very brief anecdote was repeated again and again during the century as evidence in favor of contagionism. It prompted further investigations on the island that confirmed the accuracy of the account, although the nature of the "cough" remained in dispute.[133] William Cullen discussed this phenomenon with his students when he lectured on catarrhs (colds).[134]

The many "remarkable" occurrences reported to the *Transactions* also subtly combined to desacralize what had earlier appeared to be miraculous signs and portents. Preternatural births, unusual weather, plagues of insects, and other frightening events became almost pedestrian when they regularly appeared in the *Transactions*. These reports helped redefine the boundaries of normal phenomena—such events might be unusual, but they were no longer seen as unique or supernatural. Preternatural births also challenged the still-emerging understanding of the rules of generation, speciation, and hybridization.

SLOANE, RICHARD BRADLEY, AND DR. JOHN MARTYN: MAGGOTS IN THE BRAIN

Sloane had ties to nearly all the known English contagionist authors of the early eighteenth century with the exception of the naturalist Dr. Richard Brookes, who remains an enigma.[135] Richard Mead and Sloane were colleagues, friends, and occasional competitors for both patients and books. Their relationship was close and long-lived. There can be no question of Mead's Whig political sympathies; his religious views are more difficult to determine, though after his marriage he was at least an occasional conformist.[136] His closest friends were the classicist Richard Bentley and the Tory physician John Freind.[137] Mead was consistently hospitable to foreigners and visitors from overseas, and he participated in the gatherings of foreigners at the Rainbow Coffee House, which included many Huguenots, such as Matthew Maty, John Baptist Sylvestre, J. T. Desaguliers, and Pierre Des Maizeaux, and heterodox authors, such as Anthony Collins and John Toland.[138]

With Richard Bradley, the leading proponent of the theory that contagion was carried by insects, Sloane's relationship was that of patron to client.[139] Bradley's early years are obscure. He was born in about 1688 and had fathered a child by 1714. There is no evidence that he had any university education, but he must have studied botany because Sloane's friend, the apothecary James Petiver, arranged for him to visit Holland in 1714 to collect natural history specimens. Petiver had previously visited Holland himself as Sloane's agent.[140] Traveling under the pseudonym of "George Grant," possibly to avoid English creditors, Bradley spent five months in Amsterdam acting as an agent for Petiver and Sloane and visiting Leeuwenhoek, whom Petiver had been unable to see on an earlier trip.[141] Constantly short of funds, Bradley took advantage of a misapprehension among his Dutch acquaintances and began to practice medicine without benefit of a license or medical education. He wrote to Petiver for medical advice, prescriptions, and even information about how to write the common abbreviations for weights and measures. On Petiver's advice, he treated his patients with Peruvian bark, ipecacuanha, camphor, cinnamon, and other remedies.

Bradley could not afford to continue on to Paris, so he returned to London and eked out a living by writing on botany, gardening, and husbandry. He sold Sloane some drawings of exotic insects that he had made in Amsterdam, and he dedicated the second "decade" or installment of his *History of Succulent Plants* to Sloane in 1717.[142] In 1721, he brought out *A Philosophical Account of the Works of Nature*, published by subscription and dedicated to Charles Boyle, fourth Earl of Orrery, the same man who owned a copy of Marten's *New Theory of Consumptions*.[143] Many Cambridge men and Fellows of the Royal Society, including Newton and Sloane, subscribed to Bradley's *Account*. At about this time, James Brydges, first Duke of Chandos, employed Bradley as a gardener for his magnificent house at Canons and asked Desaguliers, then his chaplain, to engineer fountains and waterworks. This project was not a complete success: the Chandos papers include a note to the effect that Bradley was "mismanaging the hot-house, the physic garden, and the sums entrusted to him."[144]

In April 1721, Bradley launched *The General Treatise of Husbandry and Gardening*, the first gardening magazine published in England. This was printed by J. Peele in 12 monthly issues and contained plates engraved by John Pine (1690–1756).[145] Each number had a separate dedicatee; the third was dedicated to Sir John Anstruther, another possible owner of Marten's *New Theory of Consumptions*, and the twelfth was dedicated to Sloane.[146]

In 1722, however, Bradley suffered a setback, for he wrote to Sloane that since "the unfortunate affair at Kensington, whereby I lost all my substance, my expectations, and my friends," he had received enough public support to pay off two hundred pounds of his debt. He hoped that establishing a "garden of experiments for general use" would clear the remainder.[147] In 1723, he wrote to Sloane requesting support for his candidacy for Professor of Botany at Oxford.[148] He was unsuccessful, but in the following year, he became the first Professor of Botany at Cambridge with the support of the classicist Richard Bentley.[149] He was appointed with the understanding that he would establish a botanical garden there.

It seems that Bradley was sincere in his desire for such a garden, but he did not have the resources to create one by himself.[150] He appealed for funds in 1725, noting that he had already collected many plants for it.[151] His professorship carried no endowment; he was dependent on lecture fees for his income, and evidently he found few students.[152] This was not an unusual problem at Cambridge; it was said that Newton often lectured to his classroom walls.[153] The professorship also earned him the bitter enmity of John Martyn, who had wanted the position for himself, although he was just 25 years old at the time, unqualified, and had achieved little of note.[154] For the rest of Bradley's life, Martyn would miss no opportunity to undermine his work.

Still struggling, Bradley turned again to the booksellers and created *The Weekly Miscellany for the Improvement of Husbandry, Arts, and Sciences*, which appeared in 21 issues between July and November 1727. This work greatly impressed Cotton Mather, but it failed to resolve Bradley's financial problems, for he wrote to Sloane in early July that he could not issue the current number because he lacked money for the stamp duty.[155] Sloane sent him a guinea.[156] In September he was begging again. He had reached an agreement with the booksellers for distribution and had found a sponsor for the next number. He hoped that the *Miscellany* would enable him to pay his debts and settle at Cambridge, "where I can have a little piece of ground & read lectures, which would also be very advantageous, but especially if some great person would raise a public physic garden there . . . But I must be free of my debts before I can appear . . . I am sure I can discharge all I owe for one hundred pounds."[157]

Bradley's hopes were not unreasonable: the German botanist Johann Jacob Dillenius, who served as foreign secretary to the Royal Society from 1728 to 1747, gained his chair at Oxford through a bequest from William

Sherard, a friend of Sloane's, who in 1728 endowed the University with his herbarium, library, and three thousand pounds on condition that Dillenius serve as its first Professor of Botany. Sloane had already granted the lease of the Chelsea Physic Garden to the Society of Apothecaries in 1722, raised funds for repairs, and ensured that Philip Miller was appointed as its gardener. He could certainly have helped to found a garden in Cambridge, but he did not take the hint. Perhaps Sloane lacked faith in Bradley's management skills. A few months later, it was Sloane who needed Bradley's assistance when he sought the presidency of the Royal Society. Bradley sent him an account of his energetic canvassing and sought to dispel a rumor, presumably spread by John Martyn, that he had abandoned Sloane's cause.

Martyn continued to harass Bradley. In 1730, he became one of two editors of *The Grub-Street Journal*, a periodical set up by Alexander Pope and his allies. The other editor, Richard Russell, was an Oxford graduate who had lost his clerical living because he was a nonjuror; that is, he had refused to swear allegiance to the Hanoverian dynasty. At every opportunity, they pilloried the classical scholar Dr. Richard Bentley, the Master of Trinity College, who had helped Bradley obtain his professorship, and who also played a major role in Pope's *Dunciad*.[158] They resented Bentley both for his assistance to Bradley and for his contributions on the side of the "Moderns" in the Battle of the Books.[159]

Martyn lost no time in reviewing *Bradley's Course of Lectures upon the Materia Medica* (1730). Bradley, still pursuing his dream of a botanical garden, had written that he had finally found a suitable location that he could supply with many plants, if only the land were secured.[160] Martyn commented sarcastically, "I was charmed with the full assurance of our soon having a physic-garden . . . Which will, no doubt, be easily effected, if the money raised on this occasion be intrusted in the hands of our Professor, whose oeconomy is equal to his learning; and whose integrity will be as beneficial to us, as his knowledge."[161] He pounced on many small slips, real or imagined, in Bradley's lectures, and criticized him for failing to deliver public lectures often enough, leading many to "fly to foreign universities, for that knowledge which we might gain at home."[162] Martyn's own expertise was questionable: Dillenius, who served as President of the small botanical society Martyn had founded, remarked in 1727 that "Martyn does not know a Nettle from a Dock."[163]

Number 18 of *The Grub-Street Journal* contained an article by Martyn entitled "An Essay towards a New Theory of Physic, in a Discourse read before the Grubean Society," which was a long and detailed parody of the animalcular theory of disease.[164] The parody began by commenting, "I am about to propose a Theory of Physic different from all, which have been already invented: a theory not depending upon precarious reasoning, but supported by a great number of curious and exact observations . . . all diseases whatsoever owe their origin to animalcules."

The essay then referred to Leeuwenhoek's discovery of animalcules (spermatozoa) in male semen and in pepper water and vinegar. It claimed that all the fluids of the body contain an infinite number of animalcules. Digestive diseases appeared when food in the stomach upset or killed the animalcules there and their bodies putrefied. If the food caused convulsions in the animalcules, they irritated the fibers of the stomach and caused it to throw up both the food and the animalcules at once. When human blood was observed through a microscope, one particular animalcule could be seen,

> which is not very much unlike the sea tortoise in shape. This . . . is the general cause of fevers. It is not improbable, that this creature is more prolific in some seasons and constitutions, than in others. When . . . great numbers of these are generated, the vessels are distended, and the animalcules, pressing each other, cause a greater motion in the blood, and consequently, a quicker and stronger contraction of the arteries. The bodies of these creatures being very broad, and their heads small, they frequently attempt to pass through some minute vessels, which cannot admit them . . . a dangerous obstruction is necessarily made . . . The sediment of the urine is nothing else but these animalcules dead in great numbers which sink to the bottom of that fluid: as will appear by examining them on a good microscope. I could easily account for all the symptoms of every disease . . . and could show how even our thoughts depend on various figured animalcules. So that when we speak of maggots in the brain, we use a more litteral expression than we generally imagine. I shall . . . take another opportunity of discoursing . . . on the application of this Theory to practice; and endeavour to lay down a most certain method of curing all diseases whatsoever, by various treatment of these troublesome inmates.[165]

This parody could not have been effective if its intended audience of ordinary readers was not already familiar with the theory of *contagium vivum*.

Bradley never responded to any of Martyn's prodding. He married a wealthy woman who sold all her possessions to pay his debts, and then he fell ill and died in 1732, leaving his wife and child destitute. We know of this because she petitioned Sloane for aid, and probably succeeded.[166]

CONCLUSION: SLOANE'S CIRCLE AND CONTAGIONISM

Hans Sloane was a transitional figure who embodied a new sort of British science, one that owed little to university training or classical learning. He was born outside England and had an unconventional education from the time he left school at the age of 16 until the very end of his medical training. His training in chemistry was unusual, his training in botany exceptional. He was cosmopolitan in ways that extended far beyond the traditional grand tour of Europe. Although he traveled to Jamaica as a physician, not as a naturalist, he was one of the first Britons to conduct a successful scientific expedition overseas—and one of few trained in natural philosophy before embarking. If he had been born a century later, he might have become a scientist instead of a physician, but science as a profession did not yet exist. He was also fully committed to a Baconian or empirical approach to knowledge, not a theoretical or mathematical one—an epistemology favored by his teachers, the chemist

Nicholas Staphorst and the iconoclastic physician Thomas Sydenham. He had little or no training in mathematics and, evidently, little interest in iatro-mathematics or Newtonian physiology.

Quiet and self-effacing, a poor speaker always interested in listening to others, encyclopedic in his interests, generous, temperate, diligent, and efficient, Sloane attracted wealthy patients, married well, prospered in trade and business, and scaled the summit of British science and medicine. The baroque facade that grew around Sloane—the huge wealth, the palatial home, the enormous collections, the numerous dedications, the honors and offices, even the imposing bust showing Sloane in an enormous curly wig—all these have invited sniping and helped to hide his unassuming character. He was exceptionally ready to give others credit for his discoveries and published little, leaving the interpretation of his contributions as much to his enemies as his friends. It is easy to see him as an institution instead of as a person, and to underestimate his accomplishments.[167]

This view has persisted to the present. In a reassessment of his work, historian Maarten Ultee has noted that although "Newton has remained a household word for scientific genius, Sloane is remembered (if at all) as a collector of curiosities . . . but not as a scientist of note." Ultee quotes comments by historians including Joseph Levine who thought in 1977 that Sloane was "essentially a dilettante collector . . . certainly no philosopher"; J. L. Heilbron, who commented in 1983 that he was "a compulsive collector, who began by stuffing the Transactions with the trifles he enjoyed" and "printed tripe not only because he liked it but also because he received little else"; and Roger Lund, who claimed in 1985 that "Sloane symbolized all that was most excessive, self-aggrandizing and ridiculous in the activities of the modern 'virtuoso.'"[168]

These historians echo politically motivated claims by Sloane's Tory and conservative contemporaries, who envied his success and despised his empirical approach to natural knowledge. They have been joined by feminist historians who have minimized Sloane's role in inoculation in order to highlight the contribution of Lady Mary Wortley Montagu and by historians of imperialism and colonial expansion, who have implied that Sloane propped up European hegemony and slavery.[169] Other authors have a more favorable view. Wendy Churchill depicts Sloane as a critic, not a supporter, of conventional medical wisdom, but she believes that his unorthodox approach to medical theory was quickly overwhelmed by proponents of humoral medicine. James Delbourgo argues that it is anachronistic to depict Sloane as an apologist for slavery.[170] A new appreciation of Sloane's contribution to eighteenth-century natural philosophy has come from the rise of ecology and climate science as research fields; they have given value to the observations made by natural historians two hundred years ago and to the methodology of natural philosophers who observed and recorded the events unfolding before them instead of deploying mathematical theories or conducting experiments.

Sloane's agnosticism or even hostility toward scientific theorizing was unusually consistent. One can open just about any early-eighteenth-century

treatise and find an attack on unprofitable theories or inadequate grand systems, but usually this is only a prelude to an argument in favor of whatever
new theory the author wishes to promulgate. Sloane, however, followed
through, both by stripping such speculation from his own publications and
by opening the Royal Society and the *Philosophical Transactions*—as well as
his personal time and patronage—to an unusually broad array of aspiring par-
ticipants. He seems to have been equally open-minded about medical ideas.
It would be pointless to try to tease out how much of this was due to his
cosmopolitan background and how much to his Helmontian early medical
training, because the two were so closely related to each other. Something in
his early upbringing led him to seek out and value this sort of training, and
his training in turn steered him in certain directions.

Sloane served as an encourager, supporter, patron, and colleague to contagionist physicians, many on the fringes of the profession. His support
of foreigners and Dissenters contributed to the survival of a cosmopolitan
medical community in London that persisted in parallel with the more chau-
vinistic and conservative Anglican community. Most important, Sloane's
curiosity, cosmopolitanism, extensive correspondence network, and open-
minded approach to ethnography and folk medicine led him and his allies
in the Royal Society to investigate and promote smallpox inoculation. This
practice would ultimately contribute as much as published medical treatises
to the spread of contagionist disease theory.

AN ENGLISH TREATISE ON LIVING CONTAGION: BENJAMIN MARTEN'S *NEW THEORY OF CONSUMPTIONS*, 1720

INTRODUCTION

Although plague and smallpox were the most dramatic human diseases to appear in the early eighteenth century, the most remarkable work of that period did not concern itself primarily with either one. Instead it claimed to be about a less sensational though possibly deadlier ailment: "consumption" or phthisis.[1] In fact, the book offered a uniquely clear and comprehensive version of the theory that animate contagion was the cause of all febrile diseases. The book was entitled *A New Theory of Consumptions, More Especially of a Phthisis or Consumption of the Lungs*. Its author, Benjamin Marten, was an obscure London physician.

Marten's book, unlike earlier works by Kircher and Nedham, or works by his contemporaries, such as Bradley, not only argued that epidemic diseases were spread by living entities, but also emphasized the idea that they spread only by contact from person to person. The agents were not worms or flies that were omnipresent in the environment or wafted about in the winds; they were specific pathogens, each of which had its own life cycle and caused symptoms of specific species of disease. Thus Marten included both the *vivum* and the *contagium* to lay out (though briefly) a completely articulated theory of *contagium vivum*. More importantly, Marten did not accept a multicausal explanation for diseases; he stated that if these animalcules were not present, the specific diseases they caused could not occur.

Marten was probably on the fringes of Sloane's circle, and his work represents a fusion between the populist Helmontian tradition of viewing diseases as separate entities with their own life and the ideas about generation, reproduction, and speciation that were circulating among members of Sloane's circle of naturalists and taxonomists, such as Grew, Ray, and Sloane himself during this period. It exemplifies the close connection between a complete theory of *contagium vivum* and an ontological theory of disease that delineated diseases as distinct entities. This connection can even be found in the work of the quack M.A.C.D., who enthusiastically embraced the idea, indeed, took it to its logical if farcical extreme.

THE *NEW THEORY*

The title of Marten's book promised an "Enquiry . . . concerning the Prime, Essential, and hitherto accounted Inexplicable Cause" of consumption. This "Enquiry" constituted chapter 2 of the treatise. After an extremely clear clinical description of the symptoms of phthisis in chapter 1, and a brief romp through an assortment of authorities, including Hippocrates, Galen, Helmont, Paracelsus, Sylvius, Willis, and Morton, on its causes, Marten dismissed all their theories, revealing himself in passing to be well read. He argued that repeated observations had shown that consumption was contagious and spread through the breath. He quoted Morton's *Phthisiologia* to the effect that consumption "like a Contagious Fever, does infect those that lie with the sick Person, with a certain taint."[2] Marten attributed consumption and many other infectious diseases, including plague, venereal diseases, smallpox, measles, leprosy, all continued and intermitting fevers, and even chronic diseases such as melancholy and gout to "inimicable Animalcula, or wonderful minute Animals in our Fluids."[3]

Some people would find this theory strange, but microscopy had revealed a "World of Wonders" in the form of an infinite number of animals that were invisible to the naked eye, and "we may reasonably conclude that there are Myriads of others infinitely smaller and wholly imperceptible to our Eyes, tho' assisted with the best Glasses that can be made."[4] Moreover, innumerable species of living microscopic creatures had been found, and "there being no such thing as Equivocal [spontaneous] Generation," each such creature must emerge from an egg that was even smaller and thus capable of floating through the air. These eggs might lie dormant in the body for a long time and might pass from a mother to her unborn child.

As there were various species of animalcula, so there were various species of diseases. Marten quoted William Oliver's view that smallpox was caused by specific "seeds," and continued,

> How Distempers happen to rage in one Year, or Season of the Year, or in one Country or part of a Country . . . and how they are spread by Degrees, and are communicated from one Person, and from one Country to another, may by this Theory perhaps be more easily explain'd than by any other . . .
>   . . . For how can we better account for the regular types, the Small Pox, Malignant, and all other Continual and Intermitting Fevers, as well as many other Distempers, keep, and the peculiar Attributes and *Crises* etc. they have, than by concluding they are severally caused by innumerable *Animalcula*, or exceeding minute Animals that variously offend us according as their Species are different . . .

... It appears highly probable that Minute Animals, stimulating, wounding or gnawing the Parts they are lodged in, are the cause of these Diseases; and in a Word, there is possibly no Ulcer or Ulcerated Matter, but what is stock'd with *Animalcula,* and as these are of different Species . . . so those Ulcerations may be more or less stubborn or Inveterate . . . [5]

Marten also argued for the converse of this hypothesis: no disease could appear where there were no animalcula that were the specific cause of that particular disease, even when circumstances favored its development. In other words, he considered a specific animalcule to be both a sufficient and a necessary cause:

I have . . . observed, that sometimes Coughs which seem to threaten an immediate Consumption, cease almost of their own Accord without that Consequence, and leave the Patient in perfect Health, when other Coughs less troublesome, and to the Patients thinking less dangerous, quickly terminate in a deplorable Phthisis or consumption of the Lungs . . . The Reason of this is very plain, because if the Body be not pre-dispos'd to a Consumption . . . that is, according to my Theory, if it be entirely free from such species of *Animalcula* or very minute Animals, or their *Ova* or Eggs, that I imagine to be the Essential Cause of a *Phthisis,* that Disease, though the Cough is very violent or severe, will not happen.[6]

Marten asserted that if these creatures were omnipresent in the environment, in the air, food, or drink, then phthisis would be universal. That might happen in the case of plague, but not in the case of phthisis, which required close and frequent contact to spread. These animalcula were conveyed by parents to their children through heredity, or "communicated immediately from distemper'd Persons to sound Ones who are very conversant with them."[7] He also pointed out that an animalcular theory explained the geographic and epidemic patterns of particular diseases and that the life cycle of particular animalcula could determine the sequence of symptoms shown in certain diseases. He thought the tubercles that appeared in the lungs of consumption victims were obstructions caused by the animalcula, manifested by a dry cough. When the tubercles ruptured, the animalcula changed their shape or behavior and escaped from the ulcerations in the lung into the blood, causing more generalized symptoms, such as shivering and fever. He speculated that people might be constantly exposed to harmful animalcula or their eggs but could usually eliminate them without falling ill.[8]

Historian Charles Singer, writing in 1911, expressed disappointment in Marten's final chapter, which discussed Marten's method of treating phthisis and recommended a proprietary remedy without revealing its contents.[9] Marten conceded that his theory did not lead to any obvious therapies for consumption, because even if a remedy could destroy the animalcula themselves, it would not remove the damage they had done, any more than destroying a dog that had bitten a man would cure the wound it had made. Instead of seeking a cure based on an *a priori* theory of disease causation,

Marten argued, it was wiser to consider the therapies that had proved effective in experience and adapt the theory to explain their efficacy. Marten, unusually for an early-eighteenth-century physician, disapproved of harsh vomits, purges, and diaphoretics and warned against burdening patients with too many medicines.[10] He commended fresh air, exercise, a digestible diet, regular hours, and medicines that he compounded himself. There is little that Singer, writing before the age of antibiotics, could have offered that would have proved more efficacious.

More important, Marten's argument for therapeutic restraint had potentially important implications for the argument over animate contagion. Many physicians argued that if diseases were caused by animalcula, then they should be cured by remedies known to destroy vermin, such as sulfur and mercury. Consequently, if such remedies failed, they concluded that the disease in question was not due to animalcula. As such failures were far more common than successes, Marten's contention that removing the pathogenic organism would not necessarily undo the harm it had caused held the potential of saving the theory of animate contagion, as well as sparing hapless patients many unpleasant and dangerous remedies. He did note that insecticides such as sulfur and mercury were effective treatments for cutaneous infestations such as the itch.[11]

MARTEN'S FAMILY AND THE "ORACLE" PHYSICIANS

Until recently nothing was known of Marten's life. A lack of confidence in eighteenth-century orthography increased the uncertainty; several "Benjamin Martins" lived in eighteenth-century London, including a distiller and a distinguished optical instrument maker, to say nothing of numerous "Martyns." John Martyn, the botanist and Grub Street journalist, does not seem to be related to Benjamin. Confused cataloguing and record keeping compounded the problem. Singer was quick to dismiss the possibility that Benjamin Marten could have been related to a writer on venereal diseases named John Marten, whom Singer considered a disreputable character. There were no contemporary reminiscences to help ensure that when we find a record we have the "right" Marten. The fact that he signed his book "Benjamin Marten, M.D." helps, for there were few "Dr." Benjamin Martens, and it was uncommon for eighteenth-century authors to claim doctorates they had not actually obtained.[12]

The thorough research of Raymond Doetsch provided some intriguing hints, which have been confirmed and expanded by Dr. David Zuck.[13] In addition, a study by Harold Cook has supplied a much fuller context. It appears that Benjamin Marten was, in fact, John Marten's brother.[14] This relationship is important, because venereal diseases were among the very few illnesses that most medical authors agreed were spread by contagion. Another source provides one further small clue to Marten by listing "Benjamin Marten, M.D." as a subscriber to the second volume of Bishop Burnet's Whiggish *History of My Own Time* (1734).[15] It is probable that this is the correct Marten.

John Marten was a surgeon, and one of the most successful venereologists in London. He wrote in 1711 that he had been apprenticed to Joseph Green, treated the sick and wounded in Ireland, and had been in practice for twenty years.[16] He was also a supporter of the Dutch physician Johannes Groenevelt, a member of the "Oracle" group who was a highly trained and expert lithotomist.[17] Groenevelt had been prosecuted before the College of Physicians for malpractice in 1694 in a case that had become a *cause célèbre*.[18]

The College complaint against Groenevelt had alleged that he had improperly used cantharides (Spanish fly) to treat a patient with an ulcer in her bladder, leaving her bedridden.[19] After winning a civil suit brought by the patient in 1697, Groenevelt published a vindication of his therapy in the form of a Latin treatise on the use of cantharides. Entitled *De Tuto Cantharidum*, it appeared in the winter of 1697/8 and was dedicated to Richard Blackmore and two other physicians who had supported him during the malpractice controversy at the College. A second Latin edition entitled *Tutus Cantharidum* appeared in 1703 with a prefatory poem by the émigré Dutch physician Bernard Mandeville, who would later gain notoriety as the author of the *Fable of the Bees* and other radical works; this was Mandeville's first English publication.

In 1706, John Marten published an English translation of this work entitled *A Treatise of the Safe, Internal Use of Cantharides in the Practice of Physick. Written . . . by that Eminently Learned and Experienc'd Physitian, Dr. John Greenfield, Member of the College of Physitians in London, in his own Vindication . . . Now translated into English with his Approbation, by John Marten, Chyrurgeon. To which are added . . . Observations . . . concerning . . . CANTHARIDES . . . As also An anatomical and Chymical Account of that INSECT, with some very curious Observations . . . thereto, made by the Fire and Microscope . . .* [20]

Groenevelt came to England in his late twenties and probably had difficulty with English; he may have worked with Marten on the translation. The title page suggests that John Marten was a well-read man capable of undertaking the English translation of a Latin work and also that he was familiar with chemistry and microscopy, if he did not carry out the work himself. In his preface, Marten also thanked the anatomist Edward Tyson for suggesting some additions to Groenevelt's work, showing that the two men were well acquainted. It is also evident that John Marten favored Groenevelt's cause.

As the malpractice proceedings continued, Groenevelt's case became entangled in the heated conflict between the apothecaries and the physicians over the Dispensary. Most of Groenevelt's supporters, including Tyson, sided with the apothecaries in opposing the Dispensary, which would have strengthened the control that the College of Physicians exercised over London medical practice.[21] Perhaps the fact that John Marten's brother James was an apothecary just freed from his apprenticeship also inclined him to side with Groenevelt.

In the early years of the century, John Marten published a book of his own: *A Treatise of all the Degrees and Symptoms of the Venereal Disease in both*

Sexes, which proved extremely popular, reaching a seventh edition by 1711.[22] It contained an appendix about sexual relations that "veered into pornography."[23] The appendix was sold with the book but printed on a separate set of pages, enabling the two to be separated.[24]

Marten's success infuriated a rival venereologist, John Spinke, who initiated a pamphlet war and published many advertisements criticizing Marten.[25] In Defoe's newspaper, *A Review of the State of the British Nation*, Spinke claimed that John Marten, "The Hatton Garden Clap-preventer," had not translated Groenevelt's book, although it carried Marten's name on its title page. Spinke continued,

> This makes People suspect that the said John Martin is an imposing, cheating quack, and an ignorant pretender, and that his Letters, Stories of Cures, pretended Medical Secrets, etc., are (like his Pretentions of being the Author of the said Translations) but so many Shams and Impositions on the Publick; as to which Particulars, the said Martin is desir'd to publish the Truth of the Matter, in some one of the News-Papers, that he has impudently monopoliz'd for his own, his Brother Ben, the Chymical Soap-Boiler, and his Brother Spooner, the Taylor.[26]

Spinke apologized to Benjamin Marten shortly afterward: his book *Quackery Unmask'd: or, Reflections on the Sixth Edition of Mr. Marten's Treatise on the Venereal Disease* (1709) included a printed correction stating, "N.B. Since the Publication of this Book, I am credibly inform'd that Mr. Benj. Marten never was (as I was told) a *Soap-Boyler*; but he has by his own Industry, acquir'd a competent Knowledge of the Theory and Practice of Physick; and does on all Accounts behave himself as a Gentleman, meriting a good Character."[27]

In the body of the book Spinke described Mr. Spooner as a tailor who had married John Marten's sister.[28] He also described John Marten as a poor tailor's son who had been a surgeon's apprentice.[29] He wrote of knowing "the names of near Twenty of his pretendedly famous News-Paper Medicines."[30] Spinke may have confused Benjamin Marten the doctor with the Benjamin Martin "citizen distiller" who also lived in London at this date. However, it is clear that Spinke, at least, linked the John Marten who was his professional rival with the John Marten who was Groenevelt's translator, and also named both James and a respectable practitioner of physic, Benjamin, as John's brothers. In his own work, John had said that he had turned over all of his practice except venereology to his brother James, so it appears that the brothers worked closely together.[31]

John Marten, Benjamin's brother, has also been identified as the probable author of another notorious eighteenth-century work, *Onania; or the Heinous Sin of Self-Pollution*, which blamed masturbation for a wide range of debilitating chronic diseases.[32] The first edition appeared in about 1710, and subsequent editions, undated until the fourth of 1722, continued to proliferate. One reason for its huge success was its interactive nature: a number of London booksellers not only offered the book for sale but facilitated contact between patients and the author. Some could arrange for personal meetings; others could leave notes containing an account of their concerns—both medical and spiritual—and receive a written reply.[33] Subsequent editions of *Onania* include some of these first-person accounts and recommended medications that, conveniently, could be obtained from the same booksellers.

The books and associated activities must have been extremely lucrative: John Marten died in 1737 with an estate of ten thousand pounds, though some of that may have come from his wife.[34] His bequest of just one shilling to his brother Benjamin suggests a fissure within the family. Perhaps Benjamin wished to distance himself from the dubious activities of his brother, or perhaps John Marten had earlier assisted his brothers by supporting their education and there was an understanding that this was sufficient. If they were the sons of a poor tailor, it would have been difficult for Benjamin to study medicine without help from his successful brother. Groenevelt, another possible patron, was by then in financial straits of his own.[35]

It is evident that John and James Marten and their brother-in-law were marketing proprietary remedies and also that they were associated with the reformist doctor Johannes Groenevelt. Thus, it is very likely that when Groenevelt referred to his "assistant" Benjamin Marten in 1710, he was referring to the brother of his translator: John Marten, the venereologist. As this was the only "Benjamin Marten" likely to have been practicing medicine at this date, this was probably our author of *A New Theory of Consumptions*, published a decade later.

In 1710, Groenevelt wrote:

> Ever since it has pleased God to deprive me of, and take to himself the Children, with which he had blessed me, I strove to agree with some Ingenious young Physician, that should delight in Lithotomy, to whom I might freely and without reserve impart the fruits of so long a Practice, as well as early Studies; which at last I accomplish'd to my Hearts desire, in pitching upon the Skillful Mr. Benjamin Marten, whose Industry and Application I have assisted with the many observations, which an almost Forty Years Experience has furnish'd me with.[36]

Groenevelt added that rumors that his eyesight and strength or steadiness of hand had failed were entirely false, but anyone concerned about his age could have Marten perform the operation with Groenevelt's supervision, and he hoped that patients would seek out Marten after his death.[37] Despite his difficulties with the College of Physicians, Groenevelt was a well-trained physician who had been a pupil in Leyden before obtaining his MD from Utrecht in 1670 with a thesis on bladder stones. His biographer, historian Harold Cook, described this as "far and away the most learned" of the theses on the stone published in Utrecht and Leyden during this period. Although he was declining at this date, it is also clear that Benjamin Marten, his assistant, was intimately associated with him and his work and was trained in surgery as well as physic.

Francois dele Boë Sylvius, Groenevelt's teacher, had revolutionized the understanding of consumption, or phthisis. He was the first medical writer to note that tubercles were typically found in the lungs of consumptive patients and to describe their suppuration into abscesses and cavities.[38] As noted in chapter 2, Groenevelt was also a friend of the Dutch naturalist Stephen Blankaart, who believed that venereal disease was caused by microscopic venomous animals. The fact that the Marten brothers consistently spelled their own surname with an "e" hints that they themselves may have had Dutch origins.

In 1715, Groenevelt's name appeared as the "author" of an introduction to physic originally written by an anatomist and chemist from Louvain, Franz vanden Zype or Franciscus Zypaeus, described by Harold Cook as a follower of Sylvius.[39] This work had appeared in 1683 as *Fundamenta Medicinae Reformatae Physico-Anatomica* and was apparently very successful.[40] A London edition in 1714 appeared anonymously, but a new Latin edition in the following year, also published in London, included the names of both Zypaeus and Groenevelt.[41] Cook comments that Groenevelt plagiarized the text but added an appendix.[42] An English translation that also appeared in 1715 carried the title *The Grounds of Physick*.[43] It seems unlikely that Groenevelt prepared this translation because he had previously enlisted assistance to translate his own work from Latin into English.

Chapter 4 of this work discusses contagious and noncontagious diseases in a manner that is reminiscent of Sennert's later work. It defines contagion as a corruption or effluvium that is emitted from a distempered body and communicates diseases such as plague, leprosy, pox, consumption, canine madness, and the itch from body to body. Even though all diseases emit effluvia, some were not contagious because they lacked the "seminary" of a disease that contained its entire energy in a very small bulk.[44] The plague was contagious even though it was not (originally) produced from a contagion. Contagion consisted of sharp particles that could either be produced by the body or could enter the body from outside either through the air, the pores of the skin, clothes, or contact.[45] There were harmful or infectious entities such as poisons, the fumes from caverns, putrefactions, and effluvia from dead bodies that were not contagious because they had not been conveyed from a sick person to another person.[46] The *Appendix*, which focused on diagnosis and therapy, discusses the different kinds of fevers in different terms, attributing fevers to air, diet, and manner of living without mentioning contagion as a cause.[47]

Perhaps Groenevelt had encouraged Benjamin Marten to read the work of Zypaeus. Perhaps he had acquired an interest in consumption from Sylvius and passed it on to his trusted assistant. Perhaps he had speculated about the role of animate pathogens in causing diseases, a topic that his Dutch friends had discussed. In any case, Marten's treatise combined these topics into a new synthesis.

The likelihood that Groenevelt trained the author of *A New Theory of Consumptions* also connects Marten to the reforming physicians at the Royal

Society. Groenevelt had first met Sydenham in 1682, and wrote a correspondent in Holland that "I am especially well acquainted with him."[48] Sloane would join Sydenham's household in 1684; Groenevelt may have met him there. His partner, John Pechey, was Sydenham's first English translator in 1696 and included a biography of Sydenham in his edition. Another partner, Christopher Crell, wrote verses for the beginning of Sydenham's *Processus Integri*.[49] The same year, Groenevelt served as an intermediary between members of the Royal Society who were interested in the Japanese practice of moxibustion and his childhood friend, the Dutch doctor Willem ten Rhijne, who had written a manuscript on the subject, further evidence of the Society's interest in folk medical practices from around the world. As we have seen, Bernard Mandeville, an émigré physician who sometimes consulted with Sloane, wrote the preface to one of Groenevelt's works.

In 1707, Groenevelt gave a copy of ten Rhijne's Dutch treatise on leprosy to Sloane, who may have intended to commission a translation. Groenevelt borrowed the book back for more than a year.[50] Later in his life, he became more dependent on his relationship with Sloane. In 1704, Groenevelt asked Sloane to encourage his neighbor, who had a son with bladder stones, to come and see him perform the surgery on another patient.[51] In 1714, the year before his death, Groenevelt wrote to Sloane about the "former favors" Sloane had done him, and a few months later he asked Sloane to put a stop to rumors that he was too old to operate safely.[52] Groenevelt also knew Richard Mead, who had studied in Utrecht and Leyden. In a book published in 1704, Mead mentions attending the postmortem of a child who had died of kidney disease and discussing the case with Groenevelt, who had attended the boy when he was alive.[53]

Thus, Groenevelt was connected to Sloane and other Royal Society members during the period when he was sharing his practice with his assistant, Benjamin Marten. Perhaps Benjamin Marten, like his brother John, also had contact with Edward Tyson, one of the most skilled anatomists of his day and a dogged supporter of Groenevelt in his battle with the Royal College of Physicians.[54]

Groenevelt was suspected of religious as well as medical heterodoxy. Among his partners in the clinic was the Polish Socinian Christopher Crell Spinowski, later Christopher Crell.[55] Perhaps because of this association, Groenevelt himself was accused of Socinianism by other members of his congregation in the Dutch Reformed Church of Austin Friars. He left the congregation after a dispute involving the minister, who had defended him, and whom, in turn, Groenevelt defended.[56] Cook has written that the Censors of the College of Physicians suspected Groenevelt because "they did not consider him to possess the right sort of character. He was certainly not English. Not only did he therefore possess the wrong medical ideas, but he possessed the wrong religion, and his friends were among the London dissenters, experimentalists, and general troublemakers."[57]

If we have the right "Benjamin Marten," therefore, his mentor was a Dutch Nonconformist who struggled to survive in London despite his superb education. A friend of Sydenham's, he had ignited explosions within the medical establishment and had turned to Sydenham's protégé Hans Sloane in times of trouble. Marten's book, then, probably emerged from within a circle of medical reformers and religious Dissenters who associated with Sloane and held a liminal position in English medicine, neither beyond the pale nor securely within the fortress.

The many references Marten gives to other works raise the question of how he found them. The British Library was a generation away. The College of Physicians' Library was probably restricted to members.[58] It is possible that he was wealthy enough to gather his own library. He purchased at least four books by subscription, but none was in medicine or science.[59] Like many others, he may have haunted used-book stalls and auctions or asked friends to borrow books for him. He may also have had access either at first or second hand (through a friend) to the libraries amassed by Sloane or Mead, or possibly to a working library owned either individually or collectively by the Oracle physicians; Pechey in particular must have had access to a large library in order to support his work as an active translator, compiler, and author.

MARTEN'S MARKETING: SECRECY, PUBLICITY, AND PROFESSIONAL STATUS

In the context of early-eighteenth-century practice, the fact that Benjamin Marten signed his preface with Theobald's Row, presumably his professional address, combined with his reference to his own secret remedies, would have raised eyebrows among his colleagues, though it probably was not sufficient to brand his work as mere puffery.[60]

Marten himself was defensive about this in the preface to his second edition:

> The giving my own Medicines . . . may probably occasion some smart Animad-versions from false Wits, and common Dealers in Scandal . . . but as whatever such may say or write, is not worth a wise Man's Notice, so I shall take Care not to expose my Folly that way.
> This I'm well satisfied of, that those Persons who labour under Consumptions, and think proper to apply to me for Cure, will be very well pleas'd with having Remedies from my own Hands, on the goodness and Efficacy of which, I, as well as they, can fully depend.[61]

Marten surely felt he deserved a modicum of self-promotion after slogging patiently and discerningly through whole libraries of old medical treatises, but his reference to his own secret medicines may have compromised his chance to gain a serious professional audience for his ideas, damaging rather than enhancing his reputation. Like other notorious eighteenth-century quacks, such as William Brodum and Samuel Solomon, he bought his Aberdeen MD. Furthermore, his brother was a venereologist who wrote titillating works and peddled dubious remedies.[62] On the other hand, if he was Groenevelt's assistant, he was well trained in a traditional way (unlike most of the quacks) and had practiced as a surgeon; the purchase of an MD by surgeons seeking to move up the professional ladder was a widely accepted practice.[63]

Standards of professional conduct were changing during the eighteenth century. In the late seventeenth century a group of reformers including Everard Maynwaringe had called for physicians to learn about the ingredients of medicines and to prepare their own remedies to shield patients from rapacious apothecaries.[64] By the early eighteenth century, however, respectable physicians were not supposed to act as apothecaries or druggists, preparing or selling remedies themselves, although they might gain indirectly from their pharmaceutical discoveries or inventions by publishing accounts of their successful treatments and hoping that enhanced reputations would send more patients to their doors. Although the College set an example for other parts of England, there was variation in what constituted permissible or socially acceptable behavior. It was still not unusual for reputable physicians to lend their names to secret remedies, or even to have their names appropriated for remedies without their permission.[65]

In the early years of the eighteenth century, personal advertising was often the province of such medical pretenders as "Dr." Ward and of surgeons claiming to treat shameful disorders such as venereal disease.[66] Ward was a dry-salter with no training in medicine who advertised his "pill and drop" for many diseases, including venereal disease, and created three private charity infirmaries in London under his supervision.[67] Although promoting secret remedies in this manner left a physician open to criticism, (especially from his rivals), respectable physicians in the early decades of the century still gained fame and fortune in this way, and managed to keep their reputations.

In 1696, Dr. William Cockburn, a Licentiate of the College of Physicians, developed his secret remedy for dysentery while serving as physician to the Navy, tested it on a hundred sailors, patented it, peddled it to the Navy for forty years, and received a letter of gratitude from the King for his national service.[68] He also published a book on gonorrhea in 1713, promoting a secret remedy for that ailment. Cockburn was criticized for his profiteering by Thomas Dover and by a "cabal" of academic physicians, including John Freind.[69] In 1730, Cockburn retaliated in *The Danger of Improving Physick*, claiming that "the most learned physicians are always most subject to obloquy, on account of their superior knowledge and discoveries."

Physicians also sought indirect ways to promote their own wares. Even so respectable a physician as James Jurin, President of the Royal College of Physicians in 1750, freely recommended his own "*Lixivium lithontripticum*" for the stone. Jurin kept the formula secret but did not compound it himself, instead sending customers to an apothecary. Critics still suggested that he was little removed from quackery.[70]

Jacob de Castro Sarmento, a Sephardic Jew, included a puff for his "Waters of England" in Part 1 of his book: *Materia Medica: Physico-Historico-Mechanica, Reyno Mineral* (1735). He prepared his patent medicine and exported it to Portugal for the treatment of fevers, especially intermittent fevers (malaria).

The main ingredient seems to have been quinine. This remedy, not surprisingly, continued to be popular for decades after his death.[71] Although he was accused by Meyer Low Schomberg, his bitter enemy, of being an "ass and a fool" and of prescribing inappropriate and dangerous treatments, de Castro Sarmento was a respectable practitioner and scientist who was both a Licentiate of the College and an FRS.[72]

Robert James, an intimate friend of the lexicographer Dr. Samuel Johnson, had a bachelor's degree from St. Johns College, Oxford, and an MD by royal decree from Cambridge.[73] His relentless promotion of his dangerous and expensive "Dr. James's Powders"—a preparation of antimony—was notorious but did not prevent him from being made a Licentiate of the College in 1745. He sold hundreds of thousands of doses through the bookseller John Newberry, who arranged for what we might call "product placements" in a variety of publications.[74]

Hans Sloane's behavior is more typical of the emerging professional standards among "respectable" practitioners of the first half of the century and reveals the contemporary ambivalence toward any commercial peddling of remedies by physicians. Although he was certainly well educated by his travels on the Continent, Sloane never matriculated at a university or earned a bachelor's degree. His only strictly medical treatise was a slight tract on "Sore Eyes," which described a secret "liniment" that Sloane claimed he had purchased from an apothecary's assistant under a solemn vow of secrecy. Late in his life, after treating some five hundred patients with this secret remedy, Sloane decided to break his promise and publish the recipe, with the weak excuse that he had modified the formula by adding viper's fat to the formula. In fact, the main ingredients of this remedy had long been in use.[75]

In his pamphlet, Sloane commented, "I cannot charge myself with making the least mystery of my practice. For in consultations . . . I have always been very free and open; far from following the example of some physicians of good morals and great reputation, who have on many occasions thought proper to conceal part of their own acquired knowledge."[76] The blurred boundaries of acceptable practice are evident in Sloane's ambivalent emphasis on the "good morals and great reputation" of physicians who concealed their knowledge and on his claim that he himself would not engage in similar activities—although he had in fact kept his remedy a secret for a long time.[77]

Within the context of urban medical practice in the first decades of the eighteenth century, Marten's promotion of his own remedies would not have seemed completely out of place despite his doctorate. There were many practitioners, some of whom had received medical instruction superior to anything offered at Oxford or Cambridge, who could not obtain a College license because their training had been too practical or because they were too foreign or simply because they could not afford the fees. Some of them were even Fellows of the Royal Society. Avoiding College oversight as much as possible, these doctors worked as general practitioners, dispensing medicines, offering therapeutic and surgical advice, even performing minor surgical procedures for a growing urban market of ordinary people who could not afford the fees of a College physician or a team of medical and surgical consultants. In order to keep their fees affordable, they relied on volume and the sale of ancillary products; this made advertising essential.

There were two sorts of practice that were especially likely to encourage such "crossover" practices: group practice "dispensaries," such as the Oracle group, and venereology.[78] It was common for venereologists and other fringe practitioners to advertise by publishing not only actual advertisements but also pamphlets touting their skill and discretion. It was less common for such practitioners to publish serious medical treatises aimed at an educated audience; most of their works, even when longer than a half column, were blatantly self-promoting.[79] The Oracle physicians, all of whom were Licentiates, reached higher, although they also engaged in marketing. In the late seventeenth century, the Oracle physician John Pechey included a note about "our cathartic pills" sold at his own house in Basing Lane at the end of his medical compendium *Promptuarium Praxeos Medica* (1693), and he placed another advertisement for "Pechey's pills" at the end of his translation of Sydenham, which appeared in 11 editions between 1696 and 1740.[80]

The College of Physicians strongly disapproved of Pechey's activities, fining him repeatedly in 1688 for advertisements that included his new address and fees, and later for affixing his name above his door. His excuse was that he was not the only offender. He was still publishing newspaper advertisements in 1700. Another member of the Oracle practice, Richard Browne, advertised his purging pills, which were sold in many locations, including his own home. Like Pechey, Browne was cited by the Censors of the College of Physicians for advertising his medications and consulting with unlicensed practitioners.[81]

The radical physician Bernard Mandeville defended this practice in his dialogue *A Treatise of the Hypochondriack and Hysterick Passions* published in 1707. Mandeville's spokesman, Philipirio, was an unabashed empiric who eschewed traditional university training to learn from experience.[82] He compounded his own drugs to ensure their purity and to lower the cost to his patients. Mandeville himself "lets the reader know that he can be reached through the bookseller, although he vigorously denies that this publicity constitutes quackery."[83] Mandeville's later work, *The Fable of the Bees* (1714), defending consumption and materialism, may have grown from his experience in a heavily and often irrationally regulated profession that tried to control medical entrepreneurs.

Although neither Browne nor Pechey held an MD, and although Mandeville, who held a Leyden MD, never obtained a license from the College of Physicians, they were still learned men who wrote serious, well-researched, and competent medical works despite their stated hostility to medical Scholasticism. Contemporaries may have disliked their commercialism, but they were not in the same social category as uneducated hucksters who treated the desperate and credulous.[84]

In addition, as the work of Nedham, Gideon Harvey, and Maynwaringe shows, there was an established genre of serious quasi-popular self-help literature in English, including works on medicinal plants and preparations, pharmacopoeias, family medicine, pediatrics, women's diseases, and the like, intended to aid those who were hoping to solve their medical problems without seeing a practitioner, or who wished to become medically literate.[85] Mandeville's 1707 dialogue on hypochondria and hysteria falls into this category. These were more respectable than quack pamphlets, but by the eighteenth century, such works did not normally mention secret remedies. They were often written by younger physicians hoping to increase their reputation or by those who were seeking reform in both social mores and the practice of medicine. Often both motives were at work. Physicians at the summit of the profession did not compile such works—they were too busy in lucrative practices—but an ordinary physician could do so without sinking himself beyond reproach professionally.

Marten's work falls in this category: a serious, literate, and carefully researched work based on extensive clinical experience and written in plain and lucid English. It seems to be written more for a lay than a medical audience, but that does not mean that it was not read by other physicians. Marten's reference to a secret remedy may have raised eyebrows, but it probably would not have made him an outcast.

MARTEN'S READERS: DISTRIBUTION OF THE *NEW THEORY*

The contemporary impact of Marten's *New Theory of Consumptions* is difficult to gauge. By Singer's day, it had disappeared from view, and it may have been disregarded or overlooked even in Marten's own time.[86] Few contemporary authors cited Marten, and Singer could find no explicit references in any other professional publications, even in works on phthisis. The two possible indirect references Singer found both dismissed the work in a few slighting sentences. On the other hand, the book was well advertised, and in the preface to the second edition in 1722, Marten writes that his work had been approved by "several learned Gentlemen."[87]

There was at least one very serious reader who was unknown to Singer: Cotton Mather referred to it at great length in his unpublished work "The Angel of Bethesda." The very fact that a copy had made its way to Boston suggests that the work was not seen merely as ephemeral advertising. Mather evidently obtained the book shortly after its publication, read it with intense interest, and used it for most of a chapter on the "new theory" that animalcula caused disease, adding a summary of Bradley's claim (from his *New Improvements of Planting and Gardening*) that horticultural blights were also caused by microscopic insects.[88] The animalcular hypothesis was also mentioned, though in passing, by Mather's colleague the Reverend Benjamin Coleman in his pamphlet "Some Observations on the New Method of Receiving the Small Pox" (1721) in support of inoculation. Thus, Marten can be credited with some role in the successful introduction of smallpox inoculation into Boston and, because of the importance of the Boston example, into Europe as well. It is possible that Mather even mentioned Marten's work to a young

Benjamin Franklin, who visited him at about the time he was composing "The Angel of Bethesda."[89]

Raymond Doetsch identified one additional reader, Edward Barry, whose *Treatise on A Consumption of the Lungs* (1726) opposed Marten's theory.[90] Barry, a devoted Boerhaavian, wrote that the claim by "the author of a late hypothesis" that animals had been found in the victims of disease mistook effects for causes. In a note, Barry cited "Martin [*sic*] on consumptions."[91] Barry became a very distinguished physician, and his treatise was probably widely circulated, but evidently few readers chased down his misspelled citation.[92]

Only a few copies of Marten's book have survived: Singer was very fortunate to find a second-hand copy while browsing in a bookstall. At the time he wrote, he could only find four known copies. The publication of his article in 1911, although it appeared in the obscure periodical *Janus*, set off a minor gold rush among bibliophiles, and major research libraries obtained several additional copies. A few copies were acquired by libraries in the normal course of events before Singer's article was published. There are probably copies in private collections that have never been catalogued or in libraries that have not reported their older holdings to a major database.[93] Copies that were not listed in earlier Union catalogues reappeared in the Royal College of Surgeons and in Newcastle.

Little is known about the provenance of most of the copies, but we can gather a few hints from them about their likely readers in the eighteenth and early nineteenth centuries. At the time of writing, 28 copies are catalogued in modern libraries. Most bear no evidence of their previous owners. Among the copies of the first edition, there is one in the Christ Church Library, Oxford, that came from a bequest of Charles Boyle, fourth Earl of Orrery (1674–1731), a distant cousin of Robert Boyle. Orrery was a bibliophile who collected both scientific instruments and works on science and medicine. He was a patient of Richard Mead, who obtained his release from the Tower of London on medical grounds toward the end of his life.[94] There is no reason to believe that Orrery deliberately chose this book among the many that he collected or that he took the time to read it, but he probably purchased it at about the time that it was published.

A copy at Newcastle University once belonged to the Newcastle Infirmary, but the date that the Infirmary acquired it is unknown. The copy in the Royal College of Surgeons similarly has no provenance.[95] Of the second editions, the one in the Boston Medical Library has no provenance, but the library itself was founded by Oliver Wendell Holmes, a champion of the theory that puerperal fever was a contagious disorder.

The most interesting copy is a second edition at the Wellcome Library, which came from the Medical Society of London and was in their collection by 1803. The probable source was either the library of Dr. James Sims (1741–1820), who donated six thousand books to the Society in return for an annuity in 1802, or the library of Dr. John Coakley Lettsom (1744–1815), who founded the Society in 1773.[96] Obviously, either man must have acquired it from another owner. The presence of this copy makes it likely that the book was available to members of the Medical Society, either by loan from Lettsom or Sims or from the library. Lettsom, Sims, and many members of the Society favored contagionism but did not cite this work. By the early nineteenth century, the Medical Society also owned a copy of the first edition. This limited evidence again suggests that the book was not treated as a mere piece of ephemera, but was purchased, if not read, by both gentlemen and doctors who passed it to their heirs, sold it to others, or donated it to medical institutions.

Marten's book was the clearest and most complete exposition of a theory of *contagium vivum*, but it reflected a line of thought that was already present in the early eighteenth century, as evidenced not only by his copious quotations from earlier works, but also by several contemporary works that attributed plague and venereal diseases to minute animalcules. Daniel Le Clerc's *History of Worms*, first published in Latin, appeared in an English translation in 1721, one year after Marten's first edition.[97] In one section of this very long book, he comments on the theories of Kircher and others that worms could be found in the blood of people suffering from malignant and contagious diseases. Le Clerc took a position of qualified skepticism that probably reflected the mainstream medical response to Marten's theory:

> If Worms of any Kind or the Seeds of Worms are contain'd in the Blood . . . we may easily understand why they should be found in . . . all other Parts of the Body . . . But somebody ought first to shew, that all those Worms really do exist by Examples and Facts . . . and also give a more accurate Description . . . that we may be certain of the Thing . . . There is not one Reason for doubting of Kircher's Blood-Worms; nor are his Experiments so safe, that we can give Credit to them . . . The Silence of Leuvenhoek concerning these Worms, who was certainly much more skilful in the Management of the Microscope than Kircher, seems to be an Argument that these Kind of Worms are perhaps imaginary.[98]

THE APPROPRIATION OF ENGLISH CONTAGIONISM: THE *SYSTÈME D'UN MÉDECIN ANGLOIS*, 1726

If Marten's *New Theory* was a serious work, the same cannot be said for a small work entitled *Système d'un Médecin Anglois* published in Paris in 1726.[99] The full title, translated, is "The System of an English Doctor on the cause of all Kinds (*espèces*) of Diseases with the surprising configurations of the different sorts of small Insects, which can be seen by means of a good Microscope in the Blood and in the Urine of different Patients, and also all those who will become [ill]." This claims to be a translation by one M.A.C.D. of a manuscript entitled "Systems of an English Doctor, on the nature of God and of Souls, on the Generation of every thing, on the cause of all sorts of Maladies, and on their cure, collected and made intelligible . . . to everyone."[100]

The author explains that he has only translated the third section of this work because the first part would be contrary to the revealed truth of religion, the second part would wound the modesty of the chaste, and the fourth would not only make all men, but also all women, their own doctors.[101] In the sequel published the following year, however, the author offers a different framing story, writing that in 1715 he became acquainted with an English doctor who was passing through on his way home from Isfahan in the company of the Persian Ambassador and that this treatise was the result of their conversations.[102] The author claims that an enormous range of illnesses are caused by the biting and gnawing of tiny insects, visible only with the aid of an especially good microscope. He then provides dozens of pictures of various animalcules that cause such ailments as venereal diseases, plague, being too fat, fainting, jaundice, paleness, boils, abscesses, ulcers, whitlows, apoplexy, vertigo, aging, insanity, headache, scrofula, and so on. In fact, it seems that he enumerates every ailment he can possibly imagine, each with its own little picture, which is supposed to be an enlargement of an animal seen through his microscope.[103]

One of his drawings was a relatively accurate illustration of an itch mite ("*gale*").[104] This suggests that one of the sources for the *Système* was the book on human worms by Nicholas Andry de Boisregard.[105] Andry preceded the text of his work with a series of plates containing pictures of various sorts of worms taken from different sources. Plate 1 included a variety of insects at the top; the bottom of the plate was a degraded and reversed reproduction of a plate from the *Acta Eruditorum* of 1682. This had accompanied articles by the naturalist Michael Ettmüller and illustrated both the microscopic "crinonibus" that Ettmüller thought caused a disease known as "Morgellons" and the "sironibus" or itch mite. Andry's plate seems to be the source for the even more degraded drawing in the *Système*.[106] Later commentators, unaware of this appropriation, expressed surprise at the accuracy of this image.

M.A.C.D. is almost certainly the "certain Quack, whose name was Boile" mentioned by Jean Astruc in his *De Morbis Venereis*, published in 1736, a decade after the first volume of the *Système*.[107] According to Astruc, Boile maintained that "all Diseases were produced by *Animalcula* in the Blood, and different Diseases by different *Animalcula*; that there were other Animals, which were capital Enemies to these noxious *Animalcula*; that he was well acquainted with the several Kinds of pestiferous *Animalcula* . . . as also with their several Enemies . . . and likewise with the several Medicines." This is a good summary of the *Système* and *Suite du Système*. Astruc tells us that Boile had a trick microscope. He could, by sleight of hand, insert various slides that either exhibited animalcules swimming vigorously around or only showed empty fluid as if the animalcules had all been destroyed, and he gave demonstrations to throngs of visitors in Paris.[108] As soon as the construction of the microscope was revealed, he fled. "The Forgery being exploded, Physick was again restored to its antient Laws, and happily retrieved from Disgrace."[109]

There are many references to the *Système* by later authors, who usually explain that the entire work was an imposture.[110] A few authors who seem to have stumbled on the book without any information about it took it seriously as an anticipation of the germ theory; others comment that the uncovering of this fraud brought the entire animalcular theory into disrepute.

Despite its fraudulent intent, some aspects of this publication are worth noting. The first is that its title and framing story attribute this "discovery" to an *English* physician, suggesting that French audiences associated contagionist theories with English doctors. Second, here again, we find the connection between the idea of specific diseases and specific animate causes. M.A.C.D. insists even more firmly than serious contagionist authors that every species of disease can be connected to a distinct species of insect that causes it and that it can be cured only by a medicine that is fatal to that particular species of insect.

Third, the author, even if he was deceptive, was a competent logician. After criticizing the prevalent theory that fevers resulted from a mixture of acids and alkalis effervescing in the blood, he points out (correctly) that his system better explains how diseases are communicated, how they increase, why different remedies work in different diseases, why different diseases affect different parts of the body, and why there are quartan, tertian, and continued fevers as well as episodic diseases such as epilepsy. These variable phenomena are more easily explained by the discontinuous behavior of insects, which lay eggs, hatch, increase, and move to the most suitable parts of the body, than by the constant presence of acids and alkalis in the body.[111] He claims that a further proof for the theory comes from the efficacy of simples; these are nothing but poisons that kill specific sorts of insects. The author probably planned to sell "specific" remedies to his marks after revealing the microscopic insect that supposedly was causing their illness.

It seems likely that this episode, which was well reported at the time, made subsequent medical authors more cautious about expressing support for theories of animalcular contagion lest they be accused of being frauds or supporters of quackery or appear foolish or credulous.[112] Andry's work also attracted ridicule: French wits referred to him as "*homini verminoso.*"[113]

PIERRE DESAULT'S *DISSERTATION SUR LES MALADIES VÉNÉRIENNES*, 1733 AND 1738, AND JOHN ANDREE

Interest in theories of *contagium vivum* receded after 1726, but as late as 1733 the French physician Pierre Desault (1675–1737) of Bordeaux published a treatise on venereal disease. He argued that the "leaven" of venereal disease consists in "imperceptible Worms, which upon approaching are communicated from one Body to another, and afterwards multiply in the Person who has received them . . . it is probable all other contagious Distempers proceed from Worms, as the Hydrophobia, Scurvy, Small-Pox, Plague, etc."[114] Desault believed that these venereal worms might remain dormant in the body until old age or excesses reduced resistance and allowed them to multiply. Desault's work was reprinted in Paris in 1738, and in the same year John Andree translated it into English.[115] After that it apparently disappeared from view until French medical historians rediscovered it at the end of the nineteenth century. Hector Grasset commented in 1899 that "French medicine owes this man an ample apology."[116]

Desault extended his argument to rabies and tuberculosis, which he attributed to the formation of tubercles in the lungs, not the ulceration that followed.[117] He explained that tubercles eventually ruptured after "worms" had multiplied within them, causing a temporary alleviation of symptoms that was often mistaken for a cure; in fact the worms were increasing and spreading, and the disease had become contagious.[118] He likened this process to the formation of tubercles in scrofula, "the Phthisis may be called the Scrophulae of the Lungs."[119] Unlike Marten, Desault did not see worms as the only cause of consumption; he also blamed grief, cold, and "acid and coagulating Juices."[120] Because all these diseases were spread by "worms"; however, he believed that they could all be cured by similar methods that included mercury rubs, horseback riding, and other remedies.

Desault's translator, John Andree (1697/8–1785), was an English physician of Huguenot descent about whom little is known. Andree obtained an MD from Rheims in 1739 at the age of 42 and became a Licentiate the following year. He was a founder of the London Hospital in 1740 and served as its only physician for several years. From the beginning, he used the London Hospital as a source of clinical information for his publications, which reveal a clear and precise observer. Andree tried and failed to become a Fellow of the Royal Society in 1740.[121] He passed his interests down to his son, also John Andree (1749/50–1833), who studied at the London Hospital and became a surgeon to the Finsbury Dispensary in 1781.[122] The younger Andree was an active member of the London Medical Society and published two further works on venereal disease that clearly distinguished between gonorrhea and syphilis.[123]

As we have seen, Desault's work was not unique in blaming microscopic "worms" for diseases, but it is of interest to us because it was translated into English in 1738, during a period when supposedly no such works appeared. Historian Hector Grasset listed other European works published in 1730 and 1739, but there is no evidence that these were known in England.[124] Astruc's lectures, which included a contagionist discussion of plague and syphilis, were also published in English in 1747, but Astruc did not blame a living agent.[125]

CONCLUSION: ENGLISH CONTAGIONISM DURING THE EARLY EIGHTEENTH CENTURY

Confronted by a revival of terrifying diseases, medical theorists of the early eighteenth century developed and combined elements of classical and early modern science, new technology, and new, more focused habits of observation and description that would contribute to a fully articulated hypothesis of contagious disease transmission. One early physician, Benjamin Marten, who probably worked with a member of the Oracle group of practitioners, put all these pieces together and created a fully worked-out treatise including all the ideas that would prove to be central to contagionism. The most important consequence of these ideas was the belief that many diseases could be seen as individual entities that could be sorted into species.

Unlike many of his contemporaries who believed in panspermism or miasmatism, Marten insisted that such illnesses as smallpox, and "Malignant, and all other Continual and Intermitting Fevers, as well as many other Distempers," could *only* be transmitted by contagion. This implied that many of these illnesses could be circumscribed or prevented by such anticontagionist measures as isolation. We know his treatise found readers at the time of composition and was available to some readers later in the eighteenth century.

It might seem that the early explorations of new approaches to the problem of disease causation and transmission led directly to the fully articulated contagionism of the end of the century. However, after about 1730, British interest in *contagium vivum* apparently declined.[126] One important factor was the rise of the University of Leyden in the early years of the eighteenth century as Padua and Montpellier became less attractive and accessible to English students. Most Anglican students attended the English universities, which still focused on classical medical thought, but Dissenters who could afford the cost went to Leyden to study under the Calvinist Boerhaave, where they were joined by postgraduate Anglicans seeking further instruction.[127] Although he encouraged clinical observation, Boerhaave's physiological approach to disease was at odds with the ontological approach connected to contagionism.

Some British physicians, however, continued to support an approach to medicine that viewed diseases as individual entities with their own characteristic patterns of symptoms. Originally inspired by the monist ideas of such authors as Sennert and Van Helmont and combined with deep devotion to Baconian empiricism, this tradition, which was especially valued by British Dissenters, continued to serve as an alternative to a mechanistic or physiological medicine that reformulated traditional classical and Galenic teaching.

CHAPTER 8

SMALLPOX INOCULATION AND THE ROYAL SOCIETY, 1700–1723

INTRODUCTION: SMALLPOX

Between 1720 and 1723, the introduction of smallpox inoculation and the possible imposition of quarantines to fend off a new outbreak of plague sparked an increased interest in contagionism. One might have expected that the London College of Physicians would have served as a locus of debate and information about inoculation, but in fact the College remained silent on the issue until later in the century. It was the Royal Society that collected information on inoculation, debated it, disseminated information and technical advice, and ultimately initiated a successful campaign to put it into practice.

Smallpox had evidently increased in both incidence and virulence during the seventeenth century. Marchamont Nedham noted in the *Medela* that both smallpox and measles had been comparatively innocuous diseases until forty years earlier.[1] Nedham's disease categories are always doubtful, but he may have been referring to a severe smallpox epidemic in 1628. Several other serious epidemics hit London in the later seventeenth century, the worst coming in 1681. This was followed by even worse epidemics in the first decades of the eighteenth century, in 1710, 1714, 1716, and 1719, and finally by the terrible epidemic of 1721, which gave the final impetus to inoculation.[2] Earlier, we noted that Kircher had attributed smallpox to "animalcules and vermicules" and that Wilkins had included it in his list of contagious diseases caused by an external substance. Many early modern authors, however, did not see smallpox as the product of a simple contagion. Inoculation would change that perception forever.

INOCULATION

In his history of smallpox, Donald Hopkins writes that people in rural parts of Europe practiced some form of deliberate smallpox exposure, but it is not clear that this involved breaking or scratching the skin.[3] There is a doubtful reference in a possibly spurious verse in a Latin poem attributed to the school of Salerno in the tenth or eleventh century.[4] Perrott Williams of Haverford-west reported to James Jurin in 1723 that schoolboys and other Cambrians had practiced an "immemorial custom" called "buying the pox" that involved holding smallpox scabs closed in their hands.[5] Antonio Vallisneri reported a similar custom in Lombardy to Sloane in 1726. Writing in opposition to inoculation in 1723, Richard Blackmore suggested using a handkerchief if one wished to contract the disease, but he did not provide further details.[6] Genevieve Miller notes that "the few references to this practice in the medical literature (in the 17c) mention it as an example of the transplantation of disease, not in the present-day sense of transmitting a disease through contagion, but in the magical sense of disease transference common among primitive people."[7]

As it had in the case of rinderpest, the effort to contain these epidemics involved communication between graduates of Padua and the Fellows of the Royal Society. The Fellows began discussing inoculation as early as 1700/1 when Joseph Lister, a trader for the East India Company, sent Dr. Martin Lister, FRS, a letter about the Chinese method of inoculating patients by blowing powdered smallpox scabs into their nostrils.[8] There is no evidence that Dr. Lister shared this letter with his colleagues, but a month later Dr. Clopton Havers shared a similar letter that the Fellows did discuss.[9] The Turks had adopted inoculation in the seventeenth century, probably from trade contacts with the East; an epidemic in Constantinople in 1706 gave European travelers an opportunity to witness inoculation themselves. At about the same time, Cotton Mather, a Congregationalist minister in Boston, learned of the practice from Onesimus, a slave who may have come from Tripoli.[10]

The first detailed account of the Turkish practice was provided by Emmanuel Timoni (FRS, 1703), an Italian physician who had attended both Oxford and Padua before setting up a practice in Constantinople.[11] In 1713, Timoni sent a letter on inoculation to Dr. John Woodward, FRS, Professor of Physic at Gresham College (and Sloane's nemesis). This was translated from Latin to English and read before the Society in June 1714; it also appeared in the *Philosophical Transactions*.[12] Timoni sent several very similar accounts back to Europe; one went to the Swedish court, and another was published in Leipzig. A copy also appeared in France, and a physician from Constantinople used one in his Leyden thesis on inoculation in 1722.[13] In 1714, Timoni wrote another account for Sir Robert Sutton, Ambassador to Turkey (published in the *Philosophical Transactions* in 1723), and in 1717, he became the household physician to the next Ambassador, Edward Wortley Montagu.[14]

The reading of Timoni's first letter to Woodward sparked a discussion among the Fellows, who requested more information from the botanist William Sherard, who was then a consul in Smyrna.[15] In his history of inoculation, written in 1736 and published posthumously in 1756, Sloane recalled that he had been the one who had asked Sherard for a report; he knew

Sherard well because of their mutual interest in botany.[16] Sherard procured an account from a friend of his and of Timoni's: Giacomo Pylarini, a Greek physician serving as Venetian consul in Smyrna. Pylarini had studied law in Venice and medicine at Padua; had traveled through Constantinople, Syria, and Egypt; and had practiced medicine in Aleppo. He had first investigated inoculation in 1701.

Following his return to Venice, Pylarini wrote an account on inoculation published in 1715 as "*Nova et Tuta Variolas Excitandi per Transplantionem Methodus.*"[17] Sloane presented this to the Society in 1716, and it appeared in the *Philosophical Transactions* shortly thereafter.[18] There also seem to have been informal discussions within the Society.[19] The practice had also been described, but not recommended, by Peter Kennedy, a Scottish surgeon, in a book on external remedies published in London in 1715.[20] In 1716, two sons of the secretary to Sir Robert Sutton, the British Ambassador, arrived in London, having been inoculated in Constantinople; in fact, this practice seems to have been fairly common among the Europeans in Turkey.[21] The American minister Cotton Mather also wrote to Woodward on July 12, 1716, and expressed his determination to introduce the practice to Boston the next time smallpox appeared.

The following year, 1717, Lady Mary Wortley Montagu, wife of the new British Ambassador, wrote a friend that she intended to bring inoculation into fashion on her return to England.[22] She fulfilled her intention during the severe epidemic in 1721. Now settled in London, Lady Mary asked Charles Maitland, a Scot who had been the surgeon to the embassy in Constantinople, to inoculate her daughter Mary.[23] A delay for negotiations ensued, because Maitland wished to have physicians witness the procedure, whereas Lady Mary did not. According to letters quoted by historian Genevieve Miller, Lady Mary's dislike of physicians was so strong that she believed they would try to squelch inoculation if they believed it might damage their revenues by reducing the number of seriously ill patients.[24] Maitland, on the other hand, believed she wished to keep the experiment secret because she was afraid it might fail. However, the witnesses were at length invited and the attempt, in April 1721, was a success, as was a second inoculation of the son of Dr. James Keith, one of the attending physicians.

Following this, the Royal Family, and particularly Caroline of Anspach (1683–1737), the Princess of Wales, became interested in the practice.[25] Caroline was known for her interest in science and her open-minded approach to religion. She was a great admirer of Newton's and was especially fond of the Deist Samuel Clarke. Caroline and her husband, the future George II, had supported Leibnitz in Hanover, and they retained the Huguenot physicist John Theophilus Desaguliers to give scientific demonstrations at Hampton Court, later appointing him as their chaplain.[26]

According to one account, the Queen consulted the three royal physicians, Hans Sloane, David Hamilton, and John George Steigherthal, who recommended another experiment with inoculation using volunteers from Newgate prison.[27] Because of conflicting accounts in the original sources, it is not clear who initiated this meeting, but Miller believes that Lady Mary was not responsible.[28] Miller concludes that the "royal patronage of the inoculation experiments was a happy coalescence of the new scientific movement, exemplified by the Royal Society, with the royal power."[29] It was also evidence of the cosmopolitanism of English medicine: in addition to the Queen herself, all three physicians had been born and educated outside England.

At about the same time, Sloane wrote to Dr. Edward Tarry, who had been physician to the English traders in Aleppo, to request additional information about inoculation. According to Sloane's later account, this was because Maitland was reluctant to continue with the experiment.[30] The first separate publication on inoculation to be published in England appeared in July 1721. Entitled "A Dissertation on the Method of Inoculating the Small-Pox; with Critical Remarks . . . " the pamphlet was signed "J.C., M.D."[31] The author was Jacob de Castro Sarmento (Henrique de Castro), a Sephardic refugee from Portugal who had graduated MB from the University of Coimbra in Portugal in 1717 and fled to England with his wife, arriving in March 1721, just a few months before his pamphlet appeared.[32] De Castro Sarmento was a botanist and a friend of Sloane's who would later recommend him for an MD from Marischal College, Aberdeen.[33] In addition to his work on small-pox, he published an important book on the *Materia Medica* in 1735, which revealed a Boerhaavian approach to intermittent fevers and recommended the use of cinchona bark.[34]

Preparations for the experiment began with the selection of prisoners on July 24 and were avidly followed in the press. On August 9, 1721, Maitland inoculated three female and three male convicts with Sloane and Steigherthal supervising and about 25 medical witnesses observing. Shortly after this, Richard Mead obtained permission to try the Chinese method of inserting the smallpox matter in the nose of a prisoner, but this prisoner became more seriously ill than the others. All the convicts recovered and were pardoned. In order to investigate the question of whether inoculation really conferred immunity to subsequent attacks of the "natural" smallpox, Sloane and Steigherthal personally paid one of the convicts to travel to Hartford, where Maitland was living.[35] She was ordered to work in a hospital where the smallpox was raging and to sleep in the same bed as one of the patients for six weeks. In addition, Maitland continued to inoculate some patients.[36] On April 17, 1722, Claude Amyand successfully inoculated two royal princesses with Sloane, Steigherthal, and Maitland in attendance. Amyand inoculated his own two children on the same day.[37]

THE ROYAL SOCIETY LOOKS FOR MICROBES, 1722–1723

Evidence of the galvanizing effect these developments had on the members of the Royal Society can be found in letters from James Jurin, the Secretary of the Society, to Leeuwenhoek. Correspondence with Leeuwenhoek had languished during the tenure of Edmond Halley as Secretary, and many of his letters had not even been translated. In February of 1721/2, Jurin wrote to the aged microscopist to introduce himself and express a hope that he would continue to correspond with the Society.[38]

In May, he wrote that at a recent meeting of the Society, Sloane, then the Vice-President, had asked Jurin to encourage Leeuwenhoek to carry out microscopic investigations into the observations of Bonomo and Cestoni on the itch mite, reported in the *Philosophical Transactions* 18 years earlier. Sloane also wanted Leeuwenhoek to "observe, whether any Insects are to be found in ye Pustules of those that are ill of ye Small Pox, as some Persons have imagined."[39] Jurin explained that "the method of inoculating ye Small Pox wch has hitherto been practiced here . . . has occasion'd Peoples' turning their thoughts more particularly to ye manner of propagation of that Distemper, wch is the cause of this Enquiry."[40] Clearly, members of the Royal Society, and Sloane in particular, now found the hypothesis of microscopic animate contagion worth investigation after years of neglect.

Leeuwenhoek was not enthusiastic about this request because he did not believe in animate contagion. About the itch he commented that if it were agreed that no animalcula were to be found in the blood of men or animals, there was no way that they could enter the vesicles: "the Small Blood Vessels have no end, and . . . consequently all that matter which breaks out like Bladders on the Superficies of the Skin is the same substance . . . which comes out through the coates of the Blood Vessels; if that watery Humour is not red, it is the Serum of the Blood." He promised, however, to examine the fluid from the vesicles as soon as he could find a person suffering from the itch. He had spoken with the surgeon of the Delft orphanage, but none of the 240 orphans there had the itch. He died the following year, apparently without completing the proposed investigations.[41]

Although the British members of the society recognized that none of them was as skillful in microscopy as Leeuwenhoek had been, they did not lose interest in the subject. In 1723, Desaguliers, then the Curator of Experiments at the Royal Society and a very respected experimentalist, examined pus from his young daughter, who was then ill with smallpox, under a microscope. He reported to the Society that he could not find any animalcula as some had supposed, but he saw several small globules of no regular shape or size, some larger and others smaller than the "globules" of the blood (that is, red blood corpuscles).[42]

Repeated failures to find "animalcula" in the blood and body fluids of patients led many physicians to conclude that the "contagious matter" was more likely to be a chemical or poison than an animate agent. Thus, historians who have grumbled that early-eighteenth-century medical authors were "diverted" from theories of *contagium animatum* because they failed to take advantage of the microscope have it backwards. In fact, it was microscopic observations that undermined the theory. This situation was reinforced by the fact that smallpox became the prototypical contagious disease in the eighteenth century and its agent, a virus, cannot be seen with a conventional microscope.[43]

Historian Marc Ratcliff has shown that natural philosophers during the early Enlightenment found it impossible to repeat Leeuwenhoek's observations or follow his methods, and so they reoriented their research program to the investigation of larger entities that could be reliably and repeatedly visualized with the equipment they had. Microscopes continued to serve as essential tools for botany, entomology, and anatomy, but during the 1730s they were used for observations and experiments on entities such as small insects and seeds, not bacteria.[44] As far as communicable disease was concerned, the brief alliance between microscopy and medicine had been severed, and it would prove difficult to rebuild.

COTTON MATHER AND INOCULATION CONTROVERSIES IN BOSTON, 1721–1723

Among the correspondents of the Royal Society was Cotton Mather, a Congregationalist minister in Boston, who had first heard about inoculation when his slave, Onesimus, told Mather he had been inoculated in his native city.[45] Mather eagerly read articles from Timoni and Pylarini in the *Philosophical Transactions* and persuaded a local doctor, Zabdiel Boylston, to inoculate his own son and two slaves when an epidemic arrived in Boston in June 1721. The two men aroused fierce resistance, and for a time Boston prohibited inoculation. The opposition was so strong that Mather wrote it had made Boston "almost a Hell upon Earth, a City full of lies, and Murders, and Blasphemies."[46] At one point a grenade was thrown into his house.[47]

Mather had already prepared a preliminary report, which he sent to London in September of 1721. It appeared anonymously in London early the following year.[48] At the same time, the Reverend Daniel Neal, a Dissenting minister and historian, reprinted two other Boston treatises in favor of inoculation: the Reverend Benjamin Colman's *A Narrative of the Method and Success of Inoculating the Small Pox in New England*, and the Reverend William Cooper's *A Reply to the Objections Made against It from Principles of Conscience*.[49] Princess Caroline sent for Neal to question him about the practice before the two royal princesses were inoculated.[50] Thus, the first really extensive inoculation campaign anywhere in the British Empire took place in Boston.

In his study of this campaign, Roger Zelt has argued that political motives for promoting inoculation were more important than religious or purely medical considerations.[51] The strife Zelt describes was extremely complex, involving disputes over the way Boston was governed, the wisdom of creating a local bank that could issue paper money, and tensions between the Overseers and administration of Harvard College. Cotton Mather was a party to many of these conflicts. Zelt has found that those who volunteered for the terrifying ordeal of inoculation were disproportionately likely to be well-educated Harvard graduates who were ministers, merchants, or professionals, or the members of their households. Moreover, the inoculees had a relatively high rate of intermarriage and political, social, and economic ties.

Zelt also found that 156 of 259 identified inoculees were affiliated with a Puritan Congregational church, whereas no inoculee could be identified as an Anglican.[52] To Zelt this indicates the "relative conservatism" of the inoculated group, since the Boston "establishment" drew heavily on the ranks of Congregationalists.

However, it is clear that Boston's medical establishment opposed inoculation, which suggests that "conservative" is not the best term to use for inoculation supporters. The one local physician who held an MD, William Douglass, vigorously opposed inoculation when it was first proposed, although he eventually changed his position. In a letter to a friend, Douglass referred to Mather as "a certain credulous Preacher."[53] One local doctor argued that inoculation might spread not only smallpox but the bubonic plague.[54] More plausibly, Louise Breen sees Mather's campaign as part of his effort to steer an American "middle way" that differed both from the Newtonian and Anglican hegemony of England and from popular superstition and credulity at home.[55] Perhaps inevitably, he was attacked from both sides of the spectrum.

Mather had great difficulty persuading any local doctor to try inoculation. When he finally persuaded "Dr." Zabdiel Boylston to begin inoculation on an experimental basis, all the other local practitioners, led by Douglass, published a letter in opposition. This letter was then answered by six prominent local ministers defending Boylston, although "he has not had the honour and advantage of an Academical Education."[56] The ministers argued that inoculation no more interfered with Divine Providence than other medical therapies, such as bleeding or blistering. On the advice of the doctors, however, the Boston Selectmen banned inoculation, a move strongly supported by James Franklin's anticlerical *New England Courant*. Boylston quietly continued inoculating patients until the epidemic had waned. In all, 6 of 287 inoculees died (2.1 percent).[57] The entire epidemic had caused 5889 illnesses, with 855 deaths (14.5 percent).[58]

Mather knew that inoculation had become a political issue, but he did not see himself as the "conservative." In fact, he depicted his opponents as Tories, writing to Jurin in 1723, "It is with the utmost Indignation, that some have sometimes beheld the Practice [of inoculation], made a meer Party-business; and a Jacobite, or High-flying party, counting themselves bound in duty to their party to decry it."[59]

Boston was so deeply divided in those years that whatever Mather supported would certainly have attracted opposition, but how did he choose that particular cause? Why did he choose to support rather than oppose inoculation in the teeth of intense public and medical opposition? Mather was not reflexively opposed to mainstream medicine. In 1713, he had sought medical approval for a pamphlet on measles.[60] His choice was rooted in his own underlying medical and philosophical beliefs.

Mather's consuming interest in medicine was partly due to his involvement with German Pietism, a movement that to some extent recapitulated the outlook of the sectaries during the English Interregnum. Pietism was millenarian, anticlerical, and "enthusiastic," favoring a direct and monist experience of nature in opposition to the rationalism and dualism inherent in the new science.[61] Through correspondence, Mather learned of the medical work carried on by the Pietist congregation of August Francke at Halle.[62] He supported Halle's therapeutic philosophy, which relied on relatively mild botanical remedies in place of more "heroic" treatments such as bleeding and purging.[63] He taught physic to his own daughter, Katy, and trained her to prepare and dispense such medicines.[64]

Mather had a great interest in folk remedies, native American plants, and the plant lore of the Native Americans, much of which he recorded and sent to the Royal Society, where his work found an enthusiastic audience.[65] He was also fascinated by problems of pollination and generation; his letter of 1716 to the English apothecary James Petiver contained the first known account of plant hybridization.[66] These interests led Mather to read as widely as possible on medical philosophy, botany, and medical botany. He took a particular interest in the works of Paracelsus and Van Helmont, who had also considered the problem of individuation.[67] He was also an avid reader of Sydenham and persuaded local physicians to adopt his "cool" treatment for smallpox, which he believed had saved many lives.[68]

During the inoculation controversy, Mather was assembling materials for a major work on the body/soul problem, "The Angel of Bethesda," much of which he apparently wrote in 1724. It was his last major work, although he failed to publish it.[69] He postulated the existence of an entity he called the "*Nishmath Chajim*," or vital principle, which was composed of particles finer than light. Like Van Helmont's *Archaeus*, it served as the intermediary between the rational soul and the body, receiving "impressions" from both.[70] It provided for sensation, governed the development of embryos, and controlled vital functions such as digestion.

According to Mather, this vital principle distinguished humans from machines: if a machine became disordered it remained so, but the human body was self-healing. The principle was "like the Soul which animates the Brutal [i.e., animal] World."[71] It was also the vehicle that impressed into the body a sin or sickness that was in the "rational soul," and thus could be described as "the Seat of our Diseases, or the Source of them."[72] Physical diseases thus had a "spiritual" cause, in the literal sense of a deranged vital "spirit." Conversely, physical ailments could impair the vital principle and cause spiritual or mental diseases; for these Mather recommended only gentle physical treatments.

Thus, Mather's evangelical religious beliefs led him to admire authors such as Van Helmont and inspired him to adopt a vitalistic medical philosophy. His Helmontian view of illness as a living entity prompted him to accept a theory of living contagion that in turn provided a rationale for inoculation. Although the prospect of an active step to avert disease and benefit the entire community surely appealed to Mather, who also wrote *Bonifacius, Essays to Do Good*, his support for inoculation depended on motives and ideas much broader than local politics.[73]

Mather was also deeply impressed by Benjamin Marten's *New Theory of Consumptions*, which he cites together with Borel, Mayerne, and Andry in support of the theory of animate contagion.[74] Mather was not the only Bostonian to entertain ideas of animate contagion. In his own treatise in support of inoculation, the Reverend Benjamin Colman also wrote of another author's view that all venomous particles entered the body through the pores of the skin, adding that

> I carry'd on the *Hypothesis* in my own mind a little further, by that *Modern Observation* which has been made of a Multitude of *Animalcula* on every Pust[u]le of the Small-Pox: *These* they tell us their Glasses have discover'd; and that our common Infection is by swarms of these. If so, Metho't these living Minute Particles, these Animated Atoms if I might so speak, may much more easily find their way into our Pores & so into our bodies, than if they were Inanimate.

Colman theorized that inoculation resulted in a less virulent illness than natural smallpox because in inoculation fewer virulent particles or animalcules entered immediately into "such parts of hazard and distress" as the nostrils, throat, and inner parts.[75]

INOCULATION CONTROVERSIES IN ENGLAND, 1722–1723

In England, inoculation aroused opposition, though more slowly than in Boston. In August of 1722, Sloane wrote his friend Richardson:

> The inoculation for the small-pox has gone on here in Town with success, till the hot weather put a stop to it. Many Physicians, Surgeons, Apothecaries, and Divines seem to oppose it with greater warmth than, in my opinion, is consistent with sound reason or the general good of mankind . . . The *Cortex Peruvianus* met the same usage at its first entrance into Europe . . . and, perhaps, the same reasons at bottom may hinder the use of the one and the other.[76]

The opposition came on two fronts. First, High Church Anglican ministers argued that it was impious for man to attempt to control diseases that God sent to punish evildoers and to test believers. "I shall not scruple to call that a Diabolical Operation which . . . tends in this case to anticipate and banish Providence out of the World, and promotes the encrease of Vice and Immorality," wrote the Reverend Edmund Massey.[77] He believed that healthy persons had no right to put their lives or health in jeopardy by deliberate "engraftment of a corrupted Body into a sound one."[78] These religious opponents did not contest the idea that inoculation introduced a specific disease into the body: they argued that doing so was dangerous or wrong.[79]

It was on different grounds that many physicians opposed the practice. Sir Richard Blackmore, a follower of Sydenham, published one of the earliest treatises against inoculation in 1723, two years after a book on the plague.[80] Blackmore, a "staunch whig" but also a pious Anglican, had graduated MA from Oxford and tutored there before obtaining a medical degree from Padua.[81] He became a Fellow of the College of Physicians in 1687 under the charter of James II, rising to Elect and Censor, and had been one of the dissident Fellows who were allied both with Edward Tyson in defending Johannes Groenevelt and with the apothecaries in opposing the Dispensary. He had also been a physician to William III and Queen Anne.[82] He wrote numerous poetical works of dubious literary value but manifest political intent, including *A Satyr against Wit* (1700), which attacked the Dispensary; *Eliza* (1705), which likened the Tory John Radcliffe to Rodrigo Lopez, a Jewish physician accused of trying to poison Queen Elizabeth I; *The Creation* (1712), which supported Locke against the Epicureans, Hobbes, and Spinoza; and *Redemption* (1722), which condemned Arians and Socinians.[83]

Blackmore's *Treatise on Smallpox* expanded on his view of febrile diseases. After a prefatory attack on schools, traditional learning, country bumpkins, and old-fashioned doctors ("I am doubtful whether the whole Faculty might not be spared without any Damage to Mankind") he discussed the proper education for an enlightened physician.[84] He recommended a smattering of chemistry and botany but felt that anatomy was unnecessary. Neither the Ancients nor the Moderns were useful; Aristotle and Hippocrates had become so much waste paper: "the trumpery and riffraff of old libraries."[85] Physicians should know that fevers were simply tumults in the blood. Acute fevers originated in the blood and humors; secondary fevers arose from some antecedent disorder, usually in the bowels. Primary fevers could be divided into simple, inflammatory, and malignant. Simple fevers came from an irregular exaltation of the fiery ingredients in the blood. Inflammatory fevers occurred when matter from the blood was cast into the solids of the body and could not break through into the pores of the skin. If the matter lodged in the joints, it became rheumatism; if in the skin, it was erysipelas; if in the glands, it became the small spots of smallpox, measles, and other exanthemata. Malignant fevers disunited some parts of the blood, permitting putrefaction.

The second part of this treatise dealt with the new practice of inoculation. It began with an attack on quacks, fanatical chemists, and wild new religious sects. Blackmore opposed inoculation on the grounds that it was probably ineffective, since it cleared the blood of particles of smallpox but left behind other putrid particles that would later cause even more dangerous fevers. Moreover, it might spread the disease to others, or communicate other serious diseases. Thus Blackmore's opposition was based on his belief that the initial source of diseases lay within the body itself, but he did not deny the possibility of contagion.

William Wagstaffe, FRS, FRCP (MD Oxford), a physician to St. Bartholomew's Hospital, published a *Letter to Dr. Freind; Shewing the Danger and Uncertainty of Inoculating the Small Pox* in June of 1722. Miller notes that this book "became the handbook of conservative thought on the problem. It was immediately translated into French and was widely quoted, even forty years later when the inoculation controversy raged in France."[86]

Wagstaffe argued that using matter from a person who was ill of one disease might easily cause a different disease in a second person. For example, gonorrhea could cause syphilis, leprosy could cause the itch, and smallpox might cause chicken pox. The severity of a disease depended on the state of the blood at the time a person became ill, not on the nature of the infection that was introduced. There was thus no reason to believe that people who had experienced the smallpox once, even if they had had the disease in a "regular manner," could not catch it again. To put it another way, if one does not characterize smallpox as a specific disease, but merely thinks of it as a nonspecific skin ailment that has been made more severe by a patient's lack of resistance, then it is plausible to view subsequent attacks of (other) skin diseases as further episodes of "smallpox." These later episodes could then prove that the initial attack did not confer permanent immunity.

This argument was difficult to counter. Anticontagionism undermined the very idea of specific diseases. Thus, the starting assumption of the medical anticontagionists in itself rendered inoculation meaningless. If all ailments were essentially a manifestation of the same internal disorder, then in order to prevent any one disease, one must prevent every disease. Even if the hypothesis of disease specificity were adopted, however, it would not in itself provide an argument for believing that an experience with a disease should confer permanent immunity. After all, both influenza and malaria frequently recur.[87] The "parasitic model" of diseases, which likened their agents to itch mites, provided no reason for assuming that the parasites could not reinfest their victims.

Defenders of inoculation argued that experience had shown that one attack of natural smallpox would protect against reinfection and that the process by which inoculation worked was the same as the process by which one naturally contracted the disease. For example, the author of *A Letter to Mr. Massey* wrote:

> It may be demanded, That I not only shew, that Persons who have been inoculated have safely recover'd, but also that by this Means they are secured from all Danger of the Return of this Distemper in the natural way . . . Tho' many have undergone this Operation, yet there is not an Instance alledged [*sic*], of the Return of this Distemper . . . And not only Experience, but Reason also induces us to believe this an effectual Preservative . . . the greater Number of those who have been invaded by the Small-Pox, have receiv'd it by Infection from others: That is, the contagious Particles have insinuated themselves thro' certain Pores, and by mixing with the Mass of Blood have corrupted it, which Corruption is thrown out in Pustules upon the Body; and this Process is found . . . to render Men secure from the same Distemper for the future. And in what does Inoculation differ from this? Is it not a Conveying the like Matter into the Blood; Matter of the same Kind?[88]

This "Reason" however, was only persuasive if one first granted that the natural smallpox was contracted by "contagious particles." If inoculation could be proven to be identical to the process that transmitted the natural disease (the entry of contagious particles into the blood), then inoculation would confer the same immunity as the natural disease. However, many anti-inoculationists denied that "natural" smallpox arose from contagious particles, and so they did not agree that inoculation and "natural" smallpox were in fact the same disease.

The fact that smallpox and chicken pox were often seen as different "forms" of the same disorder also caused uncertainty. When one followed the other, the same disease did seem to recur. Furthermore, as smallpox was known by its severity, if inoculation conferred a milder case than the "natural" disease, it was easy to argue that it was not in fact the same illness. Indeed, this question remained open, although Thomas Fuller clearly differentiated the two diseases in 1730.[89]

Maitland grappled with this problem in defending his practice against criticism:

> As to the Objection . . . denying the Small Pox to be hereby rais'd: did the Eruptions come the natural and common way, I acknowledge it to be highly reasonable: for the only way we have to distinguish between the genuine Small Pox of the several kinds, and the spurious, or what they call the Chicken Pox, is not only by observing the Nature and Periods of the Pustules, but also by the first Invasion, and subsequent Progress of the Fever, and of the usual Symptoms attending it. The known Periods of the Small Pox are different, according to their different kinds; and this difference ariseth, not only from the various Dispositions of the Juices and Habits of the Bodies affected; but likewise from the different Degrees and Qualities of the External Infection, howsoever communicated, as it is more or less subtil, malignant and epidemic.
>
> Now in those rais'd by Inoculation, the Periods, tho' a little different from the common and natural, are, in their kind, as certain and regular as the other: a Difference indeed is observ'd of sometimes two or three Days, as to the Time of Eruption, but that is no more than what is common in the natural Sort. And thus, too, it is with Respect to the other Stages of the Disease . . . the Process is certain, and the Prognostick infallible.[90]

The controversy over inoculation, therefore, forced both defenders and opponents to confront the question of what in practice defined a disease; was it merely the severity of the illness, or were there other aspects that could be used to define a particular syndrome as a separate "disease," such as the staging of symptoms? In what sense could an illness suffered by one person be said to be the "same as" or "different from" an illness suffered by another person or by the same person at another time? This issue would reappear even more urgently at the end of the century as doctors wrestled with the question of the relationship between cowpox and smallpox.

Experience with smallpox had persuaded most people that it did not normally recur. For example, nurses who had survived smallpox did not catch the disease from their patients, even with close and frequent contact. Similarly, even when children shared the same bed, survivors did not catch it again from their siblings. The pro-inoculation forces argued that "constant experience" showed that natural smallpox did not recur and that there was no reason to believe that inoculated smallpox would behave differently.[91]

A theory had developed to account for this natural immunity: it held that children were born with a "seed," "egg," or substance in their bodies that permitted smallpox to take hold. An attack of the disease depleted this substance, so repeated exposure could not incur repeated illnesses. This idea originated in the work of the Arab medical writer Rhazes (al-Razi), who believed that babies were born with watery blood that had to be transformed into dryer, more adult blood by boiling up, losing its moisture, and throwing its sediment out through the pores in a process analogous to the fermentation of wine.[92] His successors thought smallpox arose from the putrefaction of blood in children's bodies due to undigested food or remnants of their mother's menstrual blood.[93] These ideas implied that attacks of smallpox were inevitable at some point in every person's life. They characterized smallpox as a specific disease, but the source of the specificity came from inside, not from outside the body.

Initially, this argument buttressed the case for inoculation: if the disease was inevitable, then it could not be considered a divine punishment. Whereas conservative ministers argued that it was impious to contract smallpox deliberately because doing so posed an unacceptable risking of life in the same category as jumping from a window, the inoculators argued that the risk of smallpox would be encountered at some point whatever the person did, so choosing the most propitious time to contract the disease was not an impious "suicidal" gamble.

However, there was considerable evidence that smallpox had not been known in ancient Europe at all but had been introduced from the East. If that were the case, then there could not be any inborn physiological cause for the disease. In 1704, the physician William Oliver, FRS, had argued that it was imported from Arabia.[94] Oliver connected this argument to his belief in the specific nature of the disease. "Thus the Seed once sown . . . has propagated its Poison in all Ages since, and when it will be worn out God knows. I call it a Seed . . . because I find diseases keep regular Types, and have particular Attributes that distinguish them one from the other, as the Seeds of Plants do their particular Species."[95] Oliver would be quoted by the sharp-eyed Benjamin Marten.[96]

As Cotton Mather wrote in "The Angel of Bethesda": "the Enquirers after Causes have suspected the Original of this Malady to be some Venom connate with every Man . . . But this old Notion loses much of its Authority, by our considering that it is a new Distemper . . . 'Tis evident unto us that the Ancients were unacquainted with it."[97] These arguments gradually chipped away at the theory of an "innate seed," although there was still no satisfactory explanation for immunity to reinfections.[98]

Accumulating evidence for the efficacy of inoculation demolished the argument that it was so dangerous that it was impious—at least as far as the health of the individual was concerned. The chief theoretical objection to inoculation became its danger to the community. Once it was granted that inoculation conferred "true" smallpox, and that smallpox was a specific contagious disease, there was still good reason to avoid introducing it into communities with susceptible individuals where it might spread uncontrollably. Responsible contagionists took this danger seriously, and some opposed general inoculation. This issue became the center of what John Coakley Lettsom's biographer has called "the great inoculation controversy," which began in 1775 and caused deep divisions among medical reformers.[99] At the end of the century, John Haygarth finally developed an effective set of procedures for minimizing this risk.

It was also difficult for authors to explain why inoculation caused a milder case than smallpox contracted in other ways. In the early eighteenth century, doctors believed it was because inoculation was carried out at an opportune time when the patient was best able to resist, with the result that they devised elaborate (and expensive) regimens to fortify patients before inoculation, although Colman had suggested that the site of inoculation might be a factor. Later in the century they also assembled data to determine the best age for the procedure.[100]

A few doctors experimented with inoculation for other diseases. For example, in 1759, Francis Home tried inoculating children with measles but found that his patients became as ill as if they had contracted the disease naturally.[101] In 1755, a Yorkshire gentleman inoculated eight of his calves with the cattle distemper and wrote *The London Magazine* that his efforts had been successful, though Dutch experiments did not succeed as well.[102] In 1757, Daniel Layard also recommended inoculation for rinderpest in his *Essay on the Nature, Causes, and Cure of the Contagious Distemper among the Horned Cattle*. The next year he repeated his advice before the Royal Society, comparing the disease to smallpox, but his effort was hindered by doubts about whether the initial exposure rendered surviving cattle immune.[103]

Although these efforts did not bear fruit, the success of inoculation unquestionably changed medical thought. It showed beyond serious doubt that it was possible to take a tiny quantity of material from one patient, put it in a container, transport it into another place and environment, and physically introduce a disease with a virtually identical and very distinctive course in a second person, a course that was not greatly affected by the health or diet of that person. For the first time, disease was transformed in practice from an event into an object.

THE NATURE OF SMALLPOX

Doctors were fortunate indeed to begin their experiments on deliberate disease transmission with a disease that is uniquely suited to this technique. Smallpox inoculation works only because it is a viral disease without an insect vector and so spreads only from person to person; because the vesicles contain the virus; because a single episode confers lifetime immunity; and finally because the virus so easily becomes attenuated, causing a milder case of the disease after inoculation than occurs in naturally acquired cases.[104] It is not difficult to think of diseases where inoculation would prove disastrous. The fact that physicians were able to identify this disease as a candidate is a testament to the value of folk medicine and to medical respect for empiricism; the fact that once they had succeeded they did not rush to inoculate their patients with such diseases as malaria, tuberculosis, syphilis, typhus, scarlet fever, influenza, or the plague is a testament to the value of professional medical traditions that emphasized the importance of caution when experimenting on patients.[105]

Historians continue to argue about the practical value of inoculation during the eighteenth century. Some have claimed that it had no discernible effect on mortality rates in most countries. Whatever its impact on the number of deaths, however, this was ultimately less important than its contribution to the development of understanding about the very nature of disease. As Miller wrote, "Smallpox inoculation . . . was to provide the capital to the column of evidence supporting the doctrines of contagion and specificity . . . [It was] clear evidence that the active substance was a *venenum sui generis*, specific for this disease."[106] She concluded that by the 1760s, both the proponents and the opponents of inoculation throughout Europe "adhered to the pure contagionist theory" of smallpox.[107]

The introduction of inoculation rested on an ontological view of disease, or at least of smallpox, as a separate entity. The fact that this disease had its own name in many languages, and that these names translate directly from one European language to another, is evidence that lay opinion had already conceptualized it as a separate thing. The effect of Sloane's correspondence and the reports he received depended on the fact that he and his correspondents shared a common understanding of this disease as a separate entity and trusted their ability to communicate about it unambiguously. If, for example, Timoni had not understood that this disease was a separate "sort" of disease or if he had been unable to diagnose smallpox when he saw it or heard about it, then he might have been describing inoculation for chickenpox or measles, "rashes" or "fevers," "the spotted disease," "humors in the process of concoction," "wind," "fire," "possession," or "sin."[108]

Because the people who practiced inoculation in Africa, China, and Turkey conceptualized disease in different ways, Western doctors depended on observers conversant with both cultures to translate not only the name but the whole idea of a disease—matching the practices they actually observed with the Western name for a particular set of symptoms. To some extent the international nature of medical education during this period facilitated this confidence in accurate communication, but not all observers were university educated; in fact one of them was a slave. Whatever the education of these intermediaries, their London audience accepted their ability as interpreters.[109]

Similarly, the many statistical comparisons of the mortality of "natural" versus "inoculated" smallpox rested on acceptance of the idea that even an observer as far away as Boston was describing epidemics of the same disease that haunted London or Wales—and that a convict might suffer the "same" illness as a princess. Even when they argued over the interpretation of statistics, the lay opponents of inoculation tacitly accepted this premise—and by doing so conceded the most critical issue. In this way, smallpox was kind to the doctors by being so distinctive, but they also benefited from a culture that had embedded the concept of "smallpox" as a separate disease so deeply.

It is easy to see how important this is if we consider how difficult it is now to attempt retrospective diagnosis of the diseases that decimated the classical world. These were named at the time, but the names have not proved stable over the centuries. Until very recently, scholars had no idea whether the "plague" of Galen was really the plague (*Yersinia pestis*). Even an event as recent as the "sweating sickness" of the sixteenth century remains undefined in modern terms.

THE SMALLPOX CONTROVERSY IN ITS SOCIAL/POLITICAL CONTEXT

Miller concludes that "if any one single individual is to be singled out as the most influential agent in promoting . . . inoculation in England, Sloane should be seriously considered."[110] She adds, "Although in his own 'Account' he preferred to credit Caroline with the initiative to have experiments performed, one sees Sloane's hand in every move that was made . . . It seems undeniable that Sloane was maneuvering the whole affair, using the . . . interest of Caroline and her husband . . . as the most powerful weapon he could find to introduce a revolutionary medical practice."[111]

In seeking information about inoculation, Sloane took advantage of his botanical friends, such as Sherard and Petiver. It was Sloane who nudged the experiment along by writing to Tarry when Maitland became reluctant to proceed, Sloane who advised the Queen when she decided to inoculate her own children, and Sloane who originated the effort to obtain microscopic observations on mites and smallpox from Leeuwenhoek. Sloane's intense involvement and interest in the progress of inoculation are evident in his letters to his friend Dr. Richardson of Bierley, which describe the progress of the Newgate convicts and the further spread of inoculation.[112] It is also clear that Sloane saw a connection between the success of inoculation and the possibility of animate contagion when he suggested that other diseases might spread in the same way as scabies.

It was the effort of Fellows of the Royal Society as a group that created a favorable climate for the introduction of inoculation, whereas the Royal College of Physicians remained silent, even though some physicians were members of both societies and Sloane had become President of the College in 1719. The *Annals* of the College make no mention of inoculation, whereas in 1722/3 alone the *Philosophical Transactions* published nine favorable articles and no critical articles. Sloane, Steigherthal, Hamilton, and Amyand, the medical attendants to the Court, were all FRS.[113] In addition, the Royal Society created six new Fellows who "conspicuously supported inoculation."[114] Between 1721 and 1729, 18 authors published 27 pamphlets or articles in favor of inoculation. Eleven of these authors, who wrote 20 of the articles, either published them in the *Philosophical Transactions* or were FRS. On the other hand, of the 10 authors of 12 publications opposed to inoculation, only 1, William Wagstaffe, was an FRS.[115]

The silence of the College was noted at the time. Perrott Williams of Haverfordwest wrote to Jurin: "I heartily wish, the College of Physicians wou'd vouchsafe to be so publick Spirited, as a Society, to Oblige the World with their Opinion of that Practice . . . And this wou'd conduce much more to their Honour, as the Publick Guardians of Health, than by vilifying one another . . . "[116] The College, however, did not recommend inoculation until 1755.[117]

Not only did inoculation expose differences in attitudes toward medical innovation between the Royal Society and the College of Physicians, it also served as a flashpoint for disputes between Whigs and Tories. Adrian Wilson, who analyzed the politics of inoculation, concluded that "inoculation was specifically a project of the Court Whigs; opposition to the practice came exclusively from Tories. This fundamental cleavage of attitudes . . . was to persist until at least the 1740s."[118]

Wilson argued that inoculation for smallpox was the Hanoverian analogue to the Stuart practice of "touching" for scrofula, "the King's evil," but with a telling difference. The Royal Touch had been associated with God's blessing on hereditary kingship and was an act of healing directly performed by the monarch. Inoculation, on the other hand, was a natural phenomenon and was conducted by professionals with the assent of the monarch. "This role for the sovereign—involved yet marginal—precisely fitted the Walpolian Whig political role assigned to George I."[119] Clearly it also signaled a fundamental change in popular ideas about the nature of illness itself.

The inoculation controversy began in a medical atmosphere that was already poisoned by party animosity.[120] It followed closely on the heels of the bitter "smallpox wars" that had begun in 1717 with a dispute between the Whig controversialist John Woodward and the Tory John Freind over the proper use of purges for patients with smallpox. In some ways this battle between Woodward and Freind had itself recapitulated the Edinburgh disputes of the 1690s between Sydenhamians and Newtonians.[121] It has also been depicted as a skirmish in the war between "Ancients" and "Moderns."[122]

Not surprisingly, opinions often divided along party lines.[123] The most sustained opposition to inoculation in the press appeared in *Applebee's Original Weekly Journal*, a Tory paper.[124] The most successful anti-inoculation pamphleteer, Dr. William Wagstaffe, was a member of a prominent High Church family.[125] Norman Moore described him as "a high churchman and a hater of the whigs."[126] His father was the vicar of Cublington. His relative and father-in-law, Thomas Wagstaffe, was a Jacobite nonjuror. Ejected from his London rectory after the Glorious Revolution, Thomas had practiced medicine in London and was secretly consecrated Bishop of Ipswich at the behest of James II. At the turn of the century he argued with Bishop Gilbert Burnet about whether Tory nonjurors should take the Oath of Allegiance.

Thomas's son, William Wagstaffe's brother-in-law, became chaplain to Charles Edward Stuart.

Wagstaffe was an enthusiastic controversialist who published attacks on Benjamin Hoadly, Richard Steele, and Dr. John Woodward. These targets were not chosen at random: Hoadly was the principal advocate for the "broad" or Latitudinarian church; Steele was a Whig journalist and Woodward's patient and friend; and Woodward himself was a leading "Modern." Wagstaffe also edited and wrote the preface for a medical treatise, *Anthropologia Nova*, by Dr. James Drake (1667–1707, MD Cambridge, FRS, FRCP), another well-known Tory controversialist.[127] Drake attributed both smallpox and leprosy to "thin Salt Humours thrown off from the Blood and arrested by the Density of the *Cuticula*."[128] Smallpox victims became immune to future attacks because the first onset had widened the pores in the skin, enabling the humors thereafter to escape without causing eruptions.

Although inoculation began as a folk practice without any theoretical justification, some "Newtonian" physicians soon realized that it offered a golden opportunity to demonstrate the value of mathematical analysis in medicine. Mathematically trained doctors, such as James Jurin and John Arbuthnot, realized that they could calculate the probability of dying from naturally caught smallpox and compare it with the mortality of inoculated smallpox. This was exactly the sort of hypothesis-driven medicine they had championed.[129] Moreover, because it was such a well-defined disease and because its life-or-death outcomes were less debatable than whether patients with other diseases had recovered or improved, smallpox mortality offered an opportunity for the sort of clinical "experiment" that the Helmontians had earlier proposed for their chemical remedies. Jurin spearheaded the effort to collect prospective as well as retrospective data from far and wide on the numbers of people inoculated and the numbers of deaths that resulted and published regular reports in the *Philosophical Transactions*.[130] Inoculation, therefore, combined the interests and resources of Sloane and the Sydenhamians with those of Jurin, Arbuthnot, and the Newtonians; their collaboration had long-lasting consequences because it led to the development of medical statistics as a tool for decision making.

Jurin's other major effort during these same years was to revive an initiative originally proposed but not implemented by Robert Hooke in 1667 to collect meteorological observations from all over the world and create a "Natural History of the Air." This proved less successful in part because of difficulties in establishing standardized measures and measurement equipment—another reminder of how difficult it was in the eighteenth century to share observations made over a distance without a fixed terminology or calibrated equipment. Jurin had hoped that observations of weather conditions would lead to improvements in medicine, but he could not even get his contributors to agree on such issues as the date and the length of a foot.[131] The Society's project did help to establish a common thermometric scale after it agreed to send out its own thermometers, graded on a Fahrenheit scale, to some of the more important correspondents.

Later in the century reforming physicians would become more adept at deploying both the "Sydenhamian" approach of reporting detailed "natural histories" of individual diseases with a minimum of interpretation (the case history), and the "Newtonian" approach of compiling and extracting fixed categories of data about many cases to support hypotheses concerning the incidence of diseases and the success of therapies (morbidity and mortality tables). Their work depended on the emergence of a medical research community with a shared conceptual vocabulary.

CONCLUSION

Inoculation confronted every family and every physician with a life-and-death decision. The smallpox would not wait while investigators collected more evidence or compiled weighty tomes or shaded their opinions. For those who had not yet suffered from the disease, and especially for their young children, there was a stark choice: embrace inoculation or take one's chances with natural infection. Dithering too long might prove deadly. Because it required a single momentous decision based on very limited evidence, the advent of inoculation flushed many hidden predispositions into view, including religious affiliations, political loyalties, and personal allegiances.

Although the conflict over inoculation did not always neatly divide Whig from Tory, Whigs were more likely to favor the practice. Inoculation appealed to those who believed that individuals could and should seize control of their world, whether it was the government or the natural world. Opponents were more fatalistic, believing that people should acquiesce in divinely ordained arrangements, whether they concerned the monarchy or personal suffering. Anti-inoculationists revered medical tradition and were cautious about innovation; they believed that medical knowledge was found in books and that doctors should above all be learned men. Inoculationists valued medical innovation, believed that knowledge came from experience, and were willing to investigate "folk" medicine and learn from uneducated practitioners. Inoculationists were also more cosmopolitan, actively seeking new practices and ideas from overseas.

The inoculationists also had ideas in common with the Helmontians and the radical sectarians of the later seventeenth century. They were prepared to believe that inoculation might work because a specific disease was a specific "thing," not a disorder, and thus could be taken from one person and given to another. Some inoculationists believed that the cause of smallpox was, or might be, a living contagion; their inability to identify any microscopic animalcules probably contributed to the decline of interest in animate contagion and the adoption of an agnostic stance about the nature of contagious matter. Overall, however, the British experience of inoculation strengthened both contagionism and the ontological model of diseases.

CHAPTER 9

CONTAGION AND PLAGUE IN THE
EIGHTEENTH CENTURY

INTRODUCTION: THE PLAGUE

Smallpox was not the only nightmare that haunted eighteenth-century Europeans. The plague broke out in Marseilles in 1720, probably imported by a trading ship from the Levant. England's last major epidemic of plague had taken place 54 years earlier, so there were still survivors scarred by the horrors of that episode.

Bubonic plague, *Yersinia pestis*, is carried by a bacillus that infects a flea that lives on rats. Flea bites transmit the most common form of the plague to humans: the "bubonic plague," named for the "buboes" or painful swellings that appear as the victim's lymph nodes become infected. It is also possible to inhale the bacteria directly into the lungs. In this form, known as "pneumonic plague," it is contagious and kills very quickly.[1] There are also cases of "septicemic" plague, where the bloodstream becomes infected. Other mammals besides rats and humans can contract the plague; immunity varies by species. Its etiology and epidemiology are still not fully understood.[2]

Modern experts have access to descriptions of plague symptoms that are based on positive laboratory confirmations of *Yersinia pestis* in patients. They also possess a better understanding of the epidemiology of the disease. Even armed with these conceptual tools and possessing ready access to all the earlier documents that relate to the plague, they failed to agree on the identity of this illness (and many others) as long as they relied only on historical accounts.[3] This underscores the complexity of the task that confronted early modern physicians seeking to produce clear and coherent accounts of ill-defined diseases using their own observations and written records composed in several different languages to which they had only partial access.

Because epidemics of plague often seemed to come from the Near East, because its diffusion from a center outward and from place to place could be

tracked over time, and because strict quarantines sometimes seemed to curtail it, the early modern experience of plague offered some support to contagionism, but it was always accompanied by features that pointed to some
other explanation. It often spread through a city in a capricious hopscotch
pattern. Many ministers and practitioners who cared for victims remained
well, whereas people who had never been near a patient fell ill.

The reappearance of the plague in 1720 led to the republication of works
spurred by earlier epidemics and hence to the recovery of Renaissance theories of contagion. The conceptualization of the plague as a single and distinctive disease was well established but still open to debate. There were many
arguments about whether the plague was a separate disease or merely the
most serious degree of disease.[4] Doctors, traders, and city authorities often
tried to shade or mitigate the diagnosis, exploiting any doubts or ambiguities
in order to deny that the plague was present in their jurisdictions.

THE EPIDEMIC IN MARSEILLES, 1720

The epidemic that harrowed Marseilles followed the usual pattern: first offi-
cial denials, then claims that the disease was not the "true" plague but merely
a "fever," then a quickly mounting death rate accompanied by panic and
the flight of the wealthy—including physicians and surgeons. The French
government and the adjacent papal state imposed a *cordon sanitaire* around
the city and completed a wall six feet high across the Vaucluse countryside
in 1721.[5] Attempted quarantines had failed in the past, and many expected
the disease to spread across the continent. This cordon, however, was unpar-
alleled in its severity; one-quarter of the French army was dispatched, in
addition to city militias and provincial levies. The soldiers executed anyone
caught attempting to escape.[6]

To those shut up in the stricken city, the epidemic was an indescribable
calamity. In a city of between one hundred thousand and one hundred and
fifty thousand inhabitants, between one-third and one-half of the population
died. Between five hundred and one thousand people died every day when
the epidemic was at its height. To prevent starvation within the city, the government provided assistance; most of this went to buy food.[7] Doubting the
competence of the local doctors, the government also dispatched a surgeon
and two physicians, including François Chicoyneau from Montpellier.[8] They
met with the local doctors, who believed both that the disease was the plague
and that it was contagious.[9]

Following the meeting, the Montpellier physicians issued a soothing public statement to the panicked citizens that the disease was "not pestilential
[i.e., the plague] but only a common malignant fever." To the Court, however, they reported that it was "a true pestilential fever, not yet arrived at
its utmost degree of malignity."[10] They then returned to Montpellier after
keeping quarantine in Aix. Pierre Chirac, the physician to the King, believed
their report to the Crown was too pessimistic, expressing his view (based only
on their description) that "the malady, though great in itself, and extremely

dangerous, is only a common malignant fever . . . and by no means the plague . . . The unwholesome food of the lower class of people in that town is quite sufficient to produce such an effect."[11] The King then ordered Professor Anton Deidier of Montpellier and a new team to Marseilles.

Deidier had begun his career as a chemist. In addition to serving as physician to the King of France, he held the post of Royal Professor of Chemistry in the Montpellier medical school from 1697 until 1732. In his lectures on chemistry (1715), Deidier upheld the idea of five chemical "principles" (water, earth, salt, sulfur, and spirit), challenging both the Galenic theory of four elements and the corpuscular theory of Descartes; he considered the Cartesian particles useless to chemists.[12] His long tenure at Montpellier left a lasting imprint on the teaching there, sustaining and enhancing Montpellier's tradition of vitalism.[13] Deidier's thought merits further study, as he wrote many successful works. The very popularity of his works and their many reprints has, however, produced a tangled publication record.

As noted above, Deidier sent a treatise on venereal diseases to James Jurin, then acting as the Secretary of the Royal Society, and the Society's printer, Samuel Palmer, produced it as a separate Latin treatise in 1723.[14] A French translation appeared in Paris in 1725.[15] Deidier explains in the French version that the work had originated as a Montpellier thesis for the baccalaureate defense of a student in 1713.[16] Thus this work on venereal disease was originally drafted long before the plague epidemic but printed after it occurred. The London and Paris editions included a fulsome dedication to Hans Sloane; the French translation included Sloane's response.[17] This thesis attributed venereal disease to minute living worms, arguing that this theory offered a better fit for the observed phenomena than blaming corrosive acids because it acted only after a delay, spread very widely, and because the venereal "virus" (poison) could only be destroyed by mercurial rubs.

At the time of his journey to Marseilles, however, Deidier did not see the plague as a contagious disease or attribute it to animalcules. Before his departure for Marseilles, Deidier wrote the physicians there, advising them that "the sudden deaths in the present case can only arise from a too great fullness of the internal viscera" and recommending that they bleed their patients until they fainted.[18]

When he arrived in Marseilles, Deidier was horrified by the scene that met him. As he wrote to a friend:

> I can scarcely describe the frightful disorder which I found in this desolated town. On entering . . . all the doors and windows of the houses were closed, the pavement was covered from one side to the other with the sick and dying . . . in the middle of the streets and in the squares one saw nothing but half rotting corpses and old clothes lying in the mud, and the plague carts led by convicts . . . We . . . went through the town from one end to the other. It was not possible to put a foot anywhere without stepping on the dead or on the beds of the sick.[19]

Another resident wrote that "the entire city is a vast cemetery."[20]

The pestilence gradually ebbed without spreading beyond Provence.[21] The visiting doctors tended the sick, organized hospitals, and quarreled with the medical men who had remained in Marseilles. Although the Marseilles doctors stubbornly insisted that the plague was contagious, the contingent from Montpellier did not have much respect for their views and even refused to join the nightly meeting held by the physicians and surgeons of the town.[22]

The visiting physicians initially adopted Chirac's view that the epidemic was due to corrupt humors caused by the weather and bad food, that fear and grief contributed to its continuation, that it had not been brought from the Levant, and that it was not contagious.[23] They began to waver when their servants and some of the surgeons who had accompanied the party began to die, but none of the physicians succumbed, although they constantly attended the sick, sitting on their beds and often touching their buboes and sores.

Deidier himself was convinced that plague was caused by some sort of poison or venom in the bile of patients. Bad weather in the preceding harvest season had forced many people to eat wheat from the Levant mixed with barley, oats, and rye. This had caused recurrent indigestion and thickened their blood. His theory was strengthened when he cut open many of the corpses without becoming ill himself. This, he wrote, "clearly proves that the plague cannot be caught from any malignant exhalation spread in the air, or adhering to the fingers or still less to clothing which we did not change."[24]

Finding that the victims' gall bladders were filled with thick black and greenish bile, he conducted chemical experiments on it and tested its effect on dogs. When he rubbed the bile into a laceration, the dogs died with plague symptoms in three to four days, as they did when the bile was injected into their veins. He then injected bile from one dog into that of a second, and from that dog into the veins of a third. All the dogs died. However, when the dogs swallowed the bile, they were ill for a time but recovered. Six control experiments performed at Montpellier by colleagues were negative.[25]

Deidier reported these experiments to his friend, Dr. Pierre Montresse of Valence.[26] He sent his most complete account to Dr. John Woodward, who shared it with the Royal Society. By then, he had realized that he had showed that the plague was in fact contagious. He commented, "Although the experiments which I made . . . showed me beyond doubt that plague can be communicated and is transmissible from one subject to another, they did not show how the disease is carried neither the portal of entry."[27] By analogy with syphilis, he thought for a time that plague might also be carried by "another species of pestilential worm," which could have been carried by ship into Marseilles and afterwards multiplied. However, he believed that mercury destroyed all small organisms, and he discarded the idea of living contagion when he found that mercurial ointment completely failed to cure the plague.[28]

Deidier also noted that when the epidemic first began to spread, the people who became ill lived in different parts of the city and had not had contact with each other.

> When we have seen as many as 500 persons fall sick in a day in different parts of the town each shut up in his own house . . . is it reasonable to think that so many widely dispersed persons could have had contact with plague patients . . . before being shut up? . . . If the disease was not epidemic but simply contagious, it should always have spread from neighbor to neighbor [*de proche en proche*] and the dwellings nearest to the town should have been infected before the more distant ones which had no commerce with others, but this was not so.[29]

Despite these misgivings, he came to believe that the poison of plague was analogous to the poison transmitted by the bite of a rabid dog, except that in rabies the poison was in the saliva and in the plague it was in the bile.[30]

Four years later, Deidier reflected on his experiences in an inaugural address at Montpellier. His colleagues were still divided on the issue of contagion. Chicoyneau remained convinced that the illness was just a "common malignant fever" due to poor diet and only became extremely fatal when it was called the plague. The sheer terror caused by the word itself led to many deaths and the abandonment of the sick.[31]

Deidier agreed that the initial cause of the plague must have been general causes such as dirt or hunger, but his experiments had shown beyond question that the disease was transmissible from case to case and so must be contagious. Deidier noted that one surgeon had a sore, which he dressed without washing his hands, and another continued to dress plague sores with an accidental wound—both contracted the disease—whereas those who merely breathed the air around the patients remained well.[32] He concluded that plague did not spread by "pestilential atoms" in the air, but by close and prolonged contact. "I mean by this contact," he added, "to breathe for a long time and very near the burning breath which comes from the mouth of the sick . . . to wear the same shirt or sleep in the same bed-clothes . . . to touch one's own sores with hands still carrying infected sweat or blood . . . The contagion of plague is something like that of venereal disease."[33] Plague, smallpox, syphilis, and rabies all spread in this way despite puzzling features such as the relative immunity of some people in close contact with the stricken, the fact that initial victims were often scattered over a wide area with no evidence of contact between them, and the ineffectiveness of shutting early victims up in their own homes.[34]

The nature of the plague was also the topic of the day among doctors in Lyons, including Jerome Jean Pestalossi (Pestalozzi) and Jean Baptiste Goiffon, a medical botanist who was the head of the Bureau of Health in Lyons.[35] In a long "avertissement" or preface to Jean Baptiste Bertrand's *Observations on the Plague*, a compilation published in Lyons in 1721, Goiffon suggested that plague was carried by microscopic insects. The plague, he wrote, spared no one and did not depend like other diseases on a framework formed within us; it was a poison from outside the body. Like other poisons, it produced its effects indifferently in every sort of patient, whether of good or evil constitution, without distinction of age or sex.[36] This venom insinuated itself quickly into the body and resembled in its strength the venoms of certain animals or poisons that caused a rapid death.[37] The imperceptibility of the insects that carried it did not undermine the argument, for this was also true of all the other assigned causes of plague: the influence of the stars, the planets, the constellations, the vapors and exhalations of the earth, the atoms, the miasma, the corpuscles, and the ferments. It was by reasoning and not by their eyes that physicians discovered the causes of diseases because there were few that were not invisible.

The air itself could not be infected (i.e., corrupted) because in that case no one could survive the contagion of the plague and all precautions would be useless. One would have to find a canton where one could live without breathing, because one portion of air infects another and soon disease would spread throughout the universe. It did not infect the great majority of the country in one stroke but spread little by little, passing through many villages after appearing in Marseilles. It struck a few people in a single family, from which it passed successively from house to house and sometimes through a whole town, often persisting for months and years before returning to the original towns.[38]

Goiffon did not blame insects for every epidemic disease. In his own treatise on the Plague of Gevaudan, Goiffon stated that some epidemic diseases were contagious whereas others were not. Those that were both epidemic and contagious, like the plague, had an animate cause that could multiply outside the bodies of those affected. Goiffon attributed the incubation time of contagious diseases to the time required for successive generations of microscopic animalcules to multiply within the body and spread through the blood.[39]

The theory that insects caused plague epidemics was also considered but eventually rejected by Jerome Jean Pestalossi in a treatise published in Lyons in 1721. He thought the cause of plague was a very active agent, as real as it was invisible. Some believed with Kircher that the pestilential germ was a multitude of small worms, dragons (*dragonaux*), or living insects that flew through the air and attached themselves, or at least their eggs, to everything, but this theory failed to explain how plague could spread so rapidly through the entire city or province. Although the fundamental cause of the disease was unknown, its symptoms must be due to a volatile saline substance, not to insects. This salt, like gunpowder, could be capable of great movement. It spread by contact, fomites, and air. As the title of his book, *Avis de Précaution contre La Maladie Contagieuse de Marseille* (advice on guarding against the contagious disease of Marseilles) suggested, Pestalossi had no doubt that the plague was contagious.

Jean Astruc, Professor of Medicine at Montpellier, became an especially tenacious advocate of the view that plague was contagious, publishing several works in quick succession.[40] He noted that the existence of contagion had been demonstrated by works on rinderpest, and he viewed contagion as a pestilential ferment that multiplied in the blood of its victims.[41]

This sampling of views suggests that, although some French physicians insisted on a "humoral" or "Hippocratic" view of the plague, others adopted a range of contagionist theories.[42] They were well versed in the theory of animate contagion as framed by Andry de Boisregard and earlier authors, although they could not agree on the exact nature of the infectious agent. In any case, when forced to decide, the French government tacitly endorsed the contagionist position by imposing a quarantine. Its apparent success in containing the epidemic lent further support to contagionist theories throughout Europe.[43]

THE PLAGUE CONTROVERSY IN ENGLAND, 1720–1722, AND RICHARD MEAD'S *SHORT DISCOURSE*, 1720

Quarantines to control the plague were not new to Britain. They reflected a well-established lay belief that plague, whatever its original cause, could spread by contagion. Named for the traditional forty-day period of isolation, quarantines had first been practiced by Ragusa (now Dubrovnik) in 1377 and had become widespread in Europe by the 1420s.[44] By 1493, the belief that some imperceptible physical agent could spread the disease was so strong in Venice that when people received money during an epidemic, they washed the coins in vinegar to purify them.[45]

The scanty English medical literature from this period attributed plague epidemics to corrupted air, but we have hints that contagious matter was also blamed. In 1495, an English statute regulating upholsterers warned against infected bedding.[46] In 1513, London authorities revealed a fear of physical contagion when they ordered that the possessions of two plague victims be thrown into the Thames; five years later London forbade the sale or gift of infected clothes and bedding.[47] The notorious practice of "shutting up" infected houses with both sick and sound inside for forty days also began in 1518, by order of the Privy Council.[48] In insisting on sealing up houses, the government was not necessarily turning its back on miasmatism but hedging its bets by guarding against both the polluted air and the hypothetical venomous substances that might be found in the vicinity of victims. An order of 1543 required the airing of infected clothing in fields and the destruction of straw or litter from infected houses. It also barred any migration of exposed people and required householders to provide for the care of diseased residents.[49]

By the mid-sixteenth century, provincial English cities such as Liverpool, Chester, and Leicester were excluding travelers and taking similar precautions.[50] In 1580, the Privy Council took steps to reduce the crowding of London and crowded gatherings and imposed a strict quarantine against ships, people, or goods from all infected foreign ports.[51] In Penrith in the late sixteenth century, local farmers refused to bring food into the town until the inhabitants set up a "plague stone": a hollow block filled with water into which coins could be dropped and collected.[52]

The Privy Council reinforced these lay practices by prosecuting anyone who preached "the bloody error of noncontagion" or who claimed that nothing would protect sinners against the punishment of God.[53] In 1578, the Privy Council insisted that plague was not due to corrupted air but was being carried about by individuals because local governments had failed to enforce isolation.[54] The Council forbade clergymen to preach that it was useless to resist the pestilence and threatened to arrest laymen who offered the same opinion.[55] In 1603, it actually imprisoned Henoch Clapham, the author of *An Epistle Discoursing upon the Present Pestilence*, for writing that plague was "a special blow inflicted on mankind for sin," and therefore ordinary medicines were useless.[56]

The Marseilles plague epidemic of 1720 spurred the publication of a huge number of tracts and treatises in England. The most important was Richard Mead's treatise, *A Short Discourse concerning Pestilential Contagion*, commissioned by the Lords Justices during the King's absence in Hanover.[57]

It is not clear why the Justices turned to Mead for advice, although he was an extremely successful doctor. His political affiliations, his expertise on venoms and poisons, his experience in Italy, and his membership of the Royal Society may all have been factors. Born in 1673, Mead was the eleventh child of a famous London Independent divine, Matthew Meade, who had been ejected by the Act of Uniformity. During Richard Mead's childhood, his father continued to preach and suffered frequent persecution. In 1683, Meade was briefly arrested for plotting against the government, and he fled to Holland in 1684 following Monmouth's Rebellion.[58] He returned to his congregation in Stepney in 1687 and died in 1699.

Richard Mead's Quaker ties are not as well known as his ties to traditional Dissent. His uncle, William Mead, had been tried in 1670 with William Penn in a celebrated case where they pleaded their cause so well that the jury refused to find them guilty and was imprisoned for its verdict. William Mead's wife, Sarah Fell, was the stepdaughter of the Quaker leader George Fox. Dr. Richard Mead's first wife, Ruth Marsh, was the daughter of a lapsed Quaker who left her a large fortune.[59]

Richard Mead obtained his MD from Padua after studying in Utrecht and Leyden, where he had been a classmate of Boerhaave's. Both men believed in irenicism and they became lifelong friends.[60] Mead was cosmopolitan, frequenting the gatherings of the foreign journalists and philosophers who constituted the "republic of letters" at the Rainbow Coffee House. Fluent in Italian, he transmitted medical information and ideas from Padua to London.[61] He seems to have conformed to the Anglican Church after his father's death and his own marriage in 1699.[62] He practiced medicine without a license from his father's house in Stepney for 12 years, becoming an FRS in 1704 on the nomination of Hans Sloane following publication of *A Mechanical Account of Poisons* in 1702/3. On Dr. John Radcliffe's retirement in 1710, Mead took over his practice, and on his death, moved into his home.[63]

Mead became a Vice-President of the Royal Society in 1707 and a Fellow of the College of Physicians nine years later.[64] Among his patients were Queen Anne, whom he attended at her death, George I and II, Isaac Newton, Alexander Pope, Robert Walpole, and Bishop Gilbert Burnet: the Scottish author of the *History of My Own Time*.[65] A "very hearty Whig" known for his kindness and gentle manner, Mead was the closest friend of Richard Bentley, the Master of Trinity College, and a confidant of Thomas Guy, whom he advised on the design and management of what would become Guy's Hospital.[66] Despite his political sympathies, he befriended some Tory physicians, such as Arbuthnot and Freind, rescuing the latter from prison after the Atterbury plot. Indeed, rescuing Jacobites from prison seems to have been a hobby of his: after the Jacobite uprising in 1715, he saved Archibald Pitcairne's son Andrew from hanging, and he obtained the release of Bentley's enemy Lord Orrery from the Tower on medical grounds in 1721.[67]

Mead often attended medical consultations with Sloane, who shared his addiction to book collecting.[68] It was once suggested that Mead would be Sloane's choice for a second in a duel with the quarrelsome John Woodward.[69] In fact, it was Mead who dueled with Woodward after Dr. John Freind had dedicated a work on smallpox to Mead, embroiling him in the "smallpox wars."[70] Mead's close friend, the Scottish Tory "wit" and Scriblerian Dr. John Arbuthnot, helped lead the charge against Woodward.[71] Although on this occasion Mead ended up on the side of the Tories, this was evidently due more to Woodward's determination to turn all of his potential allies into rivals than to any Tory sympathies of Mead's.[72]

The prompt publication of his book on plague in 1720 meant that it anticipated much of the French literature on the subject, but Mead's views were similar to those of Pestalossi. Mead also had access to the recommendations drafted in 1631 by Theodore de Mayerne, the Huguenot physician to James I. As a graduate of Padua, Mead was familiar with Italian practices as well as Italian publications.[73] In his first sentences he uncompromisingly described the plague as a contagious disease and promised that his book would explain how to curtail the spread of such diseases by examining the nature of contagion itself. Contagion had three causes: "the *Air, Diseased Persons, and Goods transported from infected Places.*" Wrestling with the problem of index cases, he concluded that the air became contaminated more easily when the weather was hot and damp. To this must be added the effect of "the Stinks of stagnating Waters in hot Weather, putrid Exhalations from the Earth; and above all, the Corruption of dead Carcasses lying unburied."[74] If the air was not corrupted, the plague could not take hold; for this reason the plague always originated in the hot countries of the east and south and was imported into England by commerce.

The corrupted air was the first condition, but there was also a second factor: the "great Quantity of active Particles upon the several *Glands* of the Body, particularly upon those of the Mouth and Skin" of plague victims.[75] These particles could always infect those who were very close to a sick person, but when the air also became corrupted it could unite with them to form a more durable infectious matter that could spread to a considerable distance away from the patient.

Mead argued that corruption of the air must be necessary for the plague to reach its full force, because otherwise it would be impossible to explain how the plague could ever cease in a given place before it had killed every inhabitant; on his hypothesis this was easily explained by a change in the weather. On the other hand, he argued, anticipating Goiffon, the air alone was never enough to cause the infection "without the Concurrence of something emitted from *Infected* Persons" because experience had shown that strict quarantines could prevent the plague from spreading.[76] People became ill by breathing in the contagious particles, which mixed with saliva and were swallowed, or mixed with the blood in the lungs.

He dismissed the theory that the disease was carried by the eggs of insects that then hatched to cause fresh outbreaks, since it was not founded on any observation. Instead, he believed that the contagious substance was "an active Substance, perhaps in the Nature of a *Salt*, generated chiefly from the Corruption of a *Humane Body*."[77] This suggestion drew on Mead's earlier research on poisons, which had included mineral poisons, venomous animals, and rabies, because Mead thought that contagions were the same as poisons. He had examined viper's venom under a microscope and saw "saline particles . . . crystals of an incredible tenuity and sharpness."[78] He had also carried out a number of chemical tests that did not reveal effervescence, acidity, or alkalinity, implicitly challenging the chemical models of Sylvius and Willis.

Mead recommended a maritime quarantine for every ship that had experienced sickness on board. All merchandise, especially hemp, cotton, wool, silk, flax, linen, hair, skins, books, and paper, was to be opened and aired separately for forty days. If a ship had come from a port that was experiencing a severe outbreak of plague, it was best to burn all the goods and even the ship itself, a drastic suggestion that proved unpopular with the merchant community.[79]

Although he supported strict quarantines, Mead strongly opposed shutting up houses when a resident was stricken. He thought this concentrated poisonous particles within the house, where they killed the inhabitants one after another and remained virulent indefinitely, ready to infect anyone who entered. Moreover, the fear of being shut up led families to conceal illness as long as possible. Instead, he argued, the sick and the healthy should immediately be removed and lodged separately three or four miles out of the town. The healthy were first to be stripped, washed, and shaved. The sick were to be given every possible care and treated at public expense. The goods in the houses and possibly the houses themselves were to be burned. Furthermore, the overseers of the poor were to take care that no one remained in close and nasty dwellings, but provide them instead with clean, sweet lodgings. The city streets should be kept free of filth, carrion, and similar nuisances.[80]

If the plague should nevertheless take hold, Mead opposed the lighting of fires and the firing of guns, two traditional measures intended to disinfect the atmosphere. Instead, he recommended strewing houses with "cooling herbs" and washing them with water and vinegar, adding, "as *Nastiness* is a great Source of *Infection,* so *Cleanliness* is the greatest Preservative: which is the true Reason, why the Poor are most obnoxious [susceptible] to Disasters of this Kind."[81]

Mead commented that after thoroughly cleaning their homes, the next question was what else people could do to protect themselves. Ideally, they would take measures to put their humors into such a sound state that the "Matter of Infection" could not alter them, but although people could help to avert the plague by living temperately and taking acidic fruits and drinks, it was more important to avoid all contact with victims and with dead bodies. He recommended that the magistrates prohibit public gatherings and ensure that all bodies were taken away at night to a distance from the towns and buried deeply. To prevent individuals from carrying the disease from town to town, he recommended a *cordon sanitaire.* Those who wished to leave could do so after a quarantine of twenty days in tents, but no goods were to be permitted. Mead felt that this would be more effective than an absolute prohibition on all travel, which, by making people desperate, led them to evade the regulations completely. Moreover, he thought that confining everyone in the affected town would concentrate the disease and thus make it more virulent, just as household confinement did.

Mead's advice received very wide publicity. In addition, the Privy Council had commissioned a report from the College of Physicians that has never been published.[82] The report was signed by Mead, John Arbuthnot, and Hans Sloane, who was then President of the College. As we have seen, these three physicians were friends and often allies. None of them was a graduate of an English university.[83] They advised dividing the City of London into six health districts, each with its own searchers, and hiring an additional fifty searchers to serve in the suburban parishes, which actually contained a larger population than the city. Victims should be removed to six barracks erected a short distance away from the city in open ground. The healthy members of their households should be removed to separate housing. The government responded by creating a Board of Health and a royal commission that included Sloane, Arbuthnot, and Mead in addition to the mayor and four aldermen, five bishops, excise officers, and nearby justices of the peace.[84]

THE DEBATE ON QUARANTINES AND CONTAGION

The Parliament that convened on December 8, following the King's return from Hanover, met shortly after the trade collapse that followed the South Sea Bubble. The government was in great disarray. Walpole had just returned from seclusion in Norfolk to manage the crisis. The magnitude of the threat posed by the plague in Marseilles and the depth of the belief that it was contagious are revealed by the steadfast determination of both the ministry and Parliament to enforce stringent quarantines, even though they annoyed merchants, raised the price of goods, burdened a struggling economy, and damaged the popularity of the government, which was already weakened by scandal. Other measures, however, failed to pass amid fears that the government was assuming too much power.

England's first Quarantine Act (as opposed to an order from the Privy Council) had passed in 1710.[85] The Parliament that convened in 1720 repealed this Act and adopted many of Mead's suggestions, approving a more stringent Act in early 1721. This applied to the entire country and required local justices of the peace to raise funds to pay for additional watchmen. It permitted the creation of a commercial *cordon sanitaire* around any infected town and even authorized courts to sentence violators to death.[86] By the date of its passage, the plague at Marseilles had ebbed, but in June, two English ships arrived from infected ports of the eastern Mediterranean and were burned with their cargoes. A second Act approved in the autumn of 1721 imposed further restrictions on trade, allowing the government to "hinder and oppose by force" any ship from an infected port that attempted to dock.[87] Britain remained committed to quarantines, even at a high political and economic cost.

Members of Parliament were not the only ones who debated the wisdom of quarantines. Mead's treatise ignited a pamphlet war. Among those who joined the fray on behalf of quarantines were Edmund Gibson, the Whig bishop of London and Walpole's chief advisor on Church affairs, who refuted the argument that quarantines were "French measures" that buttressed arbitrary government.[88] Richard Boulton (d. 1724), a medical graduate of Brasenose College, Oxford, and a client of Sloane's, wrote what historian Charles Mullett described as a "somewhat slavish acceptance of Mead's ideas." Sir John Colbatch (d. 1729), a Worcester apothecary who had become a Licentiate of the College of Physicians in 1696, also sided with Mead.[89]

Another account, very similar to Mead's, came from an obscure country practitioner named Richard Brookes. In *A History of the Most Remarkable Pestilential Disorders that Have Appeared in Europe for Three Hundred Years Last Past*, published in London in 1721, Brookes commented:

> One great Property of the Plague is, that it is Contagious; which leaves us no room to doubt of the Existence of something which has the Effects of Poison; what this is, or the Nature of it, has not been sufficiently explained by Authors. Some have thought it to be the Eggs of Insects; others a Poison, of the Nature of Arsenick; others again, something arising from the Putrefaction of a Human Body. However, this is certain, that it is something very fine and subtle, and that it will lie lodged in Clothes, or such like things, for several Years together.[90]

Some authors, Brookes continued, said that the plague must consist of a putrefaction or stagnation of the air itself, but it was not difficult to see that particles might exist in the air without the corruption of the air itself.[91] He interspersed his comments with a collection of remedies culled from other authors.

Among Mead's opponents was Joseph Browne, a colorful controversialist who probably graduated MB from Jesus College, Cambridge, was twice imprisoned for libel under Queen Anne, opposed Harvey's theory of circulation, and was in the habit of dedicating his works to distinguished people without their permission. Browne wrote two separate works on the plague. His *Practical Treatise on the Plague* (1720) had a preface addressed to Mead, and his *Antidotaria: or a Collection of Antidotes against the Plague* (1721) was dedicated to the College of Physicians.[92] Browne attributed the disease to polluted air. Browne accused Mead of misrepresenting the work of Nathaniel Hodges, whose 1672 work *Loimologia: or an Historical Account of the Plague in London in 1665* was reprinted in 1720. Hodges had believed that the disease originated in a poisonous "aura" that arose from a corruption of the nitrous spirit in the air, but he thought it was transmitted by contagion.[93]

Another opponent was the anonymous author of *Dr. Mead's Short Discourse Explain'd or His Account of Pestilential Contagion, and Preventing, Exploded.*[94] This was a scathing account that left virtually no paragraph of Mead's free from assault. For example, the author quotes Mead's suggestion that "when the sick Families are gone, (Whither?) all the Goods of the Houses should be burnt; nay, the Houses themselves . . . A very good Advice; and, I hope, the City of London will erect another Monument for the Doctor, after they have burnt their City." The author attributed plague to contaminated air and recommended purifying the atmosphere in houses by burning scented pastilles and shavings of cork.

Another anonymous work, *Some Observations concerning the Plague* (Dublin: 1721), took the opposite position, chiding Mead for placing too much emphasis on the air as a vehicle for epidemics and claiming that the contagious agent must be something heavier than Mead's "volatile salts."[95] A London physician, Dr. George Pye, implied that Mead contributed to the spread of plague by causing unjustified terror, which led people to contract the disease. In fact, the plague was not always the same disease and did not always stem from the same cause.[96] It arose from the epidemic constitution of the air and a poor diet, and it was not infectious. In 1722, a "Dr. John Pringle" published *A Rational Inquiry into the Nature of the Plague. Shewing That as the Air Only is Capable of Producing, or Communicating It; the Method of Prevention Now Practis'd in France, Is Not Only Inhumane, but Useless, and Even Pernicious.*[97]

Richard Blackmore's *Discourse upon the Plague* blamed noxious airborne particles or vapors. These arose from "internal Vicious Humours, or Pestilential Air," which was often bred in the bowels of the earth.[98] Its destructive quality resulted from the great minuteness and refinement of these vapors or particles and their opposition to the active principles of the blood. Their size enabled them to penetrate the blood and cause a separation of its parts that resulted in putrefaction. A similar process caused all fevers, which did not differ in their nature but only in their degree. Contagion came from the effluvia of an infected body, which converted to its own nature any noxious vapors in the air. Blackmore was especially concerned with the problem presented by the origin of the plague, arguing that it could not always be due to contagion because then the disease would stretch back indefinitely in a continuous chain of illness. Although he believed that it sometimes came as a direct visitation from God, he did not seem to find this a satisfactory explanation for the origins of all epidemics.[99]

RICHARD BRADLEY'S THEORY OF INSECT-BORNE DISEASES

If Blackmore presented a conservative view of the plague, seeking to combine a limited sphere for contagion with iatromechanical and humoral theories, Richard Bradley offered a contrasting view in his treatise of 1721, *The Plague at Marseilles Considered*, dedicated to Isaac Newton as President of the Royal Society.[100] Bradley believed that plague was carried by insects. In the summer, he argued, a vast variety of the smaller kinds of insects floated in the air. Each kind of insect sought its proper "nidus" to hatch and was attracted by "certain Effluvia" that arose from the body that was suitable to it.

As a botanist, Bradley drew on the analogy with plant diseases and infestations. Bradley had visited Holland, where he had met Leeuwenhoek, and he used a microscope in his work. His *New Improvements of Planting and Gardening* attributed spring blights to insects that either overwintered from eggs laid the previous year or blew in on the east wind.[101] He also blamed "insects," by which he meant microscopic animalcules, for putrefaction, wheat smut, and human diseases.[102] He continued this line of argument in *The Plague of Marseilles*, writing:

> In the Blight of Trees, we find, such Insects as are appointed to destroy a Cherry Tree, will not injure a Tree of another Kind, and again, unless the leaves of some Trees are bruised by Hail, or otherwise Distemper'd, no Insect will invade them; so in Animals it may be, that by ill Diet the Habit of their Body, may be so altered, that their very Breath may entice those poisonous Insects to follow their way, till they can lodge themselves in the Stomach of the Animal, and thereby occasion Death.[103]

Once infested, diseased people then flung out eggs of these insects "in Parcels" in the process of respiration. These eggs could infect those standing by the patient, and because of their extraordinary smallness they could travel through the air some distance. These insects might be carried long distances by the east winds from Tartary.

Charles Singer described Bradley's work on plague as the "best attempt to solve the problem of the nature of infection of any writer previous to Pasteur," but it is not appreciably different from the nearly simultaneous work of Goiffon, who also attributed the plague to minute insects.[104] Singer was impressed by Bradley's emphasis on the specificity of the plague, but Bradley laid his primary stress on the specificity of the host of the disease, not of the disease itself. However, it is interesting that Bradley explicitly referred to Ramazzini's work on rinderpest.

Bradley's theory inspired the anonymous author of *A New Discovery of the Nature of the Plague . . . Contrary to the Opinion of Dr. Meade, [sic]*

Dr. Browne, and others, who give for the First Causes of the Plague . . . Air, Diet, and Disease. This very rare work, whose author may have lacked medical training, attributed the disease to "a subtle, active poisonous Body or Insect, very minute . . . living on and subsisting by the virulent Matter in the body . . . the Air being no more to it, than it is to Birds."[105]

In Defoe's *Journal of the Plague Year* (1722), a complex mixture of history and fiction, the narrator refers to Bradley's theory, though he ultimately rejects it. The narrator said that he had observed that when a servant fell ill of the plague and the family either sent the servant away from the house or left the house themselves, they survived. On the other hand, when a family was shut up in a house with one person who was ill, they all died. This, the narrator continued, made him certain that the plague had spread by infection:

> That is to say, by certain steams or fumes, which the physicians call effluvia, by the breath, or by the sweat, or by the stench of the sores of the sick persons, or some other way, perhaps, beyond even the reach of the physicians themselves, which effluvia affected the sound who came within certain distances of the sick . . . and so those newly infected persons communicated it in the same manner to others . . . Some people, now the contagion is over, talk of its being an immediate stroke from Heaven . . . which I look upon with contempt . . . likewise the opinion of others, who talk of infection being carried on by the air only, by carrying with it vast numbers of insects and invisible creatures, who enter the body with the breath, or even at the pores with the air, and there generate or emit most acute poisons, or poisonous ovae or eggs, which mingle themselves with the blood, and so infect the body.[106]

Bradley's work on plague was popular, reaching a third edition. He followed it with two more works on the subject: *The Virtue and Use of Coffee, with Regard to the Plague*, also published in 1721, and *Precautions against Infection*, which probably appeared in 1722. This work included a long quotation from a letter to Bradley from "J. Phillips," suggesting that the plague was transmitted by insects in the same manner as the itch was spread by mites. In addition, Bradley's *General Treatise of Husbandry and Gardening* ascribed color breaks in plants to plant disease, likening it to inoculation, and argued that: "There is the same possibility of ingraffing [*sic*] Distempers, and vitiating the Juices of Vegetables as of poisoning or infecting the Blood in Animal Bodies."[107]

During the period of the plague at Marseilles, many reputable medical authors agreed that plague was contagious. Governments throughout Europe shared this view or they would not have persisted in imposing politically unpopular and economically ruinous quarantines that were difficult and expensive to administer, greatly impeded trade with the Levant, and sometimes brought entire cities close to starvation.[108] In England, contagionist measures were introduced by a Whig government on the advice of a Whig physician and defended by a Whig Bishop, but the party lines were not as clearly drawn as they would be over inoculation.

The English debate often focused on whether the *contagium* itself was animate or chemical, on the size of the animate agent, or on the nature of the "ferment," not on whether plague was contagious at all. In this context, neither the work of Bradley nor that of Mead was exceptional, although Mead's elegant account was extremely widely read throughout the century. Mead thought of the pathogenic agent as a chemical, whereas Bradley characterized it as an insect, but both men also emphasized the contribution of other factors to the prevalence of epidemics. Mead emphasized the presence of hot, moist air, contaminated by various pollutants, such as rotting corpses. Bradley emphasized the resistance or predisposition of the patient.

THE PLAGUE CONTROVERSY AFTER MARSEILLES

The debate over plague cooled after about 1722, when it became evident that Britain would escape the outbreak at Marseilles, but it did not die out entirely. Plague remained a threat in many parts of the world where Britain had economic or military interests, and the British government, reacting to its prevalence, continued to annoy traders and travelers with a strict quarantine policy. Although the two quarantine acts lapsed in 1723 and were not renewed, a similar act passed in 1728.[109] This expired in 1730, but a new act passed in 1733, in response to a fresh outbreak in Europe. In 1753, Parliament passed an act governing trade with the Levant that required all vessels coming from a country where plague existed to undergo quarantine before entering British waters. In the following years the orders issued under this act were regularly strengthened. The British government maintained a strong quarantine policy throughout the rest of the century. An even more stringent act passed in 1788, and it was not until after the turn of the century that the rules were eased.[110]

The outbreak of a serious epidemic in Messina in 1743 spurred another wave of debate about quarantines. Critics of quarantines were never lacking; among the most furious and persistent was the obstetrician Sir Richard Manningham, author of *The Plague No Contagious Disease* (London: 1744) and *A Discourse concerning the Plague and Pestilential Fevers: Plainly Proving That the General Productive Causes of All Plagues of Pestilence, Are from Some Fault in the Air: or from Ill and Unwholesome Diet* (London: 1758).

Manningham is difficult to place on the political/religious spectrum of the mid-eighteenth century. In his book on man-midwifery in the eighteenth century, Adrian Wilson argues that he was a "Court Whig," because he was knighted by the Hanoverians in 1721/2 and opposed the use of obstetrical forceps. Forceps were improved by members of the Tory Chamberlen family in the seventeenth century, who incurred Whig resentment by sharing the secret only with a small handful of like-minded colleagues. Wilson argues that until the middle of the eighteenth century, Whig physicians favored the technique of manual version promoted by the Dutch obstetrician Hendrik van Deventer.[111]

Manningham's distrust of the notoriously unsatisfactory forceps is not a reliable guide to his political views. His criticism of forceps appeared in

1739 when the inadequacies of Chamberlen's design were well known. His "Newtonian" iatromathematical medical theories do not definitively place him in either camp, but his family background yields a different story.[112] Manningham's father, the Reverend Thomas Manningham, a dedicated Tory "High Flyer" and strong supporter of the Tory radical Henry Sacheverall, rose under Queen Anne to become Bishop of Chichester, where he tried to purge the chapter of Whigs. Three of his four sons became Anglican clergymen.[113] It is likely that the fourth son, Dr. Richard Manningham, was also a devoted Anglican, although we can only guess at his political affiliation. Medically, he was more conservative than Blackmore.

In *The Plague No Contagious Disease*, Manningham argued that the "Moderns" had misunderstood the distinction between contagion and infection. When Hippocrates spoke of "infection," he meant "a pollution, or infection" caused by a contaminated atmosphere. On the other hand, when modern physicians spoke of contagion, "they speak of this *Infection* passing from one Thing, or one Person into another; as *Fracastorius*, the first Person I think, that supposed the Plague to be a *contagious* Disease expresses it."[114]

Manningham attributed the success of Fracastoro's theory to papal politics: contagionism offered Paul III an excuse to move the Council of Trent away from an outbreak of typhus. The professors of physic from the states dominated by the Pope espoused the new opinion, "a Doctrine *newly* broached in *arbitrary* States and Governments."[115] Manningham considered quarantines "un-English," the products of popery and Continental despotism.[116]

Manningham cited an analogy of Dr. Radcliffe's: if two men went out in the rain both would be "infected" with the wet, but the dampness did not pass by contagion from one to another.[117] Only a few diseases, he noted, were considered contagious; they included leprosy, consumption, sore eyes, and the itch.[118] If the plague were contagious, how could the first person have been infected? Moreover, the steep increases in the number of plague victims during epidemics showed that it could not have been communicated from one person to another. After a rapid summary of the literature, he claimed that the epidemic in Marseilles had convinced French physicians that the plague was not contagious.

Manningham then reviewed the nature of febrile diseases. He defined fever as "a preternatural Motion of the Blood" that occurred when blood became vitiated in quantity, motion, or quality. Because there were an infinite variety of fevers, the only reliable method was to name fevers for their symptoms. The universal cause of fevers was "a lentor": a viscosity or thickening of the blood that impeded its free circulation through the capillaries. All fevers stemmed from the lentor's greater or lesser coherence and its degree of solution in the blood. Fevers were thus a matter of degree, not of kind; they represented a spectrum of severity with infinite gradations, not a fixed number of species. This solved the question of how the plague arose for the first time—it was merely a severe case of viscosity.[119] Manningham, like many others, thus tied his anticontagionism to a unitary fever theory: the denial that fevers could be different entities.

Quarantines, Manningham argued, created the plague instead of preventing it, because they raised the price of food. These "artificial famines" forced the poor to eat unhealthy food, making their blood slimy and viscid. The "disposition of the air," low spirits, immoderate passions, too luxurious a diet, and "poor hygiene" also caused plague. To cure it, Manningham recommended fires, gunpowder explosions, bleeding, and sudorifics. Manningham was still insisting that plague was not contagious in 1758.

Theophilus Lobb's *Letters relating to the Plague and Other Contagious Distempers* (London: 1745) proffered a contagionist account.[120] Lobb, an English Dissenter with a Scottish degree, blamed the plague on light volatile infectious particles that spread only a short distance through the air before becoming diluted. These particles might lodge in clothing or goods but were lighter than dust, and so tended to rise. "It does not seem probable that the Sick can infect the Sound at any great Distance."[121] He recommended that suspect goods be buried or boiled rather than burnt, since burning released the particles as smoke. He also recommended that plague patients be kept one to a bed in the top floor of pesthouses and that each room have just one bed. Lobb was not sure whether the infectious particles were animate or inanimate matter, but the air was simply the medium.

Lobb warned that seemingly healthy people could transmit infectious effluvia to others. He thought that attendants should lodge below their patients and wash their hands with vinegar and water before and after they touched the sick. He also recommended the cleansing of floors with vinegar and boiling lemon juice and the use of special gowns. Lobb may have simply meant ordinary clothing designated for this purpose or he may have been thinking of earlier European physicians who had worn long waxed robes, gloves, boots, and masks with crystal eyepieces and a long beak stuffed with herbs. As the waxed clothing prevented fleas from penetrating, this attire may well have helped to ward off infection.[122]

Lobb's comments on the transmission of plague were echoed by a first-hand observer. Writing to Mead, Dr. Mordach Mackenzie wrote of the plague he saw in Constantinople (Istanbul) that

> it is brought from Cairo commonly; and that when once a house or ship is infected, it is very difficult to eradicate the *animalcula, semina, effluvia, miasmata*, or whatever name is proper for the reliques or remains of it, which by getting once into a nidus, lodge there . . . condens'd by the cold during the winter, and when rarefied by a certain degree of heat, they act upon bodies, which have a disposition, as women and children mostly, and so spread by contact only, without communicating any malignancy to the ambient air . . . Whoever kept their doors shut, run no risque, even if the plague were in the next house; and the contact was easily trac'd in all the accidents.[123]

Mackenzie was clearly more interested in the efficacy of practical measures to prevent the transmission of the disease than in the nature of the agent that carried it—a viewpoint that characterized mid-eighteenth-century contagionism in general.

A nonmedical author, Alexander Bruce, published a treatise in 1759 arguing that the plague was the same illness as jail fever, dysentery, and scurvy and that they were all due to retained perspiration in individuals who had been active but suddenly became inactive and so stopped sweating. This work was demolished by an essay in the *Critical Review* that may have been written by the novelist Tobias Smollett, the journal's editor. The reviewer stated that the communication of plague by contagion "is a truth as plain as human evidence can make it" and complained that Bruce had inappropriately confounded the plague with other illnesses.[124]

A serious epidemic in Moscow prompted publication of William Brownrigg's strongly contagionist treatise *Considerations on the Means of Preventing the Communication of Pestilential Contagion* (London: 1771).[125] Brownrigg championed the use of military cordons to quarantine cities or neighborhoods where outbreaks had occurred in place of the old practice of "shutting up" households. He argued that cordons had prevented the spread of the plague from southern France into the rest of the country and contained the plague at Messina in 1743. Venice had successfully deployed them to keep an outbreak in Bosnia from spreading to Istria and Dalmatia.[126]

In 1783, *Medical Commentaries* published Dr. Matthew Guthrie's "Observations on the Plague, Quarantines etc. in a Letter . . . to Dr. Duncan." Guthrie believed in contagion and quarantines but thought 14 days would be sufficient for the isolation of suspect individuals.[127] Another account of the Moscow epidemic came from Dr. Richard Pearson, who translated the work of the Belgian witness Dr. Charles de Mertens in 1799.[128]

William Black, in his *Historical Sketch of Medicine and Surgery* (London: 1782) felt so confident that the contagionist argument had prevailed that he sneered at the opposition:

> Strange as it may appear, in the present century, several Physicians in France, and one or two in England, published and maintained this extraordinary position, that the plague which broke out at Marseilles was not a contagious disease. Dr. Mead and Astruc exerted great pains to combat a doctrine so rash, inconsiderate, and mischievous. The authors of this ruinous hypothesis, seem to have been conceited pedants, who at the hazard of thousands of lives, obstinately hugged their own theories, and should have been chained to the gallies or locked up in a mad-house . . . It is now well known that the infection . . . can be spread to a very trifling distance by the air alone.[129]

Black, an Irish-born student of William Cullen, Professor of Medicine at Edinburgh, also wrote in favor of smallpox inoculation.[130]

At about the same time, Cullen himself was describing the plague to his students as a "disease which always arises from contagion" and recommending quarantines.[131] Lighting fires or other "general fumigations" were ineffective because neither the atmosphere as a whole nor any significant portion of it became tainted or impregnated with contagious matter. Instead, he recommended frequent changes of clothing, good ventilation, and the destruction of suspected goods.[132]

In 1789, John Howard's *Account of the Principal Lazarettos in Europe with Various Papers Relative to the Plague* appeared.[133] Howard's heroic life and equally heroic death in 1790 ensured that this work reached a very large audience.[134] Howard had prepared for his investigation before he left England in 1785 by asking "two of my medical friends, Dr. Aikin and Dr. Jebb" for a set of queries concerning the plague. "Dr. Aikin" was Dr. John Aikin, also a student of Cullen's and a very close collaborator of Howard's, who worked with him on his earlier editions of the *State of the Prisons*.[135] It is unfortunate that Howard does not provide a first name for "Dr. Jebb" as there were two active physicians with that surname in the early 1780s, but it was probably the radical Dr. John Jebb, who was extremely interested in prison construction and reform.[136] The questions these two physicians prepared show that contagion was still controversial, but they presumed an affirmative reply.

At the turn of the century, the tide of contagionism began to ebb. In 1799, the Scottish physician Gilbert Blane, another student of Cullen's, wrote that "there have been physicians paradoxical enough to maintain that the plague itself is not infectious; and their principal argument is, that numbers are exposed to it without being affected by it. But the same may be said of the small pox."[137] In a work of the same year, however, Dr. Charles Maclean was as caustic as Black had been on the other side of the argument:

> The alarm excited . . . by the . . . plague of Marseilles, having come to its height . . . Dr. Mead was requested . . . to draw up directions for its prevention. But . . . either panic had swept away all powers of reasoning, or the doctor's powers were not correspondent with his reputation. His "Discourse," since so blindly admired . . . we now find to have been . . . nothing but the fables of Fracastorius . . . as sanctioned by the Pope, and the Council of Trent.[138]

As Erwin Ackerknecht pointed out in his classic article on "anticontagionism" in which Maclean played a central role, Maclean distinguished between "epidemic" diseases such as the plague, typhus, cholera, and yellow fever, and those he believed were "truly contagious," such as smallpox and syphilis. Unlike the truly contagious diseases, epidemic diseases took many forms, appeared seasonally, and might attack the same person more than once. He attributed plague to some atmospheric influence; typhus to hunger, unemployment, and despair; yellow fever to filth; and cholera to undefined "local causes" and "winds." Echoing Manningham, he argued that it was the quarantines themselves, not the diseases they purported to control, that caused the vast majority of all epidemic illnesses.[139]

This again reveals the association between anticontagionism and a denial of disease specificity. Maclean and his allies believed that epidemics were due to the social forces that caused poverty, hunger, overcrowding, filth, and despair. Ascribing illness to an emotion such as terror or despair or blaming a social problem such as poverty grouped a wide range of symptoms, signs, and ailments together as diseases with a common cause. The contagionists whom he attacked believed that plague and typhus, like smallpox and syphilis, were caused by a specific morbid substance that could only cause a specific species of illness. Social conditions might exacerbate an epidemic, but a quarantine that successfully blocked the specific morbid poison would prevent or arrest the plague even among the poor and distressed. Despite Maclean's efforts, the British government continued to enforce quarantines through the nineteenth century. In the face of constant and occasionally bitter criticism, the government did not entirely abandon its quarantine policy until 1896, when it was replaced by ship inspections.[140]

Maclean's hostility toward both the contagionism of his elders and the medical "establishment" represented by the College of Physicians may have misled Ackerknecht, who adopted Maclean's viewpoint that his enemies must be allies. To Maclean, Ackerknecht wrote:

> The political background of the anticontagionist discussion was no less obvious. The leading anticontagionists . . . were known radicals or liberals. The leading contagionists . . . were high ranking royal military or naval officers like . . . Sir Gilbert Blane . . . Maclean thundered against the 'monopoly of the College of Physicians' . . . Quarantines were ready 'engines of despotism.'[141]

Dissenters such as Lobb, Howard, and Aikin and Irish and Scottish physicians such as Black and Blane, were indeed contagionists, but they were not conservatives who opposed reform. Aikin and Howard campaigned energetically to reform institutions, including hospitals and prisons. Aikin's radical political views cost him his livelihood and his home; Howard paid for his work with his life. Blane fought to implement James Lind's advice to prevent scurvy among seamen and improve their conditions. These men were tenacious critics of the views and practices of the College of Physicians, not its representatives as Ackerknecht had believed.[142] Contagionism had deep roots among Dissenters, but throughout the century the College saw Nonconformists and doctors from outside England as second-class citizens.

CONCLUSION

The belief that plague was a contagious disease and the corresponding efforts by governments to prevent or circumscribe it had begun in the Renaissance, but contagionist accounts of plague peaked during the eighteenth century. Both the theory and the practice of quarantine were reiterated in a new and more uncompromising form during the epidemic in Marseilles. A wave of contagionist works by authors associated with William Cullen also followed the outbreak of plague in Russia in 1770.[143] After coming under attack in the early nineteenth century, contagionist disease theory would be forever transformed by the advent of the germ theory of disease, though *Yersinia pestis* still poses unanswered questions.

Mead's *Short Discourse* helped define British contagionism throughout the century. Mead is often, and justly, characterized as a Newtonian. His first book, which gained him admission to the Royal Society, was entitled *A Mechanical Account of Poisons*, a title that speaks for itself. His next work, *De Imperio Solis et Lunae* (1704), argued that heavenly bodies (especially the sun and moon) affected air pressure and this then affected the fluids and humors in the body, causing some ailments and controlling the periodicity of others.[144] It was Newtonian in its rhetoric, though not in its dependence on unsupported hypothesis. Yet when he discussed the plague 15 years later, Mead's approach had changed. His *Discourse* displays little evidence of the hydraulic physiology favored by Newtonians.

Historian Theodore Brown noted this "strange" and "radically different" approach in the book on plague.[145] Brown explained it as a result of the increasing influence of Boerhaave on medical theory, but Mead's *Discourse* is even less Boerhaavian than it is Newtonian. Instead of the weak fibers, gluey blood, and obstructions that dominated Boerhaave's theories, we find references to putrefaction and ferments that might almost have come from Willis. Mead drew on many elements for the *Discourse*, including his encyclopedic knowledge of the classics and his familiarity with modern Italian thought, but his work also had features in common with iatrochemical authors such as Deidier.[146] Moreover, his plan to prevent the plague was similar to one created by Theodore de Mayerne, a Paracelsian Court physician, nearly a century earlier.

As we have seen, Mead conformed to the Anglican Church, but he had been a Dissenter into early adulthood. Perhaps this early education shaped the *Discourse*, even as his adult skepticism stripped it completely of the mysticism and vitalism that had accompanied earlier Helmontian thought: an editing process that made the theory more palatable to many Enlightenment readers. He thought that changes in the quality of the air accounted for the origins of pestilences but also that "the original cause of a disease, and the communication of it, are very different things."[147] Most of his work speculates about the "original cause" of diseases, but the *Discourse* focuses on the "communication" without resolving the underlying tensions between miasmatism and contagionism.

Throughout the eighteenth century, contagionists viewed plague as a separate disease entity that was distinct from other serious fevers. Anticontagionists, in contrast, rejected the idea that acute diseases constituted distinct entities because it entailed the assumption that diseases were caused by separate kinds of matter. The sorts of causes they posited—conditions such as hunger, fear, humoral imbalances, and the air itself—dovetailed with unified disease theories that characterized plague as an especially severe degree of illness, not a separate species of illness. The humoral theory of Galen and the hydraulic theory of the iatromathematicians both rested on physiological explanations for illness that could not adequately account for disease specificity.

Counterfactual arguments must always be doubtful, but it is possible that the adoption of a contagionist regimen helped to avert a British epidemic of plague during the centuries following the Great Plague of London. Across Europe, medical policies designed to secure public health may actually have saved lives—perhaps even millions of lives.[148] Even so, the way that contagionism transformed the conceptualization of acute diseases from dis-orders to separate entities may have been more consequential in the long run.

CONCLUSION

Restoration medical thought was dominated by Galenism or neo-Galenism, pushing other ideas to the fringes of professional medical discourse. The modified iatrochemistry of Willis and the modified corpuscularism of Sydenham were also influential, although Willis's influence gradually faded and Sydenham's continued to grow. All three of these approaches focused on the patient's own internal equilibrium and the relationship between the patient and the environment; none of them encouraged practitioners to delineate acute diseases as individual entities with features that persisted through time and space. On the other hand, the overall philosophy of Helmontian medical reformers did encourage the development of contagionism or, perhaps, it merely restated a lay contagionism that had been present all along.

Most medical authors, however, felt free to choose among an assortment of approaches. Indeed, many considered it praiseworthy to reconcile warring ideologies. Contagionism was by no means self-evident: it was merely one way in which conflicting evidence could be interpreted. In some cases the evidence in its favor was very strong; in others it was nonexistent, and in most it was equivocal. To construe any given disease as contagious was an intellectual choice made in a complex scientific, political, social, and religious context—a choice that was always constrained by the behavior of the disease itself.

Nevertheless, even though it often remained unstated or in the background, contagionism was still reshaping the way natural philosophers and physicians thought about how diseases might be classified and conceptualized. Even when authors such as Petty, Ray, and Wilkins did not set out to write about medicine, the Helmontian influence, the idea of contagious entities, and the ontological view of disease still colored their understanding of how illnesses were organized.

Charles-Edward Amory Winslow wondered why no "open-minded and imaginative observer" had combined the insights of Kircher, Leeuwenhoek, and Redi to develop the "modern germ-theory of disease" by 1700.[1] Winslow blamed the influence of Sydenham, alleging that "his almost complete neglect of contagion as a practical factor in the spread of epidemic disease and his major stress upon the metaphysical factor of [the] epidemic constitution held back epidemiological progress for two hundred years."[2]

As we have seen, however, many observers did put these insights together in the late seventeenth century, although they reached many different conclusions. Some argued that disease was conveyed by some sort of animate agent, but they did not therefore become contagionists. Marchamont Nedham (following Sennert and Kircher) thought minuscule animate life forms spread disease, but he saw them as omnipresent. Gideon Harvey attributed syphilis to animated "pocky bodies" that bred within the bodies of their victims, though he thought they could also be spontaneously generated. He also claimed that the plague, consumption, and leprosy were contagious illnesses. Robert Hooke, who accepted the common view that the plague was contagious, saw microscopic animals and fungi but did not blame them for diseases. William Petty later read Hooke's *Micrographia* and theorized that the plague might be carried by armies of minuscule insects, but he thought they were wafted by the winds from Africa, not conveyed from person to person. John Locke posited "seminal principles" that gradually altered the nature of their host as they grew. William Oliver thought diseases had "particular species" like the seeds of plants. Edward Tyson saw microscopic worms and blamed them for several different diseases—as he rejected spontaneous generation, he must have believed, with Ray, that they or their eggs were somehow ingested. Frederick Slare thought diseases were due to the inhalation of contagious particles and speculated about minuscule insects as a cause of rinderpest. Discussions about an animate cause of rinderpest continued into the early eighteenth century. It is difficult, therefore, to blame the influence of Sydenham for dampening speculation about this question. Indeed, it is equally likely that Sydenham's view only seems so influential in retrospect because later medical authors, especially Herman Boerhaave, found it appealing or convenient.

Sydenham was still a controversial figure at the end of the century; his influence was less significant in discouraging extended discussions of theories of living contagion than the associations between such theories and the swirling steams of vitalism, Helmontianism, alchemy, monism, occultism, populism, professional insubordination, anti-intellectualism, political resistance, and religious heterodoxy that had drifted through Britain during the Civil War and Interregnum before the Restoration Establishment and regime attempted to bottle them up once more. In general, and especially among English university graduates, by the late seventeenth century mechanist explanations were preferable to vitalist ones, and many considered the observation-based life sciences such as botany and biology to be vulgar, "unscientific," or inferior to research that employed mathematics and focused on physics or astronomy. With no terrifying epidemics, authors had little motive to ransack libraries for the works of such authors as Fracastoro and Kircher or to publish treatises on epidemic diseases.

The proliferation of conjectures about contagionism, microscopic pathogens, and disease-causing animalcules, however, may be more evident to the historian looking backward than they were to natural philosophers at the time. Most of these discussions took place in private and did not reach a wider audience, in part because there was no institution that had the desire and the means to pursue them and in part because many established physicians and members of the London College of Physicians stigmatized such theories as the conceits of poorly educated enthusiasts.

Stray comments appeared, often in letters or conversations, but there was no organized effort to follow up with more focused observations, experiments, or even publications. Moffet's observations on the itch mite were buried in a compilation that went unpublished for more than forty years; the follow-up letter from Bonomo to Redi remained unknown until Mead published it in 1702/3. Hooke shared his thoughts on plague in an unpublished letter; Charleton interviews at the pesthouse didn't happen; no one followed up Slare's suggestion that insects caused murrain in cattle; Tyson never published his thoughts, and they only survived by accident; Petty's and Locke's ideas about contagious life forms remained unpublished notes, and Petty's ideas about how to reorganize the reporting of the causes of deaths were never implemented. John Twysden and Nathaniel Hodges both looked for microscopic microbes, but neither one succeeded; we know about their efforts only because of their offhand comments. In fact Hodges' remark, made in 1665, would also have disappeared if it had not been reprinted in 1721.

In the later years of the seventeenth century, men such as Tyson, Locke, and Charleton evidently realized that the publication of contagionist ideas would not further their careers, or they simply changed their minds or developed other interests. They left the publication of works about living pathogens to populist authors who were interested in attracting patients and in selling books and remedies to as many customers as possible. These authors were less concerned about preserving their professional status and conciliating a conventionally educated elite, but they also lacked the reputation, resources, and community support to generate further investigations. At the turn of the eighteenth century, British publications that referred to some form of animate contagion or animate causation were largely confined to a small group of unconventional and isolated authors.

Despite many hypotheses about the agents of contagion, therefore, there was little sustained discussion and no consensus among the intellectual leaders of early eighteenth-century medicine and natural philosophy. Wilkins's early death and the financial and organizational problems the Royal Society experienced in the final decade of the century may also have played a role in the eclipse of this idea. When Hans Sloane tried to initiate a new investigation in 1722/3, it was frustrated first by the death of Leeuwenhoek and then by the continued inability of Desaguliers (or anyone else) to visualize any microscopic disease agents. The Temple Coffee House Botany Club dissolved, the Royal Society moved on to other research topics, and Sloane's personal patronage died with him, though the British Museum that was established by his will continued to serve as a gathering place for medically trained outsiders.[3]

Marten's work remained an anomaly in England, and few later authors referred to it. The very unconventionally educated Sloane was the most important patron of both the life sciences and contagionism, but the authors associated with him who were interested in ideas about contagion were not working together on a shared research program. The natural philosophers and medical men of the early eighteenth century did not institutionalize contagionism or the ontological approach to epidemic disease; there was no contagionist "school" with mutually supportive members.

This was a significant barrier to the adoption of contagionist theories for acute diseases because establishing the fact that a given disease has passed from one place to another by contagion or that it is the same disease as one previously observed depends on a network of observers and great confidence in a shared nomenclature. Without this, authors were unsure that they were discussing the same ailment. Nathaniel Hodges, for example, wondered whether the plague that Kircher had investigated in Rome was a different disease than the one he was seeing in London. Anticontagionists continued to believe that the plague was not a distinct disease but only a more severe degree of disease.

The rejection of spontaneous generation may actually have intensified the logical problem that medical authors confronted when they attempted to explain the origin of epidemics. If everything came only from a parent, how did the first illness (the index case) of an epidemic arise? How could the advent of new diseases be explained? Had God created all diseases in the beginning? This problem was especially acute in a providentialist culture, and providentialism was intertwined with the sort of physico-theology that Ray and his colleagues embraced.

Because the *contagium* and the *vivum* seemed to be at odds, many authors who believed in contagion found it more reasonable to assume that disease was due to a nonliving substance. This rendered the debate about the spontaneous generation of animalcules moot. The belief that contagion was a nonliving substance allowed many medical authors to continue to assume that diseases could be "generated" by some combination of factors and then might become contagious.

After briefly converging in the late seventeenth and early eighteenth century, medical philosophy and research in the life sciences again diverged as each field became more specialized. British medical men after 1730 were no longer anxious to determine what the *contagium* itself was or whether it was even a living entity. Most of those who believed in contagion were content to posit some sort of "morbid substance" without trying to specify its nature. They believed that they did not have the tools to establish what this mysterious entity was and focused instead on determining what it did. Authors elsewhere, such as Linnaeus, occasionally published works on living contagion, but to British physicians debates about *contagium vivum* did not seem to be essential components of the more important conversation about whether and how diseases were transmitted from person to person by contagion, whether and how to identify and treat separate diseases, and how to forestall illnesses and epidemics. The answers to these questions had very immediate consequences.

If theories of *contagium vivum* did not find a solid foothold in Britain, the idea of contagion itself as a phenomenon received a boost in the early years of the eighteenth century with the reappearance of the plague and the introduction of smallpox inoculation. Although neo-Galenic and traditionalist physicians continued to stress physiological models of disease, the idea that the plague was contagious had a solid grounding in lay thought and in the actual measures to prevent or circumscribe outbreaks imposed over the preceding two centuries by lay governments. Whether or not they personally subscribed to a belief in contagion, government officials needed to be seen taking action in some manner to stem the tide of fear and despair. Because they were statist interventions, quarantines and the policies that underlay them were not seen as the products of a radical mind-set. British measures against rinderpest, plague, and smallpox emerged from consultations and cooperation between government authorities and medical advisors. The quarantines and *cordons sanitaires* established during the Plague of Marseilles and the British strategy to control rinderpest were perceived by a large segment of the public to have been effective, reinforcing the belief that some diseases were indeed contagious.

Smallpox inoculation proved to be more contentious. Despite the example of the British royal family, smallpox inoculation in Britain proceeded by fits and starts in the period from 1721 to 1740. Its spread was impeded by the high cost of both the preparation process and the procedure itself and by a reduction in the incidence of natural smallpox during the 1730s.[4] It was not practiced at all in most of Europe until after the middle of the century. Nevertheless, the very public introduction of inoculation into Britain demonstrated to the satisfaction of many doctors that smallpox should be added to the short list of separate diseases that were transmitted from person to person by an unspecified contagious matter or "principle."

Marten's British successors would not insist, as he had, on the animate nature of contagion, but a growing number of physicians did come to believe that a greater number of diseases were contagious, that there were separate species of diseases that persisted across space and time, and that outbreaks of contagious diseases could be prevented if doctors and administrators worked together. By the end of the century they would create a network that sustained and institutionalized contagionist theory in a way that had not been possible in the early eighteenth century.

NOTES

PREFACE

1. A note about the formatting: Unless otherwise indicated, quotations from early modern authors retain their original italics and spelling except that, where appropriate, the letter "v" has been converted to "u" and the long letter "s" has been converted to a modern "s." "Sic" is only used to confirm very unusual or critical variant spellings.
2. Charles-Edward Amory Winslow, *The Conquest of Epidemic Disease: A Chapter in the History of Ideas* (Princeton, NJ: 1943, rpt. Madison, WI: 1980), 159–60.
3. Fielding Garrison, *An Introduction to the History of Medicine* (3rd rev. and enl. ed., Philadelphia: 1923), 314, and 317. The attribution of "exaggerated sobriety" to the age of Smollett and Hogarth is surprising.
4. The "modern germ-theory of disease" is a very problematic phrase. The phrase does not refer to a theory of "disease" but to a subset of acute, often febrile, diseases. Each word within it also poses difficulties.
5. This includes those who were raised as Dissenters or Jews who later conformed to the Church of England. Even if these were excluded, Dissenters' influence in these medical debates was disproportionate to their numbers. James E. Bradley, *Religion, Revolution and English Radicalism* (Cambridge: 1990), 93, estimates the proportion of Baptists, Congregationalists, and Presbyterians combined at about 5 percent of the population in 1715 and the Quakers at less than 1 percent. On page 89 he wrote: "the Dissenters' radical [political] ideology could influence even those laymen who had occasionally conformed."
6. Henry Cohen [Lord Cohen of Birkenhead], "The Evolution of the Concept of Disease," *Proceedings of the Royal Society of Medicine* 48 (Section of Medicine, March 1955):155–60; Lloyd G. Stevenson, "New Diseases in the Seventeenth Century," *Bulletin of the History of Medicine* 39 (1965): 1–21; and Owsei Temkin, "The Scientific Approach to Disease: Specific Entity and Individual Sickness," in same, *The Double Face of Janus* (Baltimore: 1977), 441–55.
7. There are many physiological diseases, such as high blood pressure or heart disease. This transformation separated diseases that are usually febrile, acute, contagious, and/or epidemic from others with more diffuse causes.
8. Winslow, *Conquest*, 159. He also mentioned Linnaeus.
9. Clifford Dobell, *Antony van Leeuwenhoek and His "Little Animals"* (New York: 1958); William Bulloch, *A History of Bacteriology* (Oxford: 1938, rpt. New York, 1979).
10. Bulloch, *Bacteriology*, 27.
11. E. Ashworth Underwood, "Charles Singer: A Biographical Note," introduction to Underwood ed., *Science, Medicine and History: Essays on the Evolution of*

Scientific Thought and Medical Practice written in Honour of Charles Singer, Vol. 1 (Oxford: 1953). Singer's articles led William Osler to offer him a Studentship in Pathology at Oxford on the understanding that he would focus on medical history.

12. Charles Singer, "Benjamin Marten, a Neglected Predecessor of Louis Pasteur," *Janus* 16 (1911): 81–9; and same, *The Development of the Doctrine of Contagium Vivum 1500–1750* (London: 1913). This also appeared in the *Transactions of the 17th International Congress of Medicine, 1913*, Section 23: History of Medicine (London: 1914), 87–206.

13. Singer, *Doctrine of Contagium Vivum*, 14. I thank the Multnomah County, Oregon, public library and the University of Alabama at Birmingham for providing this.

14. M. C. Buer, *Health, Wealth, and Population in the Early Days of the Industrial Revolution* (London: 1926; New York: 1968) provides a comprehensive overview of eighteenth-century British medical reform, but as an economic historian, Buer considered the effect of diseases and medical intervention on population growth, not on medical ideas. Other optimistic views of the century either focused, like Buer, on the "standard of living debate" or a related debate about the role (if any) of hospitals and public health measures in reducing mortality during the Industrial Revolution, giving only passing attention to disease theory. Mabel Craven Buer (1881–1942), Independent Lecturer in Economics at Reading University, deserves more recognition.

15. Richard Shryock, *The Development of Modern Medicine* (London, 1936, rpt. Madison, WI: 1979).

16. Peter Gay, "The Enlightenment as Medicine and as Cure," in W. H. Barber et al. eds., *The Age of the Enlightenment: Studies Presented to Theodore Besterman* (Edinburgh: 1967), 381. This claim dates to the eighteenth century.

17. Walter Pagel, *Paracelsus: An Introduction to Philosophical Medicine in the Era of the Renaissance*, 2nd ed. (Basel: 1958); Charles Webster, introduction to *Samuel Hartlib and the Advancement of Learning* (Cambridge: 1972); and same, *The Great Instauration, Science Medicine and Reform, 1626–1660* (New York: 1976); and Allen Debus, *The English Paracelsians* (London: 1965). Debus hesitated to publish this work because he thought there was little interest in Paracelsian chemistry. Paul H. Theerman and Karen Hunger Parshall, eds., *Experiencing Nature* (Dordrecht: 1997), 270.

18. Jon Arrizabalaga, John Henderson, and Roger French, *The Great Pox: The French Disease in Renaissance Europe* (New Haven: 1997) shows that Sennert combined elements of Galenism, atomism, and Paracelsianism. Sociologist Ludwik Fleck discussed the construction of syphilis as a disease entity in *Genesis and Development of a Scientific Fact* (1935), ed. Thaddeus J. Trenn and Robert K. Merton, trans. Fred Bradley and Thaddeus J. Trenn (Chicago: 1979).

19. A valuable older work is Charles F. Mullett, *The Bubonic Plague and England: An Essay in the History of Preventive Medicine* (Lexington, KY: 1956). See also Walter George Bell, *The Great Plague in London in 1665*, ed. and abr. Belinda Hollyer (London: 1924, 2001); Stephen Porter, *The Great Plague* (Thrupp, Stroud, Gloucestershire: 1999); Paul Slack, *The Impact of Plague in Tudor and Stuart England* (London: 1985, rpt. 2003); and same, "The Disappearance of Plague: An Alternative View," *Economic History Review* (1981) 34, no. 3:469–76; R. Oratz, "The Plague. Changing Notions of Contagion: London 1665–Marseilles 1720," *Synthesis* (1977) 4:4–27; and Andrew Wear, *Knowledge and Practice in English Medicine, 1550–1680* (Cambridge: 2000).

20. Margaret Pelling, "Introduction" in Margaret Pelling and Scott Mandelbrote, ed., *The Practice of Reform* (Aldershot, UK: 2005).

21. Richard Harrison Shryock, "Germ Theories in Medicine Prior to 1870: Further Comments on Continuity in Science," *Clio Medica* (1972) 7:81–109.

22. Shryock, "Germ Theories," 84.

23. William R. Le Fanu, "The Lost Half-Century in English Medicine, 1700–1750," *Bulletin of the History of Medicine* (July–August 1971) 46:319–48.

24. Le Fanu, "Lost Half-Century," 327–8.

25. Adrian Wilson, "The Politics of Medical Improvement in Early Hanoverian London," in Andrew Cunningham and Roger French, eds., *The Medical Enlightenment of the Eighteenth Century* (Cambridge: 1990), 4–39, on 10. Wilson discusses the creation of voluntary hospitals, smallpox inoculation, and man-midwifery as examples of early-eighteenth-century medical innovation. Other essays in this book also present a positive view but do not discuss contagionism.

26. Nicholas Jewson, "Medical Knowledge and the Patronage System," *Sociology* (1974) 8:369–85.

27. See also David Harley, "Rhetoric and the Social Construction of Sickness and Healing," *Social History of Medicine* (1999) 12:407–35, esp. 413. I thank Dr. Harley for a copy of this article.

28. David Wootton, *Bad Medicine: Doctors Doing Harm Since Hippocrates* (Oxford: 2006, rpt. 2007), 116. See also John Waller, *The Discovery of the Germ* (Duxford: 2002, rpt. 2004). Wootton posted responses to his claim that medicine did more harm than good until 1865, at www.badmedicine.co.uk/main.asp. Historians criticized him for implying that they were unaware of the deplorable nature of earlier medicine, which, they claimed, was uncontroversial. Instead of arguing about whether early medicine was "bad" or "good," or whether "Whiggish" and "progressivist" accounts are justified, we should first ask whether historians have portrayed the actual ideas and practice of earlier medicine adequately and accurately.

29. J. N. Hays, *The Burdens of Disease: Epidemics and Human Response in Western History* (New Brunswick: 1998, third paperback ed., 2003), 110.

30. For example, K. Codell Carter, *The Rise of Causal Concepts of Disease* (Aldershot, UK: 2003) begins in the nineteenth century, and Robert Hudson, *Disease and Its Control: The Shaping of Modern Thought* (Westport: 1983) skips from Fracastoro in 1546 to early-nineteenth-century anticontagionism in the space of one page (144–5).

31. See for example, Stephen Gaukroger, *The Collapse of Mechanism and the Rise of Sensibility: Science and the Shaping of Modernity, 1680–1760* (Oxford: 2010, rpt. 2012); and same *The Emergence of a Scientific Culture: Science and the Shaping of Modernity 1210–1685* (Oxford: 2006, rpt 2008); Jonathan Israel, *Enlightenment Contested: Philosophy, Modernity, and the Emancipation of Man 1670–1752* (Oxford: 2006); and same, *Radical Enlightenment: Philosophy and the Making of Modernity 1650–1750* (Oxford: 2001); Christoph Lüthy, John E. Murdoch, and William R. Newman, ed., *Late Medieval and Early Modern Corpuscular Matter Theories* (Leiden: 2001); works by or edited by Lawrence Principe including *New Narratives in Eighteenth-Century Chemistry* (Dordrecht: 2007), *Chymists and Chymistry: Studies in the History of Alchemy and Early Modern Chemistry* (Sagamore Beach, MA: 2007), 25, and *The Secrets of Alchemy* (Chicago and London: 2013); Peter Anstey, *John Locke and Natural Philosophy* (Oxford: 2011); John C. Powers, *Inventing Chemistry: Herman Boerhaave and the Reform of the Chemical Arts* (Chicago: 2012); Peter Hanns Reill, *Vitalizing Nature in the Enlightenment*

(Berkeley: 2005); Anna Marie Eleanor Roos, *The Salt of the Earth: Natural Philosophy, Medicine, and Chymistry in England, 1650–1750* (Leiden: 2007); and Mary Terrall, *Catching Nature in the Act: Réaumur and the Practice of Natural History in the Eighteenth Century* (Chicago: 2014).

32. See for example, Roy Porter, "Was There a Medical Enlightenment in Eighteenth-Century England?" *The British Journal of Eighteenth-Century Studies* (1982) 5, no. 1:49–63; and same, "Medical Science and Human Science in the Enlightenment," in Christopher Fox, Roy Porter, and Robert Wokler, eds., *Inventing Human Science: Eighteenth-Century Domains* (Berkeley and Los Angeles, 1995), 53–87; and same, "Introduction," in Roy Porter, ed., *The Cambridge History of Science, vol. 4: Eighteenth-Century Science* (Cambridge: 2003), 1–2.

33. Porter, "Medical Enlightenment?" 51–2. See also "Chapter 5: Medical Science" in same, ed., *The Cambridge History of Medicine* (Cambridge and New York: 1996, rpt. 2004), 171.

34. Porter, "Medical Enlightenment?" 52.

35. Porter's reference to "Evangelical Christianity" is also problematic. Does he mean "Evangelical" in the sense of membership in the Evangelical movement or that they were motivated by Christianity? Most of the Dissenters and Anglicans who were active medical reformers were not Evangelicals in the former sense.

36. Porter, "Medical Enlightenment?" 53.

37. Porter, "Medical Enlightenment?" 56. On page 58, he added, "What we really need to know for England . . . is the relation of quacks to the Enlightenment."

38. Historians' interest in replacing traditional history with social history "written from below" has generated its own history. See Leonard G. Wilson, "Medical History without the Medicine," *Journal of the History of Medicine* (1980), 35:5–7; Frank Huisman and John Harley Warner, *Locating Medical History: The Stories and their Meanings* (Baltimore: 2004, rpt. 2006); Ruth Linker, "Resuscitating the 'Great Doctor': The Career of Biography in Medical History," in Thomas Söderqvist, ed., *The History and Poetics of Scientific Biography* (Aldershot, UK: 2007), 221–40; Roger Cooter, "After Death/After-'Life': The Social History of Medicine in Post-Modernity," *Social History of Medicine* (2007) 20:441–54; Howard I . Kushner, "The Art of Medicine: Medical Historians and the History of Medicine," *The Lancet* (August 30, 2008), 372, no. 9640:710–11, doi: 10.1016/so140-6736(08)61293-3; William Bynum, "The Past Is Never Dead," (review of Jackson Mark, ed., *The Oxford Handbook of the History of Medicine*) *The Lancet* (August 20, 2011) 378, no. 9792:655–6, doi: 10.1016/S0140-6736(11)61319-6; and Howard I. Kushner and Leslie S. Leighton, "The Histories of Medicine: Toward an Applied History of Medicine," in Eleonora Belfiore and Anna Upchurch, eds., *The Humanities in the Twenty-First Century* (London: 2013), 111–37.

39. Adrian Wilson and T. G. Ashplant, "Whig History and Present-Centred History," *Historical Journal* (1988) 31, no. 1:1–16; Adrian Wilson and T. G. Ashplant, "Present-Centered History and the Problem of Historical Knowledge," *Historical Journal* (1988) 31, no. 2:253–74. For an overview of twentieth-century medical history, see John Pickstone, "A Brief History of Medical History" (London: Institute for Historical Research, 2008), online article http://www.history.ac.uk/makinghistory/resources/articles/history_of_medicine.html.

40. Partial exceptions can be found in articles by Guenter Risse, such as "Epidemics and Medicine: The Influence of Disease on Medical Thought and Practice," *Bulletin of the History of Medicine* (1979) 53:505–19; Christopher J. Lawrence, "Early

Edinburgh Medicine: Theory and Practice," in R. G. W. Anderson and A.D.C. Simpson, eds., *The Early Years of the Edinburgh Medical School* (Edinburgh: 1976); Dale C. Smith, "Medical Science, Medical Practice, and the Emerging Concept of Typhus in Mid-Eighteenth-Century Britain," in W. F. Bynum and V. Nutton, eds., *Theories of Fever from Antiquity to the Enlightenment* (London: 1981), 121–34; and William F. Bynum, "Cullen and the Study of Fevers in Britain, 1760–1820," 135–47 in the same volume.

41. James Riley, *The Eighteenth-Century Campaign to Avoid Disease* (New York: 1987), 52. He also wrote that "the elements of a germ theory were not, as Winslow supposed, at hand," 150. The essays by Lester King on the eighteenth century were unenthusiastic, especially those in *The Medical World of the Eighteenth Century* (Chicago: 1958). *The Philosophy of Medicine: The Early Eighteenth Century* (Cambridge, MA: 1978) is less severe.

42. See in addition to "Politics of Medical Improvement," (above), Adrian Wilson, "On the History of Disease-Concepts: The Case of Pleurisy," *History of Science* (2000) 38:271–319, esp. 273–5 on the value of a history of changing ideas about particular diseases. Wilson traces this view to Fleck, *Scientific Fact.* The rise of contagionism is mentioned by Anne Hardy, but her brief article merely states it as a fact, "The Medical Response to Epidemic Disease during the Long Eighteenth Century," in J. A. I. Champion, ed., *Epidemic Disease in London* (London: 1993, 2001), 65–70, online at www.history.ac.uk/ihr/Focus/Medical/epimenu.html.

43. We know, however, that some early modern hospitals either tried to segregate patients with infectious diseases or barred them entirely, suggesting a continued lay fear of contagion. Kevin P. Siena, *Venereal Disease, Hospitals and the Urban Poor: London's 'Foul Wards,' 1600–1800*, 64–74 and 106, says that belief in contagion was "waning" in hospitals by the late seventeenth century, but he refers to a Georgian "class terror" of contagious disease in "Contagion, Exclusion, and the Unique Medical World of the Eighteenth-Century Workhouse," in Jonathan Reinarz and Leonard Schwarz, eds., *Medicine and the Workhouse* (Rochester, NY: 2013), 19–39, on 27. Graham A. J. Ayliffe and Mary P. English, *Hospital Infection from Miasmas to MRSA* (Cambridge and New York: 2003), 16–17, 25–7 give other eighteenth-century examples.

44. Erwin H. Ackerknecht, "Anticontagionism between 1821 and 1867," the Fielding H. Garrison Lecture, *Bulletin of the History of Medicine* (1948) 22:562–93. This was reprinted, abridged, and with portions translated in the *International Journal of Epidemiology* (2009) 38, no. 1:7–21, doi: 10.1093/ije/dyn254.

45. Ackerknecht, "Anticontagionism," 591. One of the men Ackerknecht names was the Scot Gilbert Blane. Blane was certainly a contagionist and a high-ranking Navy officer, but he was not a defender of the status quo. He was a reformer who finally persuaded the Admiralty to provide citrus juice to sailors. See chapter 9 below.

46. Separate articles by Christopher Hamlin, Charles Rosenberg and Alexandra Minna Stern, and Howard Markel in the *International Journal of Epidemiology* (2009) 38, no. 1:22–33 on the fiftieth anniversary of Ackerknecht's article did not address the misperception that Whig reformers of the late eighteenth century were anticontagionists.

47. Margaret Pelling, *Cholera, Fever and English Medicine* (Oxford: 1978).

48. With the exception of Pelling herself, who began her career as a student of Webster's. See R. J. Morris, *Cholera 1832: The Social Response to an Epidemic* (London: 1976); Michael Durey, *The Return of the Plague: British Society and the Cholera, 1831–2* (Dublin: 1979); Richard Evans, *Death in Hamburg: Society and*

Politics in the Cholera Years 1830–1910 (Oxford: 1987, rpt. New York: 2005); Frank Snowden, *Naples in the Time of Cholera, 1884–1911* (Cambridge: 1995, rpt. 2002); Christopher Hamlin, *Public Health and Social Justice in the Age of Chadwick: Britain 1800–1854* (Cambridge: 1998); and same, *Cholera: The Biography of a Disease* (Oxford: 2009); Anne Hardy, *The Epidemic Streets: Infectious Diseases and the Rise of Preventive Medicine, 1856– 1900* (Oxford: 1993); Peter Baldwin, *Contagion and the State in Europe, 1830–1939* (Cambridge: 1999); David McLean, *Public Health and Politics in the Age of Reform: Cholera, the State and the Royal Navy in Victorian Britain* (London: 2006); and Pamela Gilbert, *Cholera and the Nation: Doctoring the Social Body in Victorian England* (Albany: 2008). Charles Rosenberg's first work on cholera in America, *The Cholera Years: The United States in 1832, 1849 and 1866,* preceded Pelling's on Britain, but neither that nor his subsequent works, including the work edited with Janet Golden, *Framing Disease* (New York: 1992), has taken issue with this idea.

49. Francis M. Lobo, "John Haygarth, Smallpox and Religious Dissent in Eighteenth-Century England," in Andrew Cunningham and Roger French, eds., *The Medical Enlightenment of the Eighteenth Century* (Cambridge and New York: 1990), 217–53.

50. Christopher C. Booth, *John Haygarth, FRS: A Physician of the Enlightenment (1740–1827)* (Philadelphia: 2005); J. Johnston Abraham, *Lettsom: His Life, Times, Friends and Descendants* (London: 1933); Leona Baumgartner, "John Howard and the Public Health Movement," *Bulletin of the History of Medicine* (1937) 5:489–508; Margaret DeLacy, "Influenza Research and the Medical Profession in Eighteenth-Century Britain," *Albion* (1993) 25, no. 1:37–66. See also chapter 9 below.

51. These articles appear in shorter form in Lise Wilkinson, *Animals and Disease: An Introduction to the History of Comparative Medicine* (Cambridge and New York: 1992). See also C. A. Spinage, *Cattle Plague, A History* (New York: 2003).

52. Catherine Wilson, *The Invisible World: Early Modern Philosophy and the Invention of the Microscope* (Princeton: 1995).

53. Wilson, *Invisible World,* 174.

54. Marc J. Ratcliff, *The Quest for the Invisible: Microscopy in the Enlightenment* (Farnham, Surrey: 2009).

55. See Bruno Latour's work including (with Jonas Salk and Steve Woolgar), *Laboratory Life: The Construction of a Scientific Fact* (Princeton: 1979, new ed., 1986), *Science in Action* (Cambridge, MA: 1987, rpt. 1999), and *The Pasteurization of France* (Cambridge: 1984); and Steven Shapin and Simon Schaffer, *Leviathan and the Air-Pump: Hobbes, Boyle and the Experimental Life* (Princeton: 1985). These emphasize the importance of both a scientific community and a mutually agreed upon epistemology. See also David S. Lux and Harold J. Cook, "Closed Circles or Open Networks? Communicating at a Distance during the Scientific Revolution," *History of Science* (1998) 36:179–211.

56. Ulrich Tröhler, *To Improve the Evidence of Medicine: The 18th Century British Origins of a Critical Approach* (Edinburgh: 2000). This is a shorter version of Tröhler's PhD thesis "Quantification in British Medicine and Surgery 1730–1830, with Special Reference to Its Introduction into Therapeutics" (London: 1978). Tröhler posted his thesis in 2004 on The James Lind Library, www.jameslindlibrary.org/articles.

57. Ulrich Tröhler, "The Introduction of Numerical Methods to Assess the Effects of Medical Interventions during the 18th Century: A Brief History," (2010),

www.jameslindlibrary.org/articles. See also Andrea A. Rusnock, *Vital Accounts: Quantifying Health and Population in Eighteenth-Century England and France* (Cambridge: 2002). Mary Poovey's *A History of the Modern Fact* (Chicago: 1998) omits vital statistics and medicine.

58. Esther Fischer-Homberger, "Eighteenth-Century Nosology and Its Survivors," *Medical History* (1970) 14:397–403, states the pessimistic case; the other side appears in A. J. Cain, "John Ray on 'Accidents,'" *Archives of Natural History* (1996), 23:343–69; and same, "Thomas Sydenham, John Ray, and Some Contemporaries on Species," *Archives of Natural History* (1999) 26:55–83; M. M. Slaughter, *Universal Languages and Scientific Taxonomy in the Seventeenth Century* (Cambridge: 1982); Rhodri Lewis, *Language, Mind and Nature: Artificial Languages in England from Bacon to Locke* (Cambridge: 2007); and John S. Wilkins, *Species: A History of the Idea* (Berkeley: 2009). See also Lennart Nordenfelt, "Identification and Classification of Diseases: Fundamental Problems in Medical Ontology and Epistemology," *Studia Philosophica Estonica* (2013) 6, no. 2:6–21, online at http://www.spe.ut.ee/ojs-2.2.2/index.php/spe/article/view/134/84; and Daniel R. Headrick, *When Information Came of Age: Technologies of Knowledge in the Age of Reason and Revolution, 1700–1850* (Oxford: 2000).

59. Egerton collected and abridged these articles as *Roots of Ecology from Antiquity to Haeckel* (Berkeley: 2012); earlier versions can be found on the Ecological Society of America website at www.esajournals.org/loi/ebul.

60. Sloane's "Account" appeared in the *Philosophical Transactions* (1756) 49 pt. 2:516–20. For the continuing stream of works on this topic, see Miller, *Inoculation*, 296–309: bibliography B. *Contemporary Printed Sources* 1. Works Exclusively on Inoculation a. *England and Her Colonies*. Jean Astruc's lectures, published in English as *Academical Lectures on Fevers . . . Read in the Royal College at Paris* (London: 1747) attributed plague to a "ferment" that multiplied in the blood of the afflicted.

61. Shryock in 1972 moved the start of the decline up to "about the 1730s."

62. But see David Boyd Haycock, "Exterminated by the Bloody Flux: Dysentery in Eighteenth-Century Naval and Military Medical Accounts," *Journal for Maritime Research* (2002); and Marjo Kaartinen's *Breast Cancer in the Eighteenth Century* (London: 2013).

63. In particular, Italian contagionism deserves further study. See Achille Monti, *The Fundamental Data of Modern Pathology*, trans. John Joseph Eyre (London: 1900), 150–3; and Luigi Belloni, *Le 'Contagium vivum' avant Pasteur*, Conferences du Palais de la Découverte, ser. D., no. 74 (Paris: 1961). Belloni comments on page 23 that after 1714, belief in living contagion was centered in Northern Italy. Beyond the perennial discussion of plague and plague quarantines, other works that mention the widespread acceptance of contagion include Mitchell Lewis Hammond, "Contagion, Honour and Urban Life in Early Modern Germany," in Claire L. Carlin, ed., *Imagining Contagion in Early Modern Europe* (Basingstoke, Hampshire: 2005), 96–104; G. Romagnoli, "L'Evoluzione del Concetto della Contagiosità e della Profilassi della Tubercolosi Polmonare attraverso I Secoli, Dall'Antichità fino alla Scoperta del Bacillo de Koch," *Giornale de Batteriologia, Virologia et Immunologia et Annali dell'Ospedale Maria Vittoria de Torino* (1968) 61:233–73; Charles Coury, *Grandeur et Déclin d'une Maladie: La Tuberculose au Cour des Ages* (Suresne: 1972); and Jacques Bernier, "*L'Interprétation de la Phtisie Pulmonaire au XVIIIe Siècle*," *Canadian Bulletin*

of Medical History (2005) 22, no. 1:35–56. Mary Lindman's *Health and Healing in Eighteenth-Century Germany* (Baltimore and London: 1996) has only incidental references to contagion (see page 293) as does Laurence Brockliss and Colin Jones, *The Medical World of Early Modern France* (Oxford: 1997) (see pages 750–1), despite frequent references to the Plague of Marseilles and the increasing use of *cordons sanitaires*.

CHAPTER 1

1. Owsei Temkin stated that the first systematic list of contagious diseases appeared in the *Book of Treasure* (ca. 900) and consisted of leprosy, scabies, smallpox, measles, "ozanea," ophthalmia, and the pestilential diseases. Ozanea is a foul discharge from the nose. A thirteenth-century work listed acute fever, consumption, epilepsy, scabies, "*ignis sacer*," anthrax, ophthalmia, and leprosy. "An Historical Analysis of the Concept of Infection," in same, *The Double Face of Janus* (Baltimore: 1977), 123–47. See also Max Meyerhof, "'The Book of Treasure,' an Early Arabic Treatise on Medicine," *Isis* (May 1930) 14, no. 1:55–76 on 61.
2. A very concise account of the various classical disease theories is Henry Cohen [Lord Cohen of Birkenhead], "The Evolution of the Concept of Disease," *Proceedings of the Royal Society of Medicine*, Section of Medicine (March 1955) 48, no. 3:155–60. Lucinda McCrae Beier, *Sufferers and Healers: The Experience of Illness in Seventeenth-Century England* (London: 1987), 249, comments that "although no coherent 'theory of infection' was articulated . . . people certainly distinguished between infectious and non-infectious ailments, feeling much safer in the presence of the latter . . . Infection was believed to pass directly from person to person."
3. Charles-Edward Amory Winslow's *The Conquest of Epidemic Disease: A Chapter in the History of Ideas* (Madison: 1943, ppb. rpt., 1980) still offers the most lucid general history of contagionism from classical times to the twentieth century. See esp. 84.
4. Winslow, *Conquest*, 85. Pre-Islamic Arabs blamed epidemics on demons but also viewed them as contagious (or infectious). This approach was rejected by some later Muslim authors who found it superstitious and inconsistent with divine omnipotence. Lawrence I. Conrad, "Epidemic Disease in Formal and Popular Thought in Early Islamic Society," in *Epidemics and Ideas: Essays on the Historical Perception of Pestilence*, eds. Terence Ranger and Paul Slack (Cambridge: 1992), 77–100.
5. This disease was probably not the illness known by that name today. For early contagionism see the essays in *Contagion: Perspectives from Pre-Modern Societies*, eds. Lawrence I. Conrad and Dominik Wujastyk (Aldershot, UK: 2000): Francois-Olivier Touati, "Contagionism and Leprosy: Myth, Ideas and Evolution in Medieval Minds and Societies," 196–7, dates the idea that leprosy was contagious to the end of the thirteenth century and its adoption to the work of Guy de Chauliac in the mid-fourteenth century. Lawrence I. Conrad, however, points out that contagion was debated by Islamic authors in the early middle ages, "A Ninth-Century Muslim Scholar's Discussion of Contagion," 163–77.
6. Winslow, *Conquest*, 85.
7. Vivian Nutton, "The Seeds of Disease: An Explanation of Contagion and Infection from the Greeks to the Renaissance," *Medical History* (1983) 27:1–34, on 9, and *Imagining Contagion in Early Modern Europe*, ed. Claire L. Carlin (Basingstoke,

Hampshire: 2005). See also Conrad and Wujastyk, *Contagion*. Jon Arrizabalaga, John Henderson, and Roger French, *The Great Pox: The French Disease in Renaissance Europe* (New Haven: 1997) shows how ideas about the transmission of a contagious disease evolved from the late fifteenth century to the work of Sennert. Annemarie Kinzelbach, "Infection, Contagion, and Public Health in Late Medieval and Early Modern German Imperial Towns," *JHMAS* (2006) 61:369–89 questions the idea of a divide between academic and popular medicine on the roles of miasma and contagion in spreading plague but shows that physicians attributed a local epidemic to humoral changes affecting only the poor. Peter Elmer, *The Miraculous Conformist: Valentine Greatrakes, the Body Politic and the Politics of Healing in Restoration Britain* (Oxford: 2013) notes on 113–14 that contemporaries also thought witchcraft operated through physical contact.

8. Charles Greene Cumston, *An Introduction to the History of Medicine: From the Time of the Pharaohs to the End of the 18th Century* (New York: 1926, rpt. 1987), 161.

9. Nutton, "Seeds," 15, citing Owsei Temkin, "The Scientific Approach to Disease: Specific Entity and Individual Sickness," in Temkin, *Double Face of Janus*. See also Walter Pagel, *Paracelsus: An Introduction to Philosophical Medicine in the Era of the Renaissance*, 2nd rev. ed. (Basel: 1982), 137–8; and same, *Joan Baptista Van Helmont*, chapter 5: "The Ontological Concept of Disease," 141–98; Johanna Geyer-Kordesch, "Fevers and Other Fundamentals: Dutch and German Medical Explanations c. 1680–1730," in *Theories of Fever from Antiquity to the Enlightenment*, eds. W. F. Bynum and V. Nutton, (*Medical History*, Supplement 1) (London: 1981), 99–120; and L. J. Rather, "Towards a Philosophical Study of the Idea of Disease," in *Historical Development of Physiological Thought*, eds. Chandler McC. Brooks and Paul F. Cranefield (New York: 1959), 351–73.

10. The most famous traditional medicine was "Venice Treacle" or Theriac. The similar London Treacle remained in the *London Pharmacopoeia* until 1788. Simplicity in prescriptions would become one of the identifying marks of the medical reformers.

11. Winslow, *Conquest*, 64–5; Ralph Hermon Major, *A History of Medicine*, 2 vols., (Springfield, IL: 1955), 1:126–7.

12. Jole Shackelford, *A Philosophical Path for Paracelsian Medicine* (Copenhagen: 2004), 156; Eric Grier Casteel, "Entrepot and Backwater: A Cultural History of the Transfer of Medical Knowledge from Leiden to Edinburgh, 1690–1740" (Ph.D. dissertation, University of California at Los Angeles: 2007), esp. 108–10 and 123; more generally see David Cantor, ed., *Reinventing Hippocrates* (Aldershot, UK: 2002).

13. Nutton, "Seeds," 13. For the concept of seeds in the Renaissance and early modern period see Hiro Hirai, *Le Concept de Semence dans les Théories de la Matière à la Renaissance: De Marsile Ficin à Pierre Gassendi* (Turnhout: 2005).

14. Walter Pagel, *New Light on William Harvey* (Basel: 1976), 21–2.

15. I use "monism" in place of other possible terms such as "neo-Platonism" as a shorthand description of this strand in medical discourse throughout the early modern and modern period.

16. Charles Webster, *The Great Instauration: Science, Medicine, and Reform, 1626–1660* (New York: 1976), 200 and 280. G. H. Williams, *The Radical Reformation* (Philadelphia: 1962), introduction, xxiii–xxix.

17. Christopher Hill, "William Harvey and the Idea of Monarchy," in *The Intellectual Revolution of the Seventeenth Century*, ed. Charles Webster (London: 1974),

170. Benjamin Franklin published a mortalist pamphlet, *A Dissertation on Liberty and Necessity, Pleasure and Pain* in 1725.

18. See John Henry, "The Matter of Souls: Medical Theory and Theology," in *The Medical Revolution of the Seventeenth Century*, eds. Roger French and Andrew Wear (Cambridge: 1989), 88.

19. Philippus Aureolus Theophrastus Bombastus von Hohenheim. See Pagel, *Paracelsus*, 17 and 69.

20. Pagel, *Paracelsus*, 96, 131–2, cf. Webster, *Instauration*, 287–8.

21. On the use of this idea by Van Helmont and Severinus see Jole Shackelford, *A Philosophical Path for Paracelsian Medicine: The Ideas, Intellectual Context, and Influence of Petrus Severinus (1540/2–1602)* (Copenhagen: 2004), 238–49, and Hirai, *Semence*. Hirai shows on 441–4 that Helmont's early work echoes that of Severinus.

22. Pagel, *Paracelsus*, 180–1.

23. Nutton, "Seeds," 9–10; Cohen, "Evolution," 4. For a more extended account of the development of atomic and corpuscular theories, see R. A. Horne, "Atomism in Ancient Medical History," *Medical History* (1963) 7:317–29, Robert Hugh Kargon, *Atomism in England from Hariot to Newton* (Oxford: 1966), and the introduction and essays in *Late Medieval and Early Modern Corpuscular Matter Theories*, eds. Christoph Lüthy, John E. Murdoch, and William R. Newman (Leiden: 2001).

24. John S. Wilkins, *Species, the History of the Idea* (Berkeley: 2009), 25–6.

25. The metaphor in the term "framing" of diseases is too prescriptive to use for this period because it implies that diseases are discrete entities that can be framed (it is impossible to frame a quality, experience, or continuous substance such as water) and that the boundary around the entity can be closed.

26. The theory that viable living spores or seeds are everywhere is called panspermia, panspermism, or panspermatism. The *Oxford English Dictionary* cites the first use of "panspermatick" in 1690; other variants of the word all date from the nineteenth century. I do not accept Nutton's claim in "Seeds," p. 1, "a belief in the theory of seeds presupposes a belief in the theory of contagious (or communicable) diseases."

27. Charles Singer and E. Ashworth Underwood, *A Short History of Medicine*, 2nd ed. (Oxford: 1962), 48–51.

28. Wilmer Cave Wright, "Introduction" to *Hieronymi Fracastorii, De Contagione et Contagiousus Morbis et Eorum Curatione, Libri III* (New York: 1930). John Herman Randall, *The School of Padua and the Emergence of Modern Science* (Padua: 1961) emphasizes the role of the Aristotelian revival in Padua. See also Jerome Bylebyl, "The School of Padua: Humanistic Medicine in the Sixteenth Century," in *Health Medicine and Mortality in the 16th Century*, ed. Charles Webster (Cambridge: 1979), 335–70. Antonio Clericuzio, "Chemical Medicine and Paracelsianism in Italy, 1550–1650," in *The Practice of Reform in Health, Medicine, and Science, 1500–2000*, eds. Margaret Pelling and Scott Mandelbrote (Aldershot, UK: 2005), 59–79 noted that the theories of Paracelsus spread slowly in Italy but permeated Italian medicine by the mid-seventeenth century.

29. Wright, "Introduction" to Fracastoro, *De Contagione*, xi–xiii. Fracastoro's *Syphilis sive Morbus Gallicus* named the disease.

30. Wright, "Introduction" to Fracastoro, *De Contagione*, xxxvii–xl. This decision pleased the Pope because it moved the council to a city where he wielded greater influence.

31. Charles and Dorothea Singer, "The Scientific Position of Girolamo Fracastoro," *Annals of Medical History* (Spring 1917) 1:1–29, 10. "Fomites" are objects that harbor infection. There is some disagreement about the first record of typhus, but most historians accept a date of 1489/90 for the first written description. Victoria A. Harden, "Typhus, Epidemic," in *The Cambridge Historical Dictionary of Disease*, ed. Kenneth F. Kiple (Cambridge and New York: 2003), 352–5, and "Typhus," in *Encyclopedia of Medical History*, ed. Roderick E. McGrew (New York and St. Louis: 1985), 350–5. See also Hans Zinsser, *Rats, Lice and History* (Boston: 1934, rpt. 1963). The subtitle includes a selection of synonyms for this disease. On Fracastoro's thought see the review by Samuel Cohn, Jr. of two works by Concetta Pennuto: a translation (into Italian) of Fracastoro's *De sympathia et antipathia rerum Liber 1* and *Sympatia, fantasia e contagio: il pensiero medico e il pensiero filosofico di Girolamo Fracastoro* in *Medical History* (October 2009) 53:616–18. See also Arrizabalaga, Henderson, and French, *The Great Pox*, 244–51.

32. Fracastoro, *De Contagione*, 83–5.

33. Fracastoro, *De Contagione*, 83.

34. Fracastoro, *De Contagione*, 35.

35. Fracastoro, *De Contagione*, 41.

36. Fracastoro, *De Contagione*, 185 and 197.

37. Fracastoro, *De Contagione*, 217.

38. Fracastoro, *De Contagione*, 183. "Lenticular" or "spotted fever" probably referred to typhus.

39. Nutton, "Seeds," 34.

40. For example, Paul Slack notes that English author Stephen Bradwell in 1636 quoted Fracastoro and described contagion as a "seminary [i.e., seminal] tincture full of a venomous quality, that being very thin and spirituous mixeth itself with the air," *The Impact of Plague on Tudor and Stuart England* (London and Boston: 1985), 27–8.

41. Also known as Girolamo Cardano. See Charles Singer, *The Development of the Doctrine of Contagium Vivum 1500–1750* (London: 1913), 4. Like most neo-Platonists, Cardan regarded all terrestrial objects as animated.

42. Also known as Gabrielle Falloppia, Falopius, Falopio, or Fallopio.

43. See Singer, *Contagium Vivum*, and Robert Jutte, *Contraception: A History*, trans. Vicky Russell (Cambridge, UK: 2008), 96, citing Falopius *De Morbo Gallico* (Padua: 1564).

44. H. Youssuef, "The History of the Condom," *Journal of the Royal Society of Medicine* (April 1993) 86, no. 4: 226–8, 226.

45. Pagel, *Paracelsus*, however, argues on 311 that there is a "fundamental ideological difference" between Paracelsus and Fernel because Fernel adds the new diseases to humoral diseases whereas Paracelsus entirely eliminates humoral causes.

46. I am paraphrasing Linda Deer Richardson, "The Generation of Disease: Occult Causes and Diseases of the Total Substance," in *The Medical Renaissance of the Sixteenth Century*, eds. Andrew Wear, Roger Kenneth French, and Iain M. Lonie (Cambridge: 1985), 175–93. See also Jean Fernel, *On the Hidden Causes of Things*, trans. and ed. John M. Forrester, introduction and annotations by John Forrester and John Henry (Leiden: 2005), esp. book 2, ch. 10–18, 533–721, and Laurence Brockliss, "Seeing and Believing: Contrasting Attitudes Towards Observational Autonomy among French Galenists in the First Half of

the Seventeenth Century," in *Medicine and the Five Senses*, ed. William F. Bynum (Cambridge: 1993), 69–84.

47. Fernel, *Hidden Causes*, trans. Forrester, 551, and see 563–5 and 577. Forrester and Henry discuss the relationship between Fracastoro's idea of "seeds" and Fernel's in the introduction, 21–4.

48. Iwao Moriyama, R. M. Loy, and A. H. T. Robb-Smith, *History of the Statistical Classification of Diseases and Causes of Death*, eds. H. M. Rosenberg and D. L. Hoyert (Hyattsville, MD: 2011), 9, online at the National Center for Health Statistics, www.cdc.gov/nchs/data/misc/classification_diseases2011.pdf. See also Arrizabalaga, Henderson, and French, *The Great Pox*, 264, "An ontological conception of disease was more available to those who maintained a belief in some hidden disease process, such as that of 'whole substance,'" and Forrester and Henry's introduction to Fernel, *Hidden Causes*, 22–3.

49. Felicis Plateri, *De febribus liber: Genera, causas, et curationes febrium* (Frankfurt: 1597), online at Hathi Trust: http://hdl.handle.net/2027/ucm.5309455207. Plater is also known as Felix Platter or Platerus. Walter Charleton prefaced his book on atomism with an extract from Fernel's *De Abditis* on atoms, Charleton, *Physiologia Epicuro-Gassendo-Charletoniana* (London: 1654).

50. Plater, *De febribus*, 92; cf. *A Golden Practice of Physick in Five Books and Three Tomes by Felix Plater*, trans. Abdiah Cole and Nicholas Culpeper (London: 1662), 201.

51. Plater, *De Febribus*, 92–4.

52. David Wootton, *Bad Medicine: Doctors Doing Harm Since Hippocrates* (Oxford: 2006, rpt. w. postscript, 2007), 127. Wootton refers to him as Platter. Wootton paraphrases Plater's argument, combining two of his works, the *De Febribus* and the *Quaestiones* (1625) published posthumously. He does not give exact page references.

53. "[Plater] does not appear in the literature (at least the English-language literature) on germ theory," Wootton, *Bad Medicine*, 128. Daniele Sennerto, *De Febribus Liber IV* (Geneva: 1647), comments on Plater's arguments on 706–11. I have not located an English version of this section.

54. Wootton, "Further Reading," *Bad Medicine*, 300–1, citing "Of Contagion, the Chief Cause of a Plague," in T. Lucretius Carus, *Of the Nature of Things, Vol. II: Containing the Fifth and Sixth Books*, trans. Thomas Creech (London: 1714), 776–81.

55. On Sennert's conversion to atomism see Christoph Lüthy, "Daniel Sennert's Slow Conversion from Hylemorphism to Atomism," *Graduate Faculty Philosophy Journal* (2005) 35, no. 4:99–121, and William R. Newman, *Atoms and Alchemy: Chymistry and the Experimental Origins of the Scientific Revolution* (Chicago: 2006), Section Two: "Daniel Sennert's Atomism and the Reform of Aristotelian Matter Theory," 85–153. See also Michael Stolberg, "Particles of the Soul: The Medical and Lutheran Context of Daniel Sennert's Atomism," *Medicina nei Secoli* (2003) 15:177–203, and Arrizabalaga, Henderson, and French, *The Great Pox*, 272–7, for Sennert's theory of syphilis.

56. Lüthy and Newman, "Daniel Sennert's Earliest Writings (1599–1600) and Their Debt to Giordano Bruno," *Bruniana and Campanelliana* (2000) 6:261–79. Tycho Brahe and Sennert both lived in Jessenius' home and must have been acquainted. Sennert was tried but acquitted on a charge of heresy for maintaining that God created the souls of animals as well as men out of nothing; his earliest

works included "one of the few references to Bruno that was printed during his lifetime and . . . may well constitute the only discussion of his doctrines in an academic context before 1600" (271).

57. Marchamont Nedham, *Medela Medicinae* (London: 1665), see esp. 112–15 and Chapter 3 below. See also Webster, "Instauration," 271.

58. An outline of Sennert's life compiled by Richard Westfall is on the Galileo Project website at http://galileo.rice.edu/Catalog/NewFiles/sennert.html. Many of Sennert's works are rare (especially his early Latin works and theses), and the bibliography has yet to be straightened out. Many dissertations by his students were credited to Sennert as well as the student author. The online Wellcome Library Catalogue, http://catalogue, wellcome.ac.uk, lists 62 titles or editions between 1611 and 1687. For Sennert's influence on the Oxford school of physiology, see Robert G. Frank Jr., *Harvey and the Oxford Physiologists* (Berkeley: 1980), 120. Walter Pagel, *The Smiling Spleen: Paracelsianism in Storm and Stress* (Basel: 1984), 86–90, emphasizes the divergences between Paracelsus and Sennert.

59. Daniel Sennert, *Institutions*, trans. N. D. B. P. (London: 1656) online from EEBO. Some Worldcat entries erroneously list B. P. N. D. as translator. Lodowick Lloyd, the publisher, also published works by Van Helmont, Jacob Boehme, Paracelsus, and Oswald Croll.

60. Sennert, *Institutions*, 58.

61. For the history of this concept, see Arrizabalaga et al., *The Great Pox*, "The French Disease Grows Old," 252–77.

62. Sennert, *Institutions*, 60–1.

63. *Dr D. Sennertus of Agues and Fevers. Their Differences, Signes and Cures. Divided into Four Books: Made English by N. D. B. M . . .* (London: 1658), Book 4, Chapter 3 [*sic*], 79. This is a translation of Book 4, Chapter 2, in Daniel Sennert, *De Febribu[s] Libri IV* (Wittenberg: 1619). I used the Google Books online edition (Geneva: 1647). Marchmont Nedham also translated portions of this chapter in the *Medela Medicinae*, see below, chapter 3.

64. Daniel Sennert, *The Sixth Book of Practical Physick: Of Occult or Hidden Diseases; in Nine Parts . . . by Daniel Sennertus*, trans. N. Culpeper and Abdiah Cole, Doctors of Physick (London: 1662, online from EEBO). Notes refer to this version. This appears to be from the final volume of Sennert's *Practicae Medicinae Liber Primus—[Sextus]* published between 1628 and 1636, but I have not seen a copy of the latter work. For the complicated status of the Culpeper-Cole translations, see chapter 2.

65. Sennert, *Practical Physick*, contents and part 1, chapter 7, 8–9.

66. Sennert, *Practical Physick*, part 1, chapter 8, 9–10.

67. Sennert, *Practical Physick*, part 3, chapter 3, ss. 1–2, p. 24. This passage is very similar to the one from *De Febribus* translated by Nedham in chapter 3 below.

68. Sennert, *Practical Physick*, part 3, chapter 3 ss. 2–3, pp. 24–6. This passage also echoes one from *De Febribus*, 660.

69. Sennert, *Practical Physick*, chapter 4, 26.

70. Sennert had extensive experience with the plague, as there were several epidemics in Wittenberg during his lifetime. He has also been credited with suggesting that cancer was contagious.

71. Sennert was the first writer to describe German measles (*roteln* or *rubella*) in 1619, but it was not firmly differentiated from measles (*rubeola*) until the nineteenth century.

72. Arrizabalaga et al. argue that "Sennert endowed the *semina* with characteristics of a living organism . . . Not only was the pox an entity, it was one that feeds on the flesh and grows at its expense." *The Great Pox*, 275–6.
73. Winslow, *Conquest*, 147. Fallopius had applied the analogy of the spontaneous generation of "worms" to syphilis and phthisis in 1564.
74. Kircher's work appeared as *Scrutinium physico-medicum contagiosae luis, quae pestis dicitur* (A medical and physical examination of the contagious pestilence that they call the plague) (Rome: 1658). See also P. Conor Reilly, *Athanasius Kircher S. J.: Master of a Hundred Arts, 1602–1680* (Wiesbaden: 1974), 88.
75. Gibbs or Gibbes (?1611–1677) studied medicine at Padua. He was the first Restoration Catholic to receive an MD from Oxford in 1671. Kircher's findings were anticipated at Kircher's request by his friend and colleague P. Gaspar Schott, S. J. in *Magia Optica* (1657). Schott wrote, "Who could believe, says Kircher, that vinegar and milk swarm with innumerable small worms . . . I pass over the . . . wormy blood of those people afflicted with fever, and the other countless things unknown to any physician up to the present time," A. B. Luckhardt, "Description of Some Early Microscopes and Their Use in the Study of Infectious Diseases," *Bulletin of the Society for Medical History of Chicago* (January, 1917) 2, no. 1:106–9, 106.
76. Singer, *Contagium Vivum*, 10.
77. Reilly, *Kircher*, 92.
78. Singer, *Contagium Vivum*, 9. See also Ralph H. Major, "Athanasius Kircher," *Annals of Medical History* (March 1939) 3rd ser., 1, no. 2:105–20; Reilly, *Kircher*, 93.
79. Winslow, *Conquest*, 149–50.
80. *Epistola Praeliminaris, Tractatui de Viva Mortis Imagine* (Frankfurt: 1650).
81. Singer, *Contagium Vivum*, 11.
82. The similarity between Kircher's thought and Sennert's has been attributed to their mutual dependence on the work of the forgotten Paduan professor Fortunio Liceti (1577–1657). See Hiro Hirai, "Kircher's Chymical Interpretation of the Creation and Spontaneous Generation," in *Chymists and Chymistry: Studies in the History of Alchemy and Early Modern Chemistry*, ed. Lawrence M. Principe (Sagamore Beach, MA: 2007), 77–87. According to Hirai, Kircher "plagiarized Liceti's discussions at great length and sometimes almost *verbatim*."
83. The only translations of the *Scrutinium* seem to be Dutch in 1669 and German in 1680.

CHAPTER 2

1. Christopher Hill, *The World Turned Upside Down: Radical Ideas During the English Revolution* (New York: 1972). For a reconsideration of the Levellers see the essays in Michael Mendle, ed., *The Putney Debates of 1647: The Army, the Levellers and the English State* (Cambridge: 2010), especially Blair Worden, "The Levellers in History and Memory, c. 1660–1960," 256–82, and J. G. A. Pocock, "The True Leveller's Standard Revisited: An Afterword," 283–91.
2. Alan Rudrum, "Research Reports, VI—Theology and Politics in Seventeenth-Century England," *The Clark Newsletter* (1988) 15:5–7 argued that the concept of universal perfection inherent in alchemy made it more compatible with Arminian views than with predestination. See also Nigel Smith, "The Charge of Atheism and the Language of Radical Speculation, 1640–1660," in *Atheism from*

the Reformation to the Enlightenment eds. Michael Hunter and David Wootton (Oxford: 1992), 131–58, esp. 135–8 on Lawrence Clarkson. Following the condemnation and burning of his monist tract "A Single Eye all Light no Darkness" (1650), Clarkson became an itinerant "professor" of astrology and physic before converting to Muggletonianism.

3. Thomas Fuchs, *The Mechanization of the Heart: Harvey and Descartes*, trans. Marjorie Grene (Rochester, NY: 2001).

4. Singer, *The Development of the Doctrine of Contagium Vivum* (London: 1913), 44–5. See also Walter Pagel, "Harvey and the 'Modern' Concept of Disease," *Bulletin of the History of Medicine* (1968) 42:496–509.

5. Singer, *Contagium Vivum*, 45.

6. Singer, *Contagium Vivum*, 47, and see Niebyl, "Venesection and the Concept of the Foreign Body," Ph.D. dissertation (New Haven: Yale University, 1970), 350.

7. P. M. Rattansi, "The Helmontian-Galenist Controversy in Restoration England," *Ambix* (1964) 12:1–23; Charles Webster, "English Medical Reformers of the Puritan Revolution: A Background to the 'Society of Chymical Physitians,'" *Ambix* (1967) 14:16–41; and Andrew Wear, "Medical Practice in Late Seventeenth- and Early Eighteenth-Century England: Continuity and Union," in *The Medical Revolution of the Seventeenth Century*, eds. Roger Kenneth French and Andrew Wear (Cambridge: 1969), 294–320.

8. See Harold J. Cook, *The Decline of the Old Medical Regime in Stuart England* (Ithaca, NY: 1986), 20, 72–4. See also George Clark, *A History of the Royal College of Physicians of London* (Oxford: 1964), vol. 1, chapter 4, "The Foundation of the College of Physicians of London," 54–64.

9. A. L. Morton, *The World of the Ranters* (London: 1979), 194. Elizabeth Furdell, *Publishing and Medicine in Early Modern England* (Rochester: 2002), chapter 7 "Medical Advertising: Publishing the Proprietary," 135–54. Patients had to buy a secret remedy from the author/owner/creator or from a designated apothecary, who often paid a kickback to the physician who recommended it. See also Allen G. Debus, "Paracelsian Medicine: Noah Biggs and the Problem of Medical Reform," in *Medicine in Seventeenth-Century England*, ed. Allen G. Debus (Berkeley: 1974), 33–48; Harold J. Cook, "Henry Stubbe and the Virtuosi-Physicians," in French and Wear, *Medical Revolution*, 246–71, on 253.

10. Paul Kléber Monod, *Solomon's Secret Arts, the Occult in the Age of Enlightenment* (New Haven: 2013), esp. chapter 2: "The Silver Age of the Astrologers," 53–81.

11. Thomas A. Horrocks, *Popular Print and Popular Medicine: Almanacs and Health Advice in Early America* (Amherst, MA: 2008) shows that astrology was popular in American almanacs into the nineteenth century. Astrology could also be rationalized by arguing that the gravity of the sun and moon might affect air pressure or fluids in the body or the soil in the same way they affected the tides. This was the argument of Richard Mead's *De Imperio Solis ac Lunae* (London: 1708).

12. Enthusiasm is "the idea that one could have direct unmediated access to religious truths," Stephen Gaukroger, *The Collapse of Mechanism and the Rise of Sensibility* (Oxford: 2010), 33.

13. Monod, *Solomon's Secret Arts*, 95.

14. Allen G. Debus, "Paracelsian Medicine"; Cook, "Henry Stubbe," 253.

15. James Riddick Partington, *A History of Chemistry*, 4 vols. (London: 1961–70), vol. 2, 212; vol. 3, 8; Walter Pagel, "The Religious and Philosophical Aspects of van Helmont's Science and Medicine," *Bulletin of the History of Medicine*, Supp. 2 (1944). Pagel does not regard van Helmont as a monist, a term he reserves

for "crude" pantheists. Instead, he uses the term "pluralist," an approach close to that of the "monadology" of Leibnitz. However, for our purposes, "monist" seems to be a more serviceable term. See 22–3.

15 Pagel, "Religious and Philosophical Aspects," 214.

16. Chandler was a Ranter who became a Quaker and joined a circle around the Mortalist author Lady Anne Conway and Helmont's son, the mystic Francis Mercury van Helmont. See Peter Elmer, "Medicine, Religion and the Puritan Revolution," in French and Wear, *Medical Revolution*, 24; Carolyn Merchant, "The Vitalism of Anne Conway: Its Impact on Leibniz's Concept of the Monad," *Journal of the History of Philosophy* (July 1979) 17, no. 3:255–69; Richard H. Popkin, *The Third Force in Seventeenth-Century Thought* (Leiden: 1992), 117 and chapter 7: "Spinoza's Relations with the Quakers in Amsterdam," 120–34; and John Henry "Medicine and Pneumatology: Henry More, Richard Baxter, and Francis Glisson's '*Treatise on The Energetic Nature of Substance*,'" *Medical History* (1987), 31:15–40. On Mortalism during this period, see Ann Thomson, *Bodies of Thought: Science, Religion and the Soul in the Early Enlightenment* (Oxford: 2008); Conway is discussed on 44.

17. Allen Debus, *Science and Education in the Seventeenth Century. The Webster-Ward Debate* (London: 1970), 26.

18. Pagel, "Religious and Philosophical Aspects," 36.

19. Allen G. Debus, "Thomas Sherley's *Philosophical Essay* (1672): Helmontian Mechanism as the Basis of a New Philosophy," *Ambix* (1980) 17, no. 2:125–35; Hiro Hirai, *Le Concept de Semence dans les Théories de la Matière à la Renaissance: De Marsile Ficin à Pierre Gassendi* (Turnhout: 2005), 456, argues that the ferment has precedence over the seed.

20. Pagel, *Joan Baptista Van Helmont* (Cambridge: 1982), 71, 82, 102.

21. Pagel, *Joan Baptista Van Helmont*, 149.

22. Pagel, *Joan Baptista Van Helmont*, 43–5; Partington, *History of Chemistry*, 236.

23. Pagel, *Joan Baptista Van Helmont*, 142; William R. Newman, "The Corpuscular Theory of J. B. Van Helmont and its Medieval Sources," *Vivarium* (1993) 31:161–91.

24. Peter Niebyl, "Galen, Van Helmont and Bloodletting," in *Science, Medicine and Society in the Renaissance*, ed. Allen Debus (New York: 1972), 2:13–23. See also Lois N. Magner, *A History of Medicine* (New York: 1992), 206. A sixteenth-century version of this challenge is mentioned in William Eamon, "'With the Rules of Life and an Enema:' Leonardo Fioravanti's Medical Primitivism," in *Renaissance and Revolution*, eds. J. V. Field and Frank A. J. L. James (Cambridge: 1993), 33.

25. George Starkey, *Nature's Explication and Helmont's Vindication* (1657) quoted in Allen Debus, *The Chemical Philosophy: Paracelsian Science and Medicine in the Sixteenth and Seventeenth Centuries*, 2 vols. rpt. in 1 vol. (New York: 2002), 501. See also William R. Newman, "Starkey, George (1628–1665)," *ODNB* online ed. (May 2008), http://www.oxforddnb.com/view/article/26315.

26. Cook, *Decline*, 150. The College was established by a royal charter from Henry VIII and was often referred to as the "King's College" or "Royal College," but the name was not ratified until 1960. See Geoffrey Davenport, "When Did the College Become Royal?" in *The Royal College of Physicians and Its Collections: An Illustrated History*, eds. Geoffrey Davenport, Ian McDonald, and Caroline Moss-Gibbons (London: 2001), 26–8.

27. For the way disagreements about terminology and classification could bedevil later trials of venesection see Lester S. King, "The Blood-Letting Controversy,

a Study in the Scientific Method," *Bulletin of the History of Medicine* (1961) 35:1–13.

28. Pagel, *Joan Baptista Van Helmont*, 91, see also Anna Marie Roos, *The Salt of the Earth: Natural Philosophy, Medicine and Chymistry in England, 1650–1750* (Leiden: 2007), 49.

29. Roos, *Salt*, 49.

30. Roos, *Salt*, 91, referring to the work of the chemist and botanist Nehemiah Grew.

31. Grew concluded that different salts accounted for the differing structures of plants. Roos, *Salt*, 91–3.

32. Roos, *Salt*, 97.

33. See http://www.contagionism.org/english_religious_laws.htm for a chart of the relevant Acts.

34. David Ogg, *England in the Reign of Charles II*, 2nd. ed. rpt. in 1 vol. (Oxford: 1972), 201. The Irish Act of Uniformity (1666) imposed the *Book of Common Prayer* in Ireland and required Episcopal ordination. John Spurr, *The Post-Reformation: Religion, Politics and Society in Britain 1603–1714* (Harlow, Essex: 2006), 156; and William Birken, "The Dissenting Tradition in English Medicine of the Seventeenth and Eighteenth Centuries," *Medical History* (April 1995) 39, no. 2:197–218, online from PubMed, http://www.ncbi.nlm.nih.gov/pmc/articles/PMC1036975/.

35. William Gibson, however, states that the Clarendon Code did not cripple the Dissenters' academies but shifted their perspective toward rational dissent and away from traditional Puritanism. See *Religion and the Enlightenment 1600–1800: Conflict and the Rise of Civic Humanism in Taunton* (Oxford: 2007). I thank Dr. Gibson for his personal communication of January 24, 2010. See also William Gibson, *The Church of England 1688–1832: Unity and Accord* (London and New York: 2001).

36. A vivid account of the repression in London is in Mark Goldie, "The Hilton Gang: Terrorising Dissent in 1680s London," *History Today* (1997) 47, no. 10, http://www.historytoday.com/mark-goldie/hilton-gang-terrorising-dissent-1680s-london. Among the persecuted ministers was Stephen Lobb, father of the physician Theophilus Lobb.

37. Spurr, *Post-Reformation*, 161.

38. There was some crossover between the religious and the political groups; some Dissenters, for example, voted Tory. See William Gibson, "Dissenters, Anglicans and Elections after the Toleration Act, 1679–1710," in *Religion, Politics and Dissent, 1660–1832: Essays in Honour of James E. Bradley*, eds. Robert D. Cornwall and William Gibson (Farnham, Surrey: 2013), 129–46.

39. John Gascoigne, *Cambridge in the Age of the Enlightenment* (Cambridge: 1989), 76–7.

40. Thus, Dissenters could study at Cambridge without subscribing if they left before graduation. This Act was repealed in 1871.

41. In 1707, following the union with England, Scottish university faculty but not students were required to subscribe to the Presbyterian *Confession of Faith*. In Ireland, Trinity College, Dublin, was limited to Anglican students.

42. For medicine in the eighteenth-century English universities, see Gascoigne, *Cambridge* and *The History of the University of Oxford*, eds. L. S. Sutherland and L. G. Mitchell, vol. 5 (Oxford: 1986), especially G. L'E Turner, "The Physical Sciences," 659–68, and Charles Webster, "The Medical Faculty and

the Physic Garden," 683–723. See also Arthur Rook, "Medicine at Cambridge, 1660–1760," *Medical History* (1969) 13:107–22.

43. Margaret DeLacy, "Timeline for Admission to Fellowship of the College of Physicians 1660–1800" (2012), online at http://www.contagionism.org/timeline _for_admission_to_fellow.htm.

44. Cecil Wall, *A History of the Worshipful Society of Apothecaries of London abstracted . . . by the late H. Charles Cameron*, ed. E. Ashworth Underwood (London: 1963), 113; Michael Cyril William Hunter, *Establishing the New Science: The Experience of the Early Royal Society* (Woodbridge, Suffolk: 1989), 276.

45. Clark, *College of Physicians*, 1:340, 348. For the physicians admitted in 1680 see Munk, *Roll*, vol. 1, 406–15. Several of these physicians had Oxford or Cambridge degrees.

46. Wall, *Society of Apothecaries*, 355 n. 21.

47. Mandate degrees were awarded by the monarch at special ceremonies (by "*Comitia Regia*") or by a petition to the Crown that was approved by Cambridge University (by "*Literae Regia*"). In 1728, George II conferred 32 Cambridge MD degrees on physicians and an additional 5 degrees on members of the nobility. Most recipients were not members of the University; 18 had studied at Leyden. E. Ashworth Underwood, *Boerhaave's Men at Leyden and After* (Edinburgh: 1977), 66–9, note "On the Royal College of Physicians of London and on Some Degrees of the University of Cambridge."

48. The charter was dated March 11, 1686/7. Lister had an AB from Cambridge and studied in Montpellier, but his MD was an honorary degree from Oxford. He became Vice-President of the Royal Society in 1685. J. D. Woodley, "Lister, Martin (bap. 1639, d. 1712), *ODNB*, online ed. (October 2008), http://www .oxforddnb.com/view/article/16763. Robinson also studied in Europe, including Montpellier, but he held a regular MD from Cambridge. For Sloane, see chapter 6.

49. A friend of John Locke's but a pious Anglican, Blackmore had recently married Mary Adams, a member of the Whig Verney family. See below in this chapter.

50. Campbell R. Hone, *The Life of Dr. John Radcliffe 1652–1714: Benefactor of the University of Oxford* (London: 1750), 57.

51. Hone, *Radcliffe*, 57. See Robert L. Martensen, "Radcliffe, John (bap. 1650, d. 1714) *ODNB* (Oxford: 2004), online ed. (March 2010), http://www .oxforddnb.com/view/article/22985; and William MacMichael, *The Gold-Headed Cane*, 2nd ed. (New York: 1926), chapter 1, 1–50. Burnett's brother, Sir Thomas, was physician to Queen Anne.

52. Wall, *Society of Apothecaries*, 124–5.

53. Morton, (bap. 1637, d. 1698) a Nonconformist minister, was ejected from his pulpit at the Restoration and studied medicine abroad. Tyson lived in his house. See R. R. Trail, "Richard Morton (1637–1698)," *Medical History* (1970) 14:166–74. For Benjamin Marten's use of Morton's *Phthsiologia* see below, chapter 7. For Tyson, see below, chapter 5.

54. Clark, *College of Physicians*, I: 357. Obadiah Grew was imprisoned in 1682 for defying the Five Mile Act after teaching in the Presbyterian Academy in Coventry. Richard Cromwell had three surviving daughters. One married Thomas Gibson, whose heir and nephew Edmund Gibson became Bishop of London and supported inoculation. Another daughter married John Mortimer, the father of Cromwell Mortimer, Hans Sloane's intimate friend and assistant.

55. Gilbert Burnet (1643–1717), became Bishop of Salisbury, his brother Sir Thomas Burnet (1638–1704, MD Montpellier, 1659), would be a leader of the Edinburgh "Sydenhamians." Thomas's son, also Thomas Burnet, graduated MD at Leiden in 1691.

56. Cook, *Decline*.

57. Cook, *Decline*, 235.

58. Clark, *College of Physicians*, I: 475. Blackmore had an AB from Oxford and an MD from Padua.

59. Birken, "Dissenting Tradition," 199. A licensed provincial practitioner could practice with a bachelor's degree in either arts or medicine or with no degree at all, Rook, "Medicine at Cambridge," 108–9.

60. Birken, "Dissenting Tradition," 202.

61. William Gibson, *Church of England*.

62. Spurr writes that the Dissenters were united only in their resentment at being lumped together under the Clarendon Code, *The Post-Reformation*, 150.

63. Gary DeKrey, *London and the Restoration, 1659–1683* (Cambridge: 2005), 120–1.

64. See for example John Fothergill's comment on Presbyterians to James Pemberton, Lea Hall, Cheshire, May 10, 1766, in *Chain of Friendship: Selected Letters of Dr. John Fothergill of London, 1735–1780*, eds. Betsy C. Corner and Christopher C. Booth (Cambridge, MA: 1971), 262–4.

65. Jacob Selwood, *Diversity and Difference in Early Modern London* (Farnham, Surrey and Burlington, VT: 2010), esp. the introduction and chapter 1, 1–50. An overview is on the Old Bailey Online website at http://www.oldbaileyonline .org/static/Population-history-of-london.jsp and a bibliography on the Old Bailey Online website at http://www.oldbaileyonline.org/static/Communities Bibliography.jsp, both retrieved March 20, 2014.

66. One notable exception was the alchemist George Starkey, born in Bermuda and educated at Harvard.

67. Examples are John Arbuthnot, William Cockburn, James Douglas, and James Welwood.

68. Roger L. Emerson, "Medical men, politicians and the medical schools at Glasgow and Edinburgh, 1685–1803," in A. Doig et al. *William Cullen and the Eighteenth Century Medical World* (Edinburgh: 1993), 188–215. See also Roger L. Emerson, *Academic Patronage in the Scottish Enlightenment* (Edinburgh: 2008).

69. John Friesen, "Archibald Pitcairne, David Gregory and the Scottish Origins of English Tory Newtonianism, 1688–1715," *History of Science* (2003) 41:163–91, online from the SAO/NASA Astrophysics Data System (ADS), http://adsabs .harvard.edu/full/2003HisSc..41..163F.

70. The best known of these were Hector Nunez, Rodrigo Lopez, and Fernando Mendes, physician to Catherine of Braganza, who had accompanied her to England. Descendants of Mendes and his brother-in-law, Alvaro da Costa, formed the large Sephardic family of Mendes da Costa. Albert M. Hyamson, *The Sephardim of England* (London: 1951), 9, 62.

71. Jews were unintentionally prevented from taking the Oath of Abjuration by the final phrase "upon the true faith of a Christian." They were threatened with litigation after passage of the Conventicle Act, but the King issued a dispensation, followed by an Order in Council in 1674. Henry Straus Quixano Henriques, *The Jews and the English Law* (Oxford: 1908), 147–50.

72. See Hyamson, *Sephardim*, 63, 130, 199: "Jews might be debarred from voting in parliamentary elections, but by the end of the eighteenth century very few could remember a returning officer refusing any Jew the right to do so."

73. Morelli was a Marrano born in Cairo. See Munk, *Roll*, vol. 1; Clark, *College of Physicians*, 1:348; and Richard H. Popkin, *The Third Force in Seventeenth-Century Thought* (Leiden: 1992), 145–7. I thank Dr. Popkin for assistance on this.

74. A law of 1697 enabled them to act as brokers in the city without becoming Freemen. See Hyamson, *Sephardim*, 129, 142. Although six Jews received the Freedom of London between 1687 and 1738, in that year, Abraham Rathom lost a suit to succeed his father as a liveryman and Freeman because an oath on the New Testament was required. According to Hyamson, no further Jews became Freemen of London for nearly a century. However J. Burnby writes that by the last quarter of the eighteenth century, "Jewish surgeons and apothecaries were well established [in London] and taking apprentices," being permitted to swear on the Old Testament, "Jewish Apothecaries and Surgeons in Eighteenth-Century London," *Pharmaceutical Historian* (November 1991) 21, no. 4:10.

75. By 1700, there were 9 Calvinist French churches in the East End and 12 in the West End: 6 Anglican and 6 Calvinist. Old Bailey Online, http://www.oldbailey online.org/static/Huguenot.jsp, retrieved March 20, 2014.

76. Uta Janssens, *Matthieu Maty and the Journal Britannique 1750–1755* (Amsterdam: 1975), 21. I thank Dr. Elizabeth Eisenstein for this reference.

77. Janssens, *Maty*, 12, quoting Pierre Jean Grosely, *A Tour to London* (London: 1772). Janssen reverses the initials of his given names.

78. Elizabeth A. Williams, *A Cultural History of Medical Vitalism in Enlightenment Montpellier* (Aldershot, UK: 2003). See also Stephen Gaukroger, *The Collapse of Mechanism and the Rise of Sensibility* (Oxford: 2010), 212.

79. Hugh Trevor-Roper, *Europe's Physician: The Various Life of Theodore de Mayerne* (New Haven and London: 2006), 137. A few German-speaking chemists or naturalists also settled in England, notably Henry Oldenburg, Peter Stahl, Theodore Haak, and Ambrose Godfrey Hanckwitz. Samuel Hartlib settled in England in 1628. Frederick Slare was born in England.

80. Harold J. Cook, "The New Philosophy in the Low Countries," in *The Scientific Revolution in National Context*, eds. Roy Porter and Mikulas Teich (Cambridge: 1992), 115–49. See also Jonathan Israel, "Enlightenment, Radical Enlightenment and the 'Medical Revolution' of the Late Seventeenth and Eighteenth Centuries," in *Medicine and Religion in Enlightenment Europe*, eds. Andrew Cunningham and Ole Peter Grell (Aldershot, UK: 2007), 5–28. Peter Elmer notes in the same volume that studying in Leyden did not change the English Dissenters' belief in witchcraft, "Medicine, Witchcraft and the Politics of Healing," 224–41.

81. John Marshall, *John Locke, Toleration and Early Enlightenment Culture* (Cambridge: 2006), 169–70.

82. Like other Protestant Dissenters in Britain, the Dutch had their own churches. The Dutch Reformed Church in Austin Friars was established in 1568. William John Charles Moens, ed., *The Marriage, Baptismal and Burial Registers, 1571–1874 . . . of the Dutch Reformed Church, Austin Friars, London with a Short Account of the Strangers and Their Churches* (Lymington: 1884), online edn. www.archive.org/stream/cu31924029785445#page/n7/mode/2up. I have not consulted Ole Peter Grell, *Dutch Calvinists in Early Stuart London: The Dutch Church in Austin Friars, 1603–1642* (1997). Walloons seem to have joined the French Church.

83. Eric Grier Casteel, "'Entrepôt and Backwater': A Cultural History of the Transfer of Medical Knowledge from Leiden to Edinburgh, 1690–1740" (Ph.D. dissertation, University of California Los Angeles: 2007); Jonathan I. Israel, "Enlightenment, Radical Enlightenment and the 'Medical Revolution' of the late 17th and 18th Centuries," in *Medicine and Religion in Enlightenment Europe*, eds. Ole Peter Grell and Andrew Cunningham (Aldershot, UK: 2007), 5–27.

84. Also referred to as François Sylvius, François de la Boe, François Dubois, and Franz (Frans) dele Boe. *Sylvius* is a translation of *de le Boë*, which means "of the forest." See Harold J. Cook, *Trials of an Ordinary Doctor, Johannes Groenevelt in Seventeenth-Century London* (Baltimore: 1994), 64; Lester King, *The Road to Medical Enlightenment, 1650–1695* (New York: 1970), 93–112; Richard Westfall, "Sylvius, Franciscus dele Bo," The Galileo Project, http://galileo.rice.edu /Catalog/NewFiles/sylvius.html, retrieved April 20, 2014.

85. Harold Cook, *Matters of Exchange: Commerce, Medicine and Science in the Dutch Golden Age* (New Haven: 2007), 296.

86. Henry Leicester, *Development of Biochemical Concepts from Ancient to Modern Times* (Cambridge, MA: 1974), 101–4. Sylvius's student, Cornelis Bontekoe, demonstrated the difference.

87. See Trail, "Richard Morton," 170: "He makes a complete break with tradition . . . with an entirely new assertion that 'the air expelled by consumptives, if drawn in by the nose and mouth by anyone . . . will produce phthisis.'" On the wider significance of his explanation for hemoptysis, see Walter Pagel, "Humoral Pathology: A Lingering Anachronism," *Bulletin of the History of Medicine* (1955) 29:299–308.

88. Audrey B. Davis, "The Virtues of the Cortex in 1680: A Letter from Charles Goodall to Mr. H.," *Medical History* (1971) 15, no. 3:293–304.

89. For the conflict in Edinburgh between the Sydenhamians, who favored a clinical approach to medicine, and the Newtonians, who favored a mechanist and theoretical approach, see Robert Peel Ritchie, *The Early Days of the Royall Colledge of Phisitians, Edinburgh* (Edinburgh: 1899). See also Andrew Cunningham, "Sydenham vs. Newton: The Edinburgh fever dispute of the 1690s," in *Theories of Fever from Antiquity to the Enlightenment, Medical History*, Supp. 1, eds. W. F. Bynum and V. Nutton (London: 1981), 71–97; Casteel, "Entrepôt and Backwater," 88–114; Friesen, "Scottish Origins of English Tory Newtonianism"; W. B. Howie, "Sir Archibald Stevenson, His Ancestry, and the Riot in the College of Physicians at Edinburgh," *Medical History* (1967) 11:269–84; Anita Guerrini, "Archibald Pitcairne and Newtonian Medicine," *Medical History* (1987) 31:70–83; and David E. Shuttleton, "'A Modest Examination': John Arbuthnot and the Scottish Newtonians," *British Journal for Eighteenth-Century Studies* (1995) 18, no. 1:47–62.

90. Casteel, "Entrepôt and Backwater," 272. Casteel writes that de Volder brought Boyle's "experiment-demonstration" method to Leiden where Boerhaave adapted it for chemical instruction and passed it on to his disciple, Charles Alston. De Volder appears often in Jonathan I. Israel, *The Radical Enlightenment* (Oxford: 2001). See esp. 251–2 on his defense of Spinoza's monism and 310–11 on his critique of Boyle and English empiricism. See also Hans Hooijmaijers and Ad Maas, "Entrepreneurs in Experiments: The Leiden Cabinet of Physics and Its Founders," in *Cabinets of Experimental Philosophy in Eighteenth-Century Europe*, eds. Jim Bennett and Sofia Talas (Leiden: 2013), 27–48.

91. Cook, *Trials*, 64–5.

92. John Edward Fletcher, *A Study of the Life and Works of Athanasius Kircher, "Germanus Incredibilis,"* ed. Elizabeth Fletcher (Leiden: 2011), 119.

93. Tulp was also known as Claes Pietersz or Pieterszoon and Tulpius. Jacob Wolff, *The Science of Cancerous Disease from Earliest Times to the Present (1907)*, trans. Barbara Ayoub (Canton, ME: 1989), 9, and 39–40. This theory would be revived in Germany by Lorenz Heister (1739) and in England at the end of the eighteenth century. See Wolff, 68; A. P. Waterson and Lise Wilkinson, *An Introduction to the History of Virology* (Cambridge: 1978), 6; and John Pearson, *Practical Observations on Cancerous Complaints: With an Account of Some Diseases which Have Been Confounded with Cancer* (London: 1793), 20–1.

94. Also referred to as Monsieur Blanchard, Stephen Blancard, Stephen Blankaert, Étienne Blankard, and Steph. or Stephani Blancardi. Harold J. Cook, "New Philosophy in the Low Countries," 13, describes this compilation, which appeared in 1680, 1683, and 1686, as "a strange combination of scientific reporting and tabloid journalism . . . not very different from the contemporary *Philosophical Transactions*." See also Edward G. Ruestow, *The Microscope in the Dutch Republic* (Cambridge and New York: 1996), 34–5. Additional references are in Ruestow's index under "Blanckaert, Steven." The best short article on Blankaart in English seems to be in Wikipedia; he deserves a longer study. He appears under the name "Steven Blankaart" in Ruben E. Verwaal's unpublished thesis, "Hippocrates Meets the Yellow Emperor: On the Reception of Chinese and Japanese Medicine in Early Modern Europe, 1650–1750" (Utrecht University Medical Center, Utrecht, November 20, 2009), online at http://dspace.library.uu.nl/handle/1874/179050, esp. 12–17, 22, 26, 27, 36, 41–2, 54. Verwaal shows how Blankaart transmitted Asian medical practices and plants (especially moxibustion, acupuncture, and camphor) from Dutch physicians in China and Japan to Europe.

95. Cook, *Trials*, 150.

96. *Verhandelinge van het Podagra en Vliengende Jicht*, cited in Verwaal, "Hippocrates," 13.

97. *Schouwburg der Ruspen, Wormen, Maden en Vliegende Dierkens daaruit voortkommende* (Amsterdam: 1688; Leipzig: 1690) (The theater of caterpillars, worms, grubs and little flying animals). David M. Damkaer, *The Copepodologist's Cabinet: A Biographical and Bibliographical History* (Philadelphia: 2002), 26. Damkaer believes Blankaart's "microscope" was a magnifying glass.

98. F. Buret, *Syphilis in the Middle Ages and in Modern Times*, vols. 2 and 3 in 1, trans. A. H. Ohmann-Dumesnil (Philadelphia: 1895), 164, online from Medical Heritage Library, https://archive.org/details/39002086312379.med.yale.edu. Buret cites Étienne Blankard [Steven Blankaart], *Traité de la Vérole*, trans. William Willis (Amsterdam: 1688). The *Traité* seems to have been published in Amsterdam in 1684 as *Venus Belegert en Ontset*. Buret wrote that this was the first time an author suggested an animate cause for venereal disease, but as we shall see, a medical student, Stephen Hamm, suggested this to Leeuwenhoek in 1677. For other early examples, see chapter 1 and chapter 3.

99. Steven Blancard [Blankaart], *The Physical Dictionary. Wherein the Terms of Anatomy, the Names and Causes of Diseases, Chyrurgical Instruments, and Their Use are Accurately Describ'd*, 4th ed. (London: 1702), italics in original. The first English edition was 1684, as a translation of the *Lexicon Medicum Graeco-Latinum* (Amsterdam: 1679). I have not been able to identify the Scotus reference. It may refer to Michael Scott or Scotus, astrologer to Frederick II.

100. Benjamin Marten, *A New Theory of Consumptions*, 2nd ed. (London: 1722), 49–50.
101. Hill, *World Turned Upside Down*, 233–4. William R. Newman and Lawrence M. Principe have argued that the distinction between alchemy and chemistry is unjustified and recommended the use of the term "chymistry" for early chemistry. See their article, "Alchemy vs Chemistry: The Etymological Origins of a Historiographic Mistake," *Early Science and Medicine* (1998) 3:32–65. For a contrasting view, see George-Florin Căllian, "*Alkimia Operativa* and *Alkimia Speculativa*: Some Modern Controversies on the Historiography of Alchemy," *Annual of Medieval Studies at CEU* (2010) 16:166–90, online from the Internet Archive, http://www.archive.org/-details/AlkimiaOperativaAndAlkimia Speculativa.SomeModernControversiesOnThe.
102. Harold J. Cook, "The Society of Chemical Physicians, the New Philosophy, and the Restoration Court," *Bulletin of the History of Medicine* (1987) 61:61–77. Cook argues that the membership of the Society is difficult to place politically. He sees the turmoil as a competition for Royal and Court patronage.
103. Elmer, "Medicine, Religion and the Puritan Revolution," 33.
104. This author should be distinguished from the younger William Oliver of Bath (1695–1764, MD Leyden, 1720, and Cambridge, 1725) who became John Pringle's father-in-law and wrote a *Practical Essay on . . . Warm Bathing in Gouty Cases*. The elder William Oliver held an MD from Rheims, although he also studied in Leyden, Prussia, and Poland. W. P. Courtney, "Oliver, William (bap. 1658, d. 1716)," *Oxford Dictionary of National Biography*, rev. S. Glaser, (Oxford: 2004), online at http://www.oxforddnb.com/view /article/20735.
105. David L. Wykes, "Quaker Schoolmasters, Toleration and the Law, 1689–1714," *Journal of Religious History* (1997) 21, no. 2:178–92, published online December 19, 2002, doi: 10.1111/j.1467–9809.1997.tb00484. See also John Chapple, *Elizabeth Gaskell: The Early Years* (Manchester: 1997), 353, and *Protestant Nonconformist Texts: vol. 2: The Eighteenth Century*, eds. Alan P. F. Sell, David J. Hall, and Ian Sellers (Farnham, Surrey: 2006), introduction to part 5, "Church, State and Society," 248.
106. Thus, burial in an Anglican churchyard is not evidence that the deceased was an Anglican. Lord Harwicke's Marriage Act of 1753 exempted Jews and Quakers from its requirement for a church marriage. See Hyamson, *Sephardim*, 26–7.
107. Earl Morse Wilbur, *A History of Unitarianism* (Cambridge, MA: 1952), vol. 2: 231–2. See also Michael Hunter, "'Aikenhead the Atheist': The Context and Consequences of Articulate Irreligion in the Late Seventeenth Century," in Hunter and Wootton, *Atheism*, 221–54.
108. Benjamin Hoadly, Bishop of Bangor, initiated the controversy with a sermon arguing for the separation of church and state. Andrew Starkie, *The Church of England and the Bangorian Controversy: 1716–1721* (Woodbridge: 2007); and John Spurr, "Latitudinarianism and the Restoration Church," *Historical Journal* (1988) 31:61–82.
109. James Bradley, *Religion, Revolution, and English Radicalism* (Cambridge: 1990), 53.
110. Elmer, "Medicine, Religion and the Puritan Revolution," 11–12, 22, 33.
111. Furdell, *Publishing*, 47.

112. George Clark, *College of Physicians*, 1:318. Among the last to have difficulties over a license to print was Richard Morton, a Nonconformist minister ejected under the Act of Uniformity and expelled from the College by James II. An admirer of Sydenham, Morton was the brother-in-law of Edward Tyson.

113. Elmer, "Medicine, Religion and the Puritan Revolution," 20–1. There is a vast literature on Culpeper. Among scholarly works see F. N. L. Poynter, "Nicholas Culpeper and His Books," *JHMAS* (1962) 17, no. 1:157–67; Charles Webster, *The Great Instauration* (New York: 1975), 267–72; and Patrick Curry, "Culpeper, Nicholas (1616–1654)," *ODNB* (Oxford: 2004), online at http://www.oxforddnb.com/view/article/6882.

114. For the tangled bibliography of works attributed to Culpeper, see Mary Rhinelander McCarl, "Publishing the Works of Nicholas Culpeper, Astrological Herbalist and Translator of Latin Medical Works in Seventeenth-Century London," *Canadian Bulletin of Medical History* (1996) 13:225–76. See also Curry, "'Culpeper,' John Symons, 'Cole, Abdiah (*fl.* 1602–1664),'" *ODNB* (Oxford: 2004), online at http://www.oxforddnb.com/view/article/5845, and Elizabeth Lane Furdell, "Cole, Peter (*d.* 1665)," *ODNB* (Oxford: 2004), online edn. January 2008 at http://www.oxforddnb.com/view/article/75231.

115. Lois G. Schwoerer, *The Ingenious Mr. Henry Care, Restoration Publicist* (Baltimore: 2001), 39.

116. Lois G. Schwoerer, "Care, Henry (1646/7–1688)," *ODNB* (Oxford: 2004), online at http://www.oxforddnb.com/view/article/4621. Care's translation of Sennert appeared as *Practical Physick* in 1676. A new edition in 1679 listed Nicholas Culpeper as a co-translator.

117. *Hieronymi Fracastorii, De Contagione et Contagiosis Morbis et Eorum Curatione, Libri III*, trans. Wilmer Cave Wright (New York: 1930), 73. Thomas Phaer had also struggled with the terminology for these fevers in 1544. See Charles Creighton, *A History of Epidemics in Britain* (Cambridge: 1891), 1:458, and, more generally, J. D. Rolleston, *The History of the Acute Exanthemata* (London: 1937).

118. David Wootton, *Bad Medicine: Doctors Doing Harm Since Hippocrates* (Oxford: 2007), 126–7.

119. Perhaps beginning with the work of Thomas Phaer (?1510–1560), author of *The Boke of Chyldren* (1544). According to Philip Schwyzer he "associated the use of the vernacular with social reform and the popularization of vital knowledge," viewing medical translation as a way "to declare that to the use of many, which ought not to be secrete for lucre of a fewe," Schwyzer, "Phaer, Thomas (1510?–1560)," *ODNB* (Oxford: 2004); online edn. October 2009 at http://www.oxforddnb.com/view/article/22085. See also Rich Bowers, introduction to *Thomas Phaer and the Boke of Children (1544)* (Tempe: 1999), 1–24, and John Ruhräh, "Thomas Phaer," *Annals of Medical History* (Winter 1919) 2:334–47. On the ambivalence of the Paracelsians and alchemists toward publicity and secrecy, see Elizabeth Eisenstein, *The Printing Press as an Agent of Change* (London: 1979) 2:562–3.

120. Charles Webster, "English Medical Reformers of the Puritan Revolution: A Background to the 'Society of Chymical Physitians,'" *Ambix* (1967) 14:16–41, 18 quoting Culpeper's *A Physicall Directory* (1649).

121. Flavio Gregori, "Blackmore, Sir Richard (1654–1729)," *ODNB* (Oxford: 2004), online edn. January 2009 at http://www.oxforddnb.com/view/article/2528, quoting Blackmore's *Treatise of Consumptions* (1724).

122. Andrew Cunningham, "Sydenham versus Newton: The Edinburgh Fever Dispute of the 1690s," in *Theories of Fever from Antiquity to the Enlightenment*, eds. W. F. Bynum and V. Nutton (*Medical History*, Supplement 1) (1981), 71–98, on 87. See also Furdell, *Publishing*, and Webster, *Instauration*, 266–73.

123. F. N. L. Poynter, "Culpeper," quoting the (old) *Dictionary of National Biography*.

124. P. M. Rattansi, "The Helmontian-Galenist Controversy in Restoration England," *Ambix* (1964) 12, no. 1:1–23, 9, quoting Noah Biggs, Chymiatrophilos Mataeotechnia Medicinae Praxes: *The Vanity of the Craft of Physic* (1651).

125. Clark, *College of Physicians*, I:361.

126. "Origins and History" on the website of the Worshipful Society of Apothecaries of London, http://www.apothecaries.org/index retrieved April 14, 2010. See also Wall, *Society of Apothecaries*; Edward Kremers and George Urdang, chapter 7: "The Development in England," in *History of Pharmacy* (Philadelphia: 1940), 88–107; T. D. Whittet, "Apothecaries and Their Lodgers: Their Part in the Development of the Sciences and of Medicine," *Journal of the Royal Society of Medicine* (Supplement 2, 1983), 76:1–32, online at http://www.ncbi.nlm.nih.gov/pmc/articles/PMC1440500/; A. C. Wootton, "Pharmacy in Great Britain," chapter 7 in *Chronicles of Pharmacy* (London: 1910) I:124–56; and Roy Porter, *Health for Sale* esp. chapter 2: "Medical Entrepreneurship in the Consumer Society," 21–59. A valuable account of the situation outside London can be found in articles by P. S. Brown, for example, "The Venders of Medicines Advertised in Eighteenth-Century Bath Newspapers," *Medical History* (1975) 19:352–69, online at http://www.ncbi.nlm.nih.gov/pmc/articles/PMC1081663/.

127. T. D. Whittet, *Clerks, Bedels and Chemical Operators of the Society of Apothecaries* (London: 1978), 52. Stephen Gaukroger, *Collapse of Mechanism*, 188, claimed that the Paracelsians emphasized disease species "so that medicines could be prescribed by an apothecary rather than a physician," and in *Emergence of a Scientific Culture*, 346, that "Paracelsians and Helmontians . . . thought of diseases as separately classifiable entities . . . and so the medicines . . . were for the illness, not the person, and could be dispensed by apothecaries." I do not believe the distinction between Galenic and Paracelsian disease theory can be mapped onto professional divisions in this way. Paracelsus himself bitterly denounced apothecaries, and apothecaries had as large a stake in Galenical remedies as did physicians. They retained their customers by stocking any and all items in high demand.

128. See Henry Benjamin Wheatley, *London, Past and Present* (London: 1891), 1, 126, online at https://archive.org/details/londonpastpresent01wheauoft, and Elizabeth L. Furdell, "The Medical Personnel at the Court of Queen Anne," *Historian* (August 23, 2007) 48, no. 3:412–29.

129. Clark, *A History of the Royal College of Physicians*, vol. 2 (Oxford: 1966), 500. Many practiced as surgeon-apothecaries, especially outside London. In 1774, the Society of Apothecaries limited membership to those working as apothecaries. In 1815, the Apothecaries Act established uniform qualifications for apothecaries throughout England and, in effect, created the "general practitioner." See Irvine Loudon, "The Apothecaries Act of 1815," Chapter 7 in *Medical Care and the General Practitioner, 1750–1850* (Oxford: 1987), 152–70.

130. Johanna Geyer-Kordesch and Fiona MacDonald, *Physicians and Surgeons in Glasgow: The History of the Royal College of Physicians and Surgeons in Glasgow,*

1599–1858 (London: 1999), 1–36, 81–5. The Glasgow surgeons were incorporated with the barbers from 1599 to 1722.

131. Frank Ellis, "The Background of the London Dispensary," *JHMAS* (1965) 20, no. 3:197–212, on 204–5. For Harvey, see below, chapter 3. The Edinburgh College of Physicians set up what was in effect a dispensary in 1682, the year after the College was founded by nominating two of its members to serve as "Physicians to the Poor." Ritchie in "Early Days," 80–3, refers to this as "The Dispensary." The Edinburgh College also opened a "Repository of medicines in 1708. It continued in various forms until the opening of the infirmary. See Guenter Risse, *New Medical Challenges during the Scottish Enlightenment* (Amsterdam: 2005), 29–30.

132. William Hartston, "Medical Dispensaries in Eighteenth Century London," [Abridged], *Proc. Royal Society for Medicine*, Section of the History of Medicine (August 1963) 56:753–8, 755.

133. Hartston, "Dispensaries," 755.

134. Richard I. Cook, *Sir Samuel Garth* (Boston: 1980), 58; Ellis, "Dispensary."

135. There are many accounts of the battle between the physicians and the apothecaries. A concise account of Rose can be found in Irvine Loudon, *Medical Care*, 22–3. Loudon believes that Rose did not effect any major change in the profession because it merely ratified the *de facto* situation. For a detailed account, see Clark, *History of the Royal College of Physicians*, vol. 2, esp. 476–79. See also Harold J. Cook, "The Rose Case Reconsidered: Physicians, Apothecaries and the Law," *JHMAS* (1990) 45, no. 4:527–55, and Roger Jones, "Apothecaries, Physicians and Surgeons," *British Journal of General Practice* (March 1, 2006) 56, no. 524:232–3.

136. R. Cook, *Garth*, 59. A short-lived "repository" that seems to have functioned as a dispensary opened in Westminster outside London in 1715/16. It seems to have suspended services a few months later, reopened in 1719, and leased a building in 1720, shortly before it became the Westminster Infirmary, serving both inpatients and outpatients. J. G. Humble and Peter Hansell, *Westminster Hospital 1716–1966* (London: 1966), 1–13.

137. Clark, *College of Physicians*, 2:482. Wesley's dispensary was in Upper Moorfields, just north of the city boundary. The London Hospital seems to have provided outpatient services to venereal patients soon after it opened in 1740. We still lack a complete picture of the full range of health services provided by metropolitan hospitals and Poor Law authorities, but see Kevin P. Siena, *Venereal Disease, Hospitals and the Urban Poor: London's "Foul Wards," 1600–1800* (Rochester, NY: 2004).

138. Cook, *Trials*, 121–4, 136–41, and same, "Pechey, John (bap. 1654, d. 1718)," *ODNB* (Oxford: 2004), online at http://www.oxforddnb.com/view/article /21737. Pechey should be distinguished from John Pechey or Pechi of Gloucestershire, an Extra-Licentiate of the College. Crell was the grandson of Johann Krell or Crellius (1590–1633) and the son of Krzystof or Christopher Crell Spinowski (Christophorus Crellius Spinovius). For the elder Crell, see Howard Hotson, "Arianism and Millenarianism: The Link Between Two Heresies from Servetus to Socinus," in *Continental Millenarians: Protestants, Catholics, Heretics*, eds. John Christian Laursen and Richard H. Popkin (vol. 4 of *Millenarianism and Messianism in Early Modern European Culture*) (Dordrecht: 2001), 25.

139. Harold J. Cook, "Browne, Richard (1647/8–1693/4?)," *ODNB* (Oxford: 2004), online at http://www.oxforddnb.com/view/article/3694.

140. Andrew Wear, *Knowledge and Practice in English Medicine, 1550–1680* (Cambridge: 2000), 464–5. See also Cook, *Trials*, 143. The Dispensary of the College of Physicians avoided this problem by treating the poor for free and dispensing medicines at cost.

141. H. Cook, *Trials*, 138–41. Cf. Brockliss and Jones, *Medical World of Early Modern France* (Oxford: 1997), 331: "Not only was [Theophraste Renaudot, a Montpellier physician] . . . suspect in the eyes of certain [Paris] Faculty doctors for his support for chemical remedies . . . but he seemed to have turned medicine into a trivial, commercial art by encouraging patients to consult by letter using simple, mass-produced images of the human body on which they could plot their symptoms." In the 1630s, Renaudot offered free medical consultations to the poor of Paris and held a weekly scientific club. See also Eisenstein, *Printing Press*, 1:246–7.

142. H. Cook, *Trials*, 141, 147.

143. For Mandeville's youth, see Rudolf Dekker, "'Private Vices, Public Virtues' Revisited: The Dutch Background of Bernard Mandeville," *History of European Ideas*, trans. Gerard T. Moran (1992) 14, no. 4:481–98. Mandeville received a PhD from Leyden in 1689 and returned home, where he and his father were implicated in an antigovernment riot in 1690. He obtained an MD from Leyden in 1691 and went to England; his father was banished from Rotterdam in 1693. Moran uses the word "Latitudinarian" for the anticonservative religious views of Mandeville and his father.

144. *In Authorem de Usu Interno Cantharidum Scribentem* (To the Author of *The Internal Use of* Cantharides). Mandeville probably wrote an anonymous satirical broadside against the mayor of Rotterdam, the "*Sanctimonious Atheist*" printed by Borstius in Dutch (Rotterdam: 1690); Dekker, "Mandeville," 485–7. See also M. M. Goldsmith, "Mandeville, Bernard (bap. 1670, d. 1733), *ODNB* (Oxford: 2004), online at http://www.oxforddnb.com/view/article/17926, and Cook, *Trials*, 199–200.

145. H. John McLachlan, *Socinianism in Seventeenth-Century England* (London: 1951), 289–90. See also chapter 1, n. 30. Bidle and Firmin were associated in the administration of Christ's Hospital, where Sloane became a physician in 1694. I have not found evidence of a connection between Sloane and the other two.

146. Stephen D. Snobelen, "Isaac Newton and Socinianism: Associations with a Greater Heresy," (June 2003), online at www.isaac-newton.org/socinian, and same, "Isaac Newton, Heretic: The Strategies of a Nicodemite," *British Journal of the History of Science* (1999) 32:381–419, online at http://www.isaac-newton .org/heretic.pdf. Although Newton was careful not to publicize his heretical views, late in his life he intimated them to Colin MacLaurin.

147. McLachlan, *Socinianism*, 290.

148. John Rurah, "John Pechey, 1655–1718," *American Journal of the Diseases of Children* (January 1930) 39:179–84. According to Matthews (see n. 151 below), who refers to him as "Peche," he was also a refugee, but this appears to be the result of confusion with the other physician of the same name, MD of Caen and Extra-Licentiate of the college.

149. Cook, "Pechey, John."

150. This included *Promptuarium Praxeos Medicae* (1693), a handbook that listed diseases and cures alphabetically, *The London Dispensatory* (1694), and *The Compleat Herbal of Physical Plants* (1694), a guide to using botanical simples and making compound medicines.
151. Leslie G. Matthews, "Philip Guide, Huguenot Refugee Doctor of Medicine (ca. 1640–1716)," *Die Veroffentlichungen der Internationalen Gesellschaft fur Geschicte der Pharmazie* (1975) *Neue Folge*, 42:109–16.
152. Guide, *Observations Anatomiques Faites sur Plusieurs Animaux au Sortir de la Machine Pneumatique* (Paris: 1674).
153. Guide, *Traité de la Nature du Mal Vénérien, Tiré de Plusieurs Expériences Physiques et des Mécaniques* (Paris: 1676). According to Matthews, Guide created his remedy by mixing corrosive sublimate with water, adding an equal amount of saliva from a salivating patient, oil of tartar, and chalk (lime) water, Matthews, "Guide," 115.
154. Guide, *An Essay concerning Nutrition in Animals Proving it Analogous to That of Plants* (London: 1699).
155. Guide, *A Kind Warning to a Multitude of Patients Daily Afflicted with Different Sorts of Fevers* (London: 1710).
156. Furdell, *Publishing*, 194.

CHAPTER 3

1. It also figured in the battle between the "Ancients," who valued traditional learning, and the "Moderns," who thought recent discoveries had surpassed classical knowledge. Richard Jones, *Ancients and Moderns: A Study of the Rise of the Scientific Movement in Seventeenth Century England* (Berkeley: 2nd ed. rpt. 1965), 111.
2. See for example Nedham, *Medela*, 131, "But you will say, if this be so, who then can be safe?"
3. J. G. A. Pocock quoted in Joseph Frank, *Cromwell's Press Agent: A Critical Biography of Marchamont Nedham, 1620–1678* (Lanham, MD: 1980), 94. I thank Glenn Burgess and Steve Pincus for replies to a question about Nedham. See also Allen G. Debus, *Chemistry and Medical Debate*, 87–102. There is a short and unsympathetic discussion of Nedham in Lester Snow King, *The Road to Medical Enlightenment* (London: 1970), 147–54. Jones, *Ancients and Moderns*, 206–10, has a more favorable view.
4. Frank, *Nedham*, 108–9.
5. Charles Harding Firth, "Needham, Marchamont," *Dictionary of National Biography* (Oxford: 1894).
6. P. M. Rattansi, "The Helmontian-Galenist Controversy in Restoration England," *Ambix* (1964) 12:1–23, on 17; and Henry Thomas, "The Society of Chymical Physitians, an Echo of the Great Plague of London, 1665," in *Science, Medicine and History*, ed. E. Ashworth Underwood, vol. 2 (London: 1953), 55–71. Peter Anstey, "John Locke and Helmontian Medicine," in *The Body as Object and Instrument of Knowledge: Embodied Empiricism in Early Modern Science*, eds. Charles T. Wolf and Ofer Gal (Dordrecht: 2010), 93–117 notes John Locke's previously unknown connections with the Society.
7. Quoted in King, *Medical Enlightenment*, 149.
8. Nedham, *Medela*, 42–3.
9. Nedham, *Medela*, 45. King objects that this was merely "going through the motions of empirical examinations," *Medical Enlightenment*, 149. John Graunt's

pioneering *Natural and Political Observations Made upon the Bills of Mortality* had appeared in 1662, just three years earlier.

10. King, *Medical Enlightenment*, 151. These arguments would be echoed by Gideon Harvey, *A New Discourse of the Small Pox and Malignant Fevers, with an Exact Discovery of the Scorvey* (London: 1685).

11. Nedham, *Medela*, 113–14 (italics omitted).

12. Nedham, *Medela*, 122.

13. Nedham, *Medela*, 170–3.

14. Nedham, *Medela*, 178.

15. Needham, *Medela*, 185.

16. Charles-Edward Amory Winslow, *The Conquest of Epidemic Disease* rpt. (Madison: 1980), 150; Brian Ford, *Single Lens: The Story of the Simple Microscope* (London: 1985), 18–20.

17. John Edward Fletcher, *A Study of the Life and Works of Athanasius Kircher*, ed. Elizabeth Fletcher (Leiden: 2011), 111. Borel's two works were sandwiched between the publication of Kircher's *Ars Magnus Lucis et Umbrae* (to which Borel refers) and the *Scrutinium*.

18. Pierre Borel (also Petro or Petrus Borello or Borell), "Observations," 31, 32, *De Vero Telescopi Inventore, cum Brevi Omnium Conspiciliorum Historia. Ubi de Eorum Confectione . . . Accessit etiam Centuria Observationum Microscospicarum* (The Hague: 1655 [1656]), 21. As noted elsewhere, the meaning of *"varioli"* was unstable during this period; it can roughly be translated as "poxes." Nedham refers to Borel in the *Medela*, 13, and Marten in chapter 2, 72, as "the learned Borellus." Borel, a friend of Descartes, and a physician to Louis XIV, also published a work defending the idea of a multiplicity of inhabited worlds and a chemical bibliography (1654) that Newton used to build his alchemical collection. Karin Figala and Ulrich Fetzold, "Alchemy in the Newtonian circle," *Renaissance and Revolution*, eds. J. V. Field and Frank A. J. L. James (Cambridge: 1993), 172–92, on 176.

19. *"In gonorrhea virulenta militis."* Gonorrhea was also an unstable term in this period. The "gonorrhea" experienced by the soldier may have been a case of pubic lice, which are visible even with the naked eye.

20. Probably Johann Heinrich Alsted (1588–1638), a Calvinist encyclopedist and admirer of Bruno.

21. It is not always clear what Borel means by "worms"—many of his observations were of animals easily seen with the naked eye. Borel, like Kircher a believer in spontaneous generation, evidently thought that the putrefying matter and bad air were generating the worms. See also Clifford Dobell, *Antony van Leeuwenhoek and his 'Little Animals'* rpt. (New York: 1960), 365. Dobell objects to a comment by Charles Singer, who suggested in a 1915 article on microscopy that Borel might have priority over Leeuwenhoek in visualizing bacteria, arguing that the passages cited by Singer were hypothetical, not actual observations.

22. (London: 1664), see 20–3. For Moffet, see chapter 5 below.

23. As noted above, an English translation by "N. D. B. M." of Sennert on agues and fevers had appeared in 1658 and one by Culpeper and Cole with the English title *The Sixth Book of Practical Physic* in 1662. Sennert wrote of contagious "seeds" or "atoms." The first full description of contagion from animals would come from Benjamin Marten in 1720.

24. Historian Marc Ratcliff has shown that Redi's experiments were not as decisive as some scholars have suggested, and it was only after Louis Joblot showed in

1711 that eggs did not appear in a vial that was boiled, immediately corked, and inspected after "a considerable time" that antispontaneism (briefly) prevailed. After decades of skepticism, the English microscopist John Turburville Needham revived the idea of spontaneous generation in the mid-eighteenth century. Ratcliff, *The Quest for the Invisible* (Farnham, Surrey: 2009), 40–1.

25. Critics are discussed by King, *Enlightenment*, 154–60 (John Twysden); Jones, *Ancients and Moderns*, 210–13 (Sprackling, Twysden); Harold Cook, *The Decline of the Old Medical Regime in Stuart London* (Ithaca and London: 1986), 147 (Sprackling); and Allen Debus, *Chemistry and Medical Debate*, 94–8 (Castle, Twysden).

26. Castle, *The Chymical Galenist: a Treatise wherein the Practice of the Ancients is Reconcil'd to the New Discoveries . . . in Which Are Some Reflections upon . . .* Medela Medicinae (London: 1667), dedication.

27. Castle, *Chymical Galenist*, 176, italics omitted.

28. See Castle, *Chymical Galenist* index, 197.

29. Castle, *Chymical Galenist*, 122–3.

30. Castle, *Chymical Galenist*, 111.

31. Cook, *Decline*, 147.

32. Sprackling, *Medela Ignorantiae*, 45–6.

33. Sprackling, *Medela Ignorantiae*, 62–3.

34. Sprackling, *Medela Ignorantiae*, 72.

35. John Twysden, *Medicina Veterum Vindicata* (The medicine of the ancients vindicated) (London: 1666), 68–9.

36. Twysden, *Medicina Veterum*, 89–94.

37. Twysden, *Medicina Veterum*, 81–2. He also discusses Nedham's (mis)use of similar views in Fernel's *De Abditis*.

38. Twysden, *Medicina Veterum*, 100. Twysden, who became a Fellow of the College in 1664, after incorporating his degree at Oxford, is best known as a mathematician.

39. "An Account of the First Rise, Progress, Symptoms and Cure of the Plague, Being the Substance of a Letter from Dr. Hodges to a Person of Quality," rpt. in *A Collection of Very Valuable and Scarce Pieces Relating to the Last Plague in the Year 1665 . . .* 2nd ed. (London: 1721), compiler not listed, 15–16. The original edition of Hodges's "Letter" appears to have been lost. The original publication date of Hodges's work is not specified and this edition only claims to contain its "substance."

40. Patrick Wallis, "Harvey, Gideon (1636/7–1702)," *ODNB* (Oxford: 2004); online ed. January 2008 at http://www.oxforddnb.com/view/article/12519.

41. Munk, *Roll*, I:11.

42. H. A. Colwell, "Gideon Harvey: Sidelights on Medical Life from the Restoration to the End of the XVII Century," *Annals of Medical History* (Fall 1921) 3, no. 3:205–37, online from the Hathi Trust. Cf. John Toland's comment on "explanatory repetitions of difficult words," in *Christianity Not Mysterious* cited in Colin Jager, *The Book of God: Secularization and Design in the Romantic Era* (Philadelphia: 2007), 44–5. See also Justin Champion, *Republican Learning: John Toland and the Crisis of Christian Culture, 1696–1722* (Manchester: 2003), 83.

43. Gideon Harvey, *Little Venus Unmask'd* (2nd ed., London: 1670), online from EEBO, 38–40. I have not been able to locate any copies of a first edition; Colwell mistakenly dates a first edition to 1671.

44. Gideon Harvey, *Great Venus Unmasked* (2nd edn. London: 1672), online from EEBO, 48. Colwell, who disliked Harvey, described this as "a dirty little book on a dirty subject."

45. Harvey, *Morbus Anglicus* (London: 1666), 3.

46. Harvey, *Morbus Anglicus*, 3–4. Richard Morton's classic work *"Phthisiologia"* (1689) maintained that consumption "like a contagious fever does infect those that lie with them," but Morton believed that the humors grew more acrid with age and predisposed the lungs to be susceptible to this disease. R. R. Trail, "Richard Morton," *Medical History* (1970) 14:166–74, 173.

47. Harvey, "Chapter XVII. Of a Uerminous Consumption," in *Morbus Anglicus*, 75–80. The worms were gut worms (*Lumbrici*); fillet worms (*Taenia*), and intestinal worms (*Ascarides*).

48. Gideon Harvey, *A Discourse of the Plague* (London: 1665), and same, *Discourse of the Smallpox, and Malignant Fevers*, and same, *A Treatise of the Small-pox and Measles* ... (London: 1696).

49. Antonio Clericuzio, "Maynwaring, Everard (b. 1627/8)," *ODNB* (Oxford: 2004), online at http://www.oxforddnb.com/view/article/18449, accessed June 20, 2011.

50. Like Nedham, he argued that scurvy was a protean disease responsible for many ailments. His *A Serious Debate, and General Concern* (2nd edn. London: 1689) was an advertisement for his Scorbutic Pills, as was *The Test and Tryal of Medicines* (London: 1690).

51. E[verard] Maynwaringe, *A Treatise of Consumptions* (London: 1668), 17, online from EEBO. This is the second edition of Tabidorum Narratio, *A Treatise of Consumptions* (London: 1667).

52. Maynwaringe, *Consumptions*, 22–3.

53. Maynwaringe, *Consumptions*, 41.

54. Maynwaringe, *Consumptions*, 84.

55. Maynwaringe, *Lues* (London: 1673), online from EEBO.

56. Fracastoro's earlier poem, *Syphilis sive Morbus Gallicus* (1530), was less definite about how the disease is communicated.

57. The evolution of ideas about syphilis and its transmission in early modern Europe (through Sennert) is traced by Jon Arrizabalaga, John Henderson, and Roger French, *The Great Pox: The French Disease in Renaissance Europe* (New Haven: 1997).

58. Maynwaringe, *Lues*, 72–3.

59. Maynwaringe, *Lues*, 74.

60. Maynwaringe, *Lues*, 75.

61. Maynwaringe, *Lues*, 77, emphasis in original.

62. E[verard] M[aynwaringe], *Inquiries into the General Catalogue of Diseases Shewing the Errors and Contradictions of that Establishment* ... (London: 1691). In Ignota Febris, *Fevers Mistaken in Doctrine and Practice* (London: 1691), he argued that fever and a quickened pulse were symptoms of diseases, not diseases in themselves.

63. Maynwaringe, *Inquiries*, 3.

64. Colwell, "Gideon Harvey: Sidelights on Medical Life," 213.

65. In a study of works on venereal disease published between 1700 and 1800, historian William Bynum noted that the participation of elite authors "gives the lie to the idea of a fringe monopoly in the treatment of what one anonymous pamphleteer actually called 'the secret disease.'" W. F. Bynum, "Treating the Wages

of Sin: Venereal Disease and Specialism in Eighteenth-Century Britain," in *Medical Fringe and Medical Orthodoxy*, eds. W. F. Bynum and Roy Porter (London: 1987), 11.

66. F. N. L. Poynter, "A Seventeenth-Century Medical Controversy: Robert Witty versus William Simpson," in *Science, Medicine and History*, ed. E. A. Underwood (London: 1953), 2:72–81. See also Peter Elmer, "Medicine, Religion and the Puritan Revolution," in *The Medical Revolution of the Seventeenth Century*, eds. Roger French and Andrew Wear (Cambridge: 1989), 30–1, and David Harley, "Honor and Property: The Structure of Professional Disputes in Eighteenth-Century English Medicine," in *The Medical Enlightenment of the Eighteenth Century*, eds. Andrew Cunningham and Roger Kenneth French (Cambridge: 1990), 143.

67. Poynter, "Controversy," 77.

68. Poynter, "Controversy," 77.

69. Christopher Hill, *The World Turned Upside Down: Radical Ideas during the English Revolution* (New York: 1972), 236–7; Allen G. Debus, "Paracelsian Doctrine in English Medicine," in *Chemistry in the Service of Medicine*, ed. F. N. L. Poynter (London: 1963), 5–26, 22. See also Andrew Wear, "The Failure of the Helmontian Revolution in Practical Medicine," chapter 9 in *Knowledge and Practice in English Medicine, 1550–1680* (Cambridge: 2000), 353–98. Debus later argued for the survival of chemical medicine into the eighteenth century in *Chemistry and Medical Debate: Van Helmont to Boerhaave* (Canton, MA: 2001).

70. See the essays in Margaret J. Osler, ed., *Rethinking the Scientific Revolution* (Cambridge: 2000), esp. Jan W. Wojcik, "Pursuing Knowledge: Robert Boyle and Isaac Newton," 183–200, and Lawrence M. Principe, "The Alchemies of Robert Boyle and Isaac Newton: Alternate Approaches and Divergent Deployments," 201–20. See also Jole Shackelford, *A Philosophical Path for Paracelsian Medicine: The Ideas, Intellectual Contest, and Influence of Petrus Severinus (1540/2–1602* (Copenhagen: 2004), 249.

71. Peter Elmer, "Medicine, Science and the Quakers: The 'Puritanism-Science' Debate Reconsidered," *Journal of the Friends Historical Society* (1981) 5:265–86. See also Jan V. Golinski, "A Noble Spectacle. Phosphorus and the Public Culture of Science in the Early Royal Society," *Isis* (1989) 80:11–39, and Paul Kléber Monod, *Solomon's Secret Arts* (New Haven: 2013), 133. Monod argues that "alchemy with its veiled heterodoxy and radical social aspirations" was at odds with (polite) culture and entered a period of intellectual torpor but was not discredited.

72. J. Andrew Mendelsohn, "Alchemy and Politics in England, 1649–1665," *Past and Present* (May 1992) 135:30–78.

73. Lawrence Principe, "A Revolution Nobody Noticed? Changes in Early Eighteenth-Century Chymistry" in same, ed., *New Narratives in Eighteenth-Century Chemistry* (Dordrecht: 2007), 1–22.

74. Anna Maria Eleanor Roos, *The Salt of the Earth: Natural Philosophy, Medicine, and Chymistry in England, 1650–1750* (Leiden: 2007), 206.

75. On Starkey, a Harvard-educated Presbyterian who moved to London in 1650, see William R. Newman, *Gehennical Fire: The Lives of George Starkey, an American Alchemist in the Scientific Revolution* (Cambridge, MA: 1994). For his influence on Boyle, see William R. Newman and Lawrence M. Principe, *Alchemy Tried in the Fire: Starkey, Boyle, and the Fate of Helmontian Chymistry* (Chicago: 2002).

76. Starkey's work *Introitus Apertus ad Occlusum Regis Palatium* (1667) had nine Latin editions and was translated into English, French, German, and Spanish, William R. Newman, "Starkey [formerly Stirk], George [pseud. Eirenaeus Philalethes] (1628–1665)," *ODNB* (Oxford: 2004), online edn. May 2008 at http://www.oxforddnb.com/view/article/26315. See also same, *Alchemy Tried in the Fire*. For the interest of the English Swedenborgians in alchemy, see August Nordenskiöld, "Spiritual Philosopher's Stone: An Address to the True Members of the New Jerusalem Church, Revealed by the Lord in the Writings of Emanuel Swedenborg" (London: 1789) rpt. on the Alchemy website at http://www.alchemywebsite.com/spiritual_stone.html, retrieved January 6, 2011, and David Ruderman, "Levison, Mordechai Gumpel Schnaber [pseud. George Levison] (1741–1797)," *ODNB* (Oxford: 2004), online at http://www.oxforddnb.com/view/article/72790. Levison was a student of John Hunter's.
77. Elizabeth A. Williams, *A Cultural History of Medical Vitalism in Enlightenment Montpellier* (Aldershot, UK: 2003). See also Stephen Gaukroger, *The Collapse of Mechanism and the Rise of Sensibility* (Oxford: 2010), 212.
78. *Officina Chymica Londinensis, sive Exacta Notitia Medicamentorum Spagyricorum* (London: 1685).
79. See William Poole, "A Fragment of the Library of Theodore Haak (1605–1690)," *eBLJ* (2007) article 6, 1–38, online at http://www.bl.uk/eblj/2007articles/article6.html.
80. Simon Werrett, *Fireworks: Pyrotechnic Arts and Sciences in European History*.
81. The Academy was carried on for some time by his son. See Figala and Petzold, "Alchemy," in Field and Frank, *Renaissance and Revolution*; Monod, *Solomon's Secret Arts*, 24, 27, 125–6, 163; and Scott Mandelbrote, "Yworth, William (*d.* 1715)" *ODNB* (Oxford: 2004), online at http://www.oxforddnb.com/view/article/40388.
82. Figala and Petzold, "Alchemy," in Field and Frank, *Renaissance and Revolution*, 177.
83. See also Elizabeth Furdell, chapter 7 "Medical Advertising: Publishing the Proprietary," in *Publishing and Medicine in Early Modern England* (Rochester: 2002), 135–54.
84. For possible sources of this expansive view of abiogenesis, see Nicholas S. Davidson, "'*Le plus beau et le plus meschant esprit que ie aye cogneu*': Science and Religion in the Writings of Giulio Cesare Vanini, 1585–1619," in *Heterodoxy in Early Modern Science and Religion*, eds. John Brooke and Ian MacLean (Oxford: 2005), 59–80, on 74–6.

CHAPTER 4

1. Robert Kargon, *Atomism in England from Hariot to Newton* (Oxford: 1966), 27–8, 92; Michael Hunter, *Science and Society in Restoration England* (Cambridge: 1981), 173–7; and Margaret J. Osler, "Baptizing Epicurean Atomism: Pierre Gassendi on the Immortality of the Soul," in *Religion, Science and Worldview*, eds. Margaret J. Osler and Paul Lawrence Farber (Cambridge: 1985), 163–84. Medieval atomism is discussed in *Late Medieval and Early Modern Corpuscular Matter Theories*, eds. Christoph Lüthy, John E. Murdoch, and William R. Newman (Leiden: 2001).
2. John Yolton, *Thinking Matter: Materialism in Eighteenth-Century Britain* (Minneapolis: 1984), 4.

3. Walter Charleton, *Physiologia Epicuro-Gassendo-Charletoniana* (London: 1654). For Charleton's career, see below.

4. Kargon, *Atomism*, 92. During the Interregnum, Charleton had also translated two of Van Helmont's works. See above, and Charles Webster, *The Great Instauration: Science, Medicine and Reform, 1626–1660* (New York: 1976), 272.

5. Richard G. Olson, *Science and Religion, 1450–1900* (Baltimore: 2004), 95–106.

6. Margaret Candee Jacob, "John Toland and the Newtonian Ideology," *Journal of the Warburg and Courtauld Institutes* (1969) 32:309–31, on 314. On the variable meaning of "Newtonian," see R. E. Schofield, "An Evolutionary Taxonomy of Eighteenth-Century Newtonianisms," *Studies in Eighteenth-Century Culture* (1998) 7:175–92.

7. Richard S. Westfall, "Newton and the Hermetic Tradition," in *Science, Medicine and Society in the Renaissance*, ed. Allen G. Debus (New York: 1972), 2:183–98; Antonio Clericuzio, "Gassendi, Charleton, and Boyle on Matter and Motion," in *Late Medieval and Early Modern Corpuscular Matter Theories*, eds. Christoph Lüthy, John E. Murdoch, and William R. Newman (Leiden: 2001), 467–82; and see for example Daniel Turner, *A Discourse concerning Fever. In Two Letters. The First, Dissuading from All Hypotheses and Theories, whether Physical or Mechanical* . . . 3rd ed. (London: 1739). Turner's own approach is best described as Sydenhamian.

8. Brent S. Sirota, "The Trinitarian Crisis in Church and State: Religious Controversy and the Making of the Postrevolutionary Church of England, 1687–1702," *Journal of British Studies* (January 2013) 52, no. 1:26–54, http://dx.doi .org/10.1017/jbr.2012.7 shows how religious claims were translated into disputes over authority and ultimately into party politics.

9. Minsoo Kang, "From the Man-Machine to the Automaton-Man: The Enlightenment Origins of the Mechanistic Imagery of Humanity," in *Vital Matters: Eighteenth-Century Views of Conception, Life, and Death*, eds. Helen Deutsch and Mary Terrall (Toronto: 2012), 148–73, esp. 158. Kang argues that popular disillusionment with government during the War of the Austrian Succession in the 1740s led to a revival of vitalism as an alternative to the mechanism that had by then become associated with an autocratic state.

10. Michael Hunter, *Science and Society in Restoration England* (Cambridge and New York: 1981), 186–7.

11. For Newton's own belief in a nonmechanical, alchemical "vegetable spirit," see Betty Jo Teeter Dobbs, *The Janus Faces of Genius: The Role of Alchemy in Newton's Thought* (Cambridge and New York: 1991).

12. Oliver, *A Practical Essay on Fevers*, 174, online from ECCO. Oliver's own theory hinged on a proper secretion of the "Animal Spirits" in the brain.

13. The roots of this theory lay in Greek Methodism. See Henry Cohen [Lord Cohen of Birkenhead], "The Evolution of the Concept of Disease," *Proceedings of the Royal Society of Medicine*, Section of Medicine (March 1955) 48, no. 3:155–60, on 158.

14. Theodore M. Brown, "The College of Physicians and the Acceptance of Iatromechanism in England, 1665–1695," *Bulletin of the History of Medicine* (1970) 44:12–30; Antonio Clericuzio, "The Internal Laboratory: The Chemical Reinterpretation of Medical Spirits in England (1650–1680)," in *Alchemy and Chemistry in the 16th and 17th Centuries*, eds. Piyo Rattansi and Antonio Clericuzio (Dordrecht: 1994), 51–84, on 73, n. 3.

15. Walter Pagel, *Joan Baptista Van Helmont* (Cambridge: 1982), 30. See also Johanna Geyer-Kordesch, "Passions and the Ghost in the Machine: Or What Not to Ask about Science in Seventeenth- and Eighteenth-century Germany," in *The Medical Revolution of the Seventeenth Century*, eds. Roger French and Andrew Wear (Cambridge: 1989), 145–63, on 150–1. For the hostility of the Paracelsian author Severinus to "geometrical" medicine, see Jole Shackelford, *A Philosophical Path for Paracelsian Medicine*, (Copenhagen: 2004), 148–9, 404.

16. Niebyl, "Science and Metaphor in the Medicine of Restoration England," *Bulletin of the History of Medicine* (1973) 47:356–74, on 360.

17. David E. Shuttleton, "'A Modest Examination,' John Arbuthnot and the Scottish Newtonians," *British Journal for Eighteenth-Century Studies* (1995) 18, no. 1:47–62, on 55. Shuttleton notes that in *The History of John Bull*, Arbuthnot equated this sort of Baconianism with Presbyterian enthusiasm. Michael Hunter, *Science and Society*, 17, describes it as the "indiscriminate collecting of information relevant to no particular hypothesis."

18. Anonymous, *The Censor Censured or the Antidote Examin'd* (London: 1704), 82.

19. H. F. Kearney, "Puritanism, Capitalism and the Scientific Revolution," in *The Intellectual Revolution of the Seventeenth Century*, ed. Charles Webster (London: 1974), 218–42, on 234–5; cf. Webster, *Instauration*, 305. In 1668, Boyle returned to a very different London.

20. Kearney, "Puritanism," 237.

21. Kearney, "Puritanism," 238.

22. Andrew Cunningham, "Thomas Sydenham: Epidemics, Experiment and the 'Good Old Cause,'" in French and Wear, *Medical Revolution*, 181–2. On chemistry and the "craftsman and scholar" thesis in the history of science, see Ursula Klein, "Apothecary-Chemists in Eighteenth-Century Germany," in *New Narratives in Eighteenth-Century Chemistry*, ed. Lawrence Principe (Dordrecht: 2007), 97–137, esp. 128–31.

23. Robert G. Frank, Jr., "The Physician as Virtuoso in Seventeenth-Century England," in *English Scientific Virtuosi in the Sixteenth and Seventeenth Centuries*, eds. Barbara Shapiro and Robert G. Frank, Jr., (Los Angeles: 1979), 57–119, on 92, and see the table in Webster, *Instauration*, 166–9.

24. Robert G. Frank, Jr., *Harvey and the Oxford Physiologists* (Berkeley: 1980), Table 3, 63–89.

25. Toby Barnard, "Petty, Sir William (1623–1687)," *ODNB* (Oxford: 2004) online edn., May 2007 at http://www.oxforddnb.com/view/article/22069.

26. J. R. Partington, *A History of Chemistry*, vol. 2 (London: 1961), 487. See also Barbara Beguin Kaplan, *Divulging of Useful Truths in Physick: The Medical Agenda of Robert Boyle* (Baltimore: 1993).

27. Allen G. Debus, "Thomas Sherley's 'Philosophical Essay' (1672): Helmontian Mechanism as the Basis of a New Philosophy," *Ambix* (July 1980) 27:124–35; W. R. Newman, "The Corpuscular Transmutational Theory of Eirenaeus Philalethes," in Rattansi and Clericuzio, *Alchemy and Chemistry*, 161–82. Philalethes was the pseudonym of the American George Starkey, a member of the Hartlib circle, who worked closely with Boyle and Newton.

28. Sydenham gained his All Souls fellowship as a reward for service in the Parliamentary Army. Webster, *Instauration*, 82–3.

29. *The Philosophical Works of the Honorable Robert Boyle* . . . ed. and abridged by Peter Shaw (London: 1725), 3:85.

30. On Willis, see Carl Zimmer, *Soul Made Flesh: The Discovery of the Brain and How It Changed the World* (New York: 2004). On the interest of members of this circle in a "vital agent," see Antonio Clericuzio, "The Internal Laboratory. The Chemical Reinterpretation of Medical Spirits in England (1650–1680)," in Rattansi and Clericuzio, *Alchemy and Chemistry*, 51–83.

31. See chapter 2 above for the creation of these honorary Fellows.

32. L. J. Rather, "Pathology at Mid-Century: A Reassessment of Thomas Willis and Thomas Sydenham," in *Medicine in Seventeenth-Century England*, ed. Allen Debus (Berkeley: 1974), 71–112, on 83. Fracastoro also likened diseases to ferments, but for him it was a metaphor whereas Willis meant it literally.

33. Thomas Willis, *De Fermentatione* (1659) trans. as "A Medical-Philosophical Discourse of Fermentation, or, Of the Intestine Motion of Particles in every Body," in *Dr. Willis's Practice of Physic . . .* trans. S[amuel] P[ordage] (London: 1681), 1. See also Alfred White Franklin, "Clinical Medicine," in Debus, *Medicine in Seventeenth-Century England*, 128–9. Franklin comments that Willis "could not escape from that compulsion to search for reasonable explanations, implanted by his Oxford education which still imposed the medical gospels of Aristotle and Galen."

34. Don G. Bates, "Thomas Willis and the Fevers Literature of the Seventeenth Century," *Medical History*, Supplement 1 (1981), 45–70, on 51, 66–7.

35. Rather, "Pathology," 83. Leeuwenhoek had seen yeast but did not realize it was alive. The living nature of yeast was demonstrated by Charles Caignard-Latour, Theodor Schwann, and F. Kuetzing independently in 1837. Maximillian Joseph Herzon, *A Text-Book on Disease-Producing Microorganisms* (Philadelphia: 1910), 20. See also Ralph Major, "Agostino Bassi and the Parasitic Theory of Disease," *Bulletin of the History of Medicine* (1944) 16:99–105.

36. We use specific yeasts for specific purposes, but that was not the case in the seventeenth century. Restoration bakers often used the barm skimmed from brewing to make bread. The Paris Faculty of Medicine recommended banning added leavens in bread-making as unhealthful, thinking that bread should be left to rise from sourdough or by itself (aided, unbeknownst to them, by wild yeasts in the air). Elizabeth David, *English Bread and Yeast Cookery* (London: 1978), 90–1, 98–100.

37. Stephen Gaukroger, *The Collapse of Mechanism and the Rise of Sensibility* (Oxford: 2010), 91 and see also 93. See also Anna Marie Roos, *The Salt of the Earth: Natural Philosophy, Medicine and Chymistry in England, 1650–1750* (Leiden: 2007), 116.

38. Rather, "Pathology," 85; Donald G. Bates, "Thomas Willis and the Epidemic Fever of 1661: A Commentary," *Bulletin of the History of Medicine* (1965) 39, no. 5:393–414.

39. See the long and informative biography by Harold J. Cook, "Sydenham, Thomas (*bap.* 1624, *d.* 1689)," in the *ODNB* (Oxford: 2004) online edn. May 2011 at http://www.oxforddnb.com/view/article/26864.

40. The entry for Sydenham in Wikipedia quotes word for word an unsigned article in the eleventh edition of the *Encyclopedia Britannica* (1911) retained from the 1887 edition: "Sydenham's nosological method is essentially the modern one." Even if "modern" refers to the dawn of the twentieth century, this is mistaken, but it has been repeated often. See also Knud Faber, *Nosography: The Evolution of Clinical Medicine in Modern Times* (New York: 1930); Arthur M Silverstein, *A History of Immunology* (Oxford: 1989), 89; William Bynum, "Nosology," in *Companion*

Encyclopedia of the History of Medicine, eds. William F. Bynum and Roy Porter (London: 2004), 1:335–6; and Lennart Nordenfelt, "Identification and Classification of Diseases: Fundamental Problems in Medical Ontology and Epistemology," *Studia Philosophica Estonica* (2013) 6, no. 2:6–21, published online May 2013.

41. Thomas Sydenham, preface to the third edition of *Medical Observations concerning the History and the Cure of Acute Diseases* in *The Works of Thomas Sydenham, M.D.*, ed. R. G. Latham, trans. from the Latin ed. of Dr. Greenhill (London: 1848, rpt. 1979, 2 vols. in 1), 1:16. For problems with this edition, see G. G. Meynell, "John Locke and the Preface to Thomas Sydenham's *Observationes Medicae*," *Medical History* (2006) 50, no. 1:93–110.

42. Sydenham, preface in *Works*, 1:13. These famous passages may have been written by Locke. Meynell, "John Locke and the Preface," describes the preface as a "mosaic." For the collaboration between the two authors, see below. Locke was an avid horticulturalist, so the metaphor would have been natural for him. The discussions of the preface below that refer to "Sydenham" should therefore be read as "Sydenham and/or Locke."

43. Sydenham, preface, in *Works*, 1:20.

44. For "Morbific causes," see the same preface, *Works*, 1:21.

45. Sydenham, "Epistle 1," in *Works*, 2:11.

46. Stephen Gaukroger is thus mistaken to view Sydenham as a Galenic physician, to claim that he "believed we had no access at all to underlying causes," and to credit him with the "invention of nosology . . . based on clinical observation," *Collapse of Mechanism*, 162. Andrew Wear, *Knowledge and Practice in English Medicine, 1550–1680*, 453, notes: "The refusal to assign causes except for the vague 'something in the air' . . . meant that Sydenham was unable to define diseases as distinct species." I argue below that his "species" was humoral and his "genus" temporal, but neither of these correspond to our taxonomic ideas.

47. Charles-Edward-Amory Winslow, *The Conquest of Epidemic Disease* (Madison: 1980), 173.

48. Andrew Cunningham, "Sydenham," 183.

49. Kenneth D. Keele, "The Sydenham-Boyle Theory of Morbific Particles," *Medical History* (1974) 18:240–8, on 246. Cf. the comment of Charles Goodall in a letter of 1680 that fevers "derive their original not from any vitious humors contained in the blood, but from a *seminium febrile* conveyed into it from the Monos-sphere"; Audrey Davis, "Some Implications of the Circulation Theory for Disease Theory and Treatment in the Seventeenth Century," *Journal of the History of Medicine* (1971) 26, no. 1:28–39, on 37.

50. See L. J. Rather, "Towards a Philosophical Study of the Idea of Disease," in *The Historical Development of Physiological Thought*, eds. C. M. Brooks and P. F. Cranefield (New York: 1959), 351–73, esp. 356–8.

51. Andrew Cunningham, "Sydenham," 174–7.

52. Lester S. King, *The Medical World of the Eighteenth Century* (Chicago: 1958), 195; and see A. J. Cain, "Logic and Memory in Linnaeus's System of Taxonomy," *Proc. Linnean Society of London*, Session 1956–7 (April 1958), 169 pts. 1 and 2:144–63, esp. 145–6, "the *differentia* is not extraneously attached to the genus; it is a particular mode in which the genus may exist."

53. Sydenham, preface, in *Works*, 1:19.

54. Sydenham, preface, in *Works*, 1:20.

55. Sydenham, *Medical Observations*, in *Works*, 1:38–9. Many historians have expressed confusion or frustration about Sydenham's theory of smallpox. Lise

Wilkinson, "The Development of the Virus Concept . . . 5. Smallpox and the Evolution of Ideas on Acute (Viral) Infections," *Medical History* (1979) 3:1–28, comments on 1–2 that "Sydenham's . . . still valid directions for the treatment of smallpox patients were not matched by clarity of thought on the subject of the aetiology of the disease."

56. See Latham, "Life of Sydenham," in *Works*, 1:liii.

57. See Andreas-Holger Maehle, *Drugs on Trial: Experimental Pharmacology and Therapeutic Innovation in the Eighteenth Century* (Amsterdam: 1999), 237. He discusses the later transformation of cinchona from a specific to a general medicine on 258 and 286–7.

58. Locke obtained an MB from Oxford, where he had studied with Willis and Boyle. He became Sydenham's pupil in 1667. Cunningham writes that Locke's influence led Sydenham to conclude by 1676 that "in medicine there can be no useful discussion at all about cause." Cunningham, "Sydenham," 184. Niebyl comments: "Locke, like the Helmontians, believed in an unbridgeable gap between the artificial and the natural," "Science and Metaphor," 371. See also James Herbert Dempster, "John Locke, Physician and Philosopher," *Annals of Medical History* (1932) n. s. 4:12–59. Kenneth Dewhurst, *John Locke (1632–1704): Physician and Philosopher: A Medical Biography with an Edition of the Medical Notes in His Journals* (London: 1963), 27, noted that Locke's early manuscripts include references to the work of Gideon Harvey.

59. J. R. Milton, "Locke, Medicine and the Mechanical Philosophy," *British Journal for the History of Philosophy* (2001) 9:221–43. For an opposing view, see Joseph Frank Payne, *Thomas Sydenham* (London: 1900), 244–9, but this appeared before many of Locke's manuscript works became available.

60. Kenneth Dewhurst, *Dr. Thomas Sydenham (1624–1689): His Life and Original Writings* (Berkeley: 1966), 62.

61. Sydenham, *Medical Observations* in *Works*, 1:101.

62. Patrick Romanell discusses "*Morbus*" in his *John Locke and Medicine* (New York: 1984) but spreads this quotation, interspersed with comments, over 58–64. I've used the quotation as it appears in Milton, "Locke," 237. For further discussion of the relationship between Locke and Sydenham, see D. L. Cowen, "Comments on Dr. Romanell's article on Locke and Sydenham," *Bulletin of the History of Medicine* (1959) 33, no. 2:173–80. For Locke's unacknowledged debt to Spinoza, see Wim Klever, "Locke's Disguised Spinozism," *Revista Conatus—Filosofia de Spinoza*, part 1 (July 2012) 6, no. 11:61–82; part 2 (December 2012) 6, no. 12:53–64.

63. Quoted in Milton, "Locke," 237.

64. Milton, "Locke," 238.

65. "Germs" as in the part of a seed that carries its identity, literally a seminal principle.

66. This entry and the attribution of authorship have generated much discussion. See Jonathan Walmsley, "*Morbus*—Locke's Early Essay on Disease," *Early Science and Medicine* (2000) 5, no. 4:367–93; Peter R. Anstey, "Robert Boyle and Locke's '*Morbus*' Entry: A Reply to J. C. Walmsley," *Early Science and Medicine* (2002) 7, no. 4:357–77; Jonathan Walmsley, "'*Morbus*,' Locke and Boyle: A Response to Peter Anstey," *Early Science and Medicine* (2002) 7, no. 4:378–97; and Peter Anstey, "John Locke and Helmontian Medicine," in *The Body as Object and Instrument of Knowledge: Embodied Empiricism in Early Modern Science*, eds. Charles T. Wolf and Ofer Gal (Dordrecht: 2010), 93–117.

67. Milton, "Locke," 240.

68. Milton, "Locke," 241.

69. Stephen Gaukroger, *Collapse of Mechanism*, esp. chapter 4, 150–86, discusses Locke's "wholesale rejection of matter theory" (163). Peter Anstey, *John Locke and Natural Philosophy* (Oxford: 2013) refers to Locke's position as "corpuscular pessimism." Gaukroger writes that Locke's position evolved; Anstey sees continuity. Both agree that Locke believed in corpuscularism but did not think it could be observed directly.

70. Robert Hooke to Robert Boyle, Gresham College, July 18, 1665, ed. by Michael Hunter et al. for the Electronic Enlightenment Project (Bodleian Libraries, Oxford: 2001). I thank the Wellcome Library for providing access to this resource.

71. Marjorie Nicholson, "The Microscope and English Imagination," *Smith College Studies in Modern Languages* (1935) 16, no. 4:1–92, on 16; and Alan Chapman, *England's Leonardo: Robert Hooke and the Seventeenth-Century Scientific Revolution* (Bristol and Philadelphia: 2004), 109. The letter came to light in the mid-eighteenth century in Thomas Birch's edition of Boyle's *Works* (London: 1744) and his *History of the Royal Society*, vol. 2 (London: 1756), which quotes the letter on 63.

72. Howard Gest, "The Discovery of Microorganisms by Robert Hooke and Antoni van Leeuwenhoek, Fellows of the Royal Society," *N&R* (2004) 58:187–201, doi: 10.1098/rsnr.2004.0055.

73. John Henry, "Charleton, Walter (1620–1707)" (Oxford: 2004), online edn. September 2010 at http://www.oxforddnb.com/veiw/article/5157.

74. Mayerne was the senior physician to Charles I; Charleton was his assistant and successor. His work included a treatise on scurvy in addition to the translations of Gassendi and Helmont noted above.

75. Thomas Birch, *History*, 2:69, March 21, 1664/5. "Dr. Bacon" was not Nathaniel Bacon; also Southwell (1598–1676), an English Jesuit in Rome, because Southwell did not practice medicine and was not in England at this time. He was probably the Matthew Bacon listed by Munk's *Roll*, vol. 1, as an MD of Padua in 1642 and an honorary Fellow of the College of Physicians in 1664.

76. George Ent, knighted in 1665, was a very close associate of William Harvey and saw Harvey's work on generation through the press. Originally a member of the Dutch church at Austin Friars, he was probably the son of a religious exile from Flanders and studied in Holland and Cambridge before obtaining an MD from Padua. He incorporated his degree at Oxford and became a Fellow (1639), Censor, and President of the College. He was also a founding member of the Royal Society.

77. Frank, *Harvey and the Oxford Physiologists*, 286.

78. Grandson of a moderate Puritan divine, Wilkins graduated from Magdalen Hall, Oxford, and became the Warden of Wadham College, Oxford. He married the younger sister of Oliver Cromwell in 1656. He conformed in 1662. A leader of the Latitudinarian party in the Church, he became Bishop of Chester in 1668. John Henry, "Wilkins, John (1614–1672," *ODNB* (Oxford: 2004), online edn. October 2009 at http://www.oxforddnb.com/view/article/29421. See also John S. Wilkins, *Species, A History of the Idea* (Berkeley: 2009), 62. Supporters of a universal language included Descartes, Kircher, Franciscus Mercurius Van Helmont, and Theodore Haak and fellow members of the Hartlib circle.

79. R. Lewis, "The Publication of John Wilkins's *Essay* (1668): Some Contextual Considerations," *N&R* (2002) 56:133–46, on 137, doi: 10.1098/rsnr.2002.0174.

80. His early death crippled efforts by his associates to continue the project.

81. Lewis, "Wilkins," 137.

82. John Wilkins, *An Essay towards a Real Character, and a Philosophical Language* (London: 1668), online from Google.

83. Wilkins, *Essay*, 219, rearranged, emphasis omitted.

84. Wilkins, *Essay*, 220.

85. Wilkins, *Essay*, 221–2, rearranged.

86. Wilkins, *Essay*, 221–4.

87. Wilkins, *Essay*, 220.

88. The single exception is the use of the synonym "French-pox" for *Lues Venerea*, where it is clear that the word is merely a common name for the disease.

89. Petty (MD Oxford, 1650) had also been the personal secretary of Thomas Hobbes, then living in exile in Paris.

90. It was there that Hooke wrote the letter quoted above about the contagion of the plague.

91. *The Collected Works of Sir William Petty in 8 volumes* (London: 1997), 87b "The Scale of Animals" (a fragment), 2:29. The introduction to this section, which contains Petty's works on philosophy, explains the provenance of this fragment, 19–20.

92. William Petty, "Observations upon the Dublin Bills of Mortality, and the State of that City," in *Several Essays in Political Arithmetick . . . To Which are Prefix'd Memoirs of the Author's Life*, 4th ed. (London: 1755), 35–50. This document is exhibited and discussed in the section on "Theories of Contagion" in the online exhibit, "Infectious Diseases at the Edward Worth Library," online at http://infectiousdiseases.edwardworthlibrary.ie/Theory-of-Contagion, retrieved April 1, 2012. See also Andrea A. Rusnock, *Vital Accounts: Quantifying Health and Population in Eighteenth-Century England and France* (Cambridge: 2002), 29–33.

93. William Petty, "A Postscript to the Stationer," in *Several Essays*, 49.

94. William Petty, "A Postscript to the Stationer," in *Several Essays*, 44–7.

95. Rusnock, *Vital Accounts*, 30. Graunt's list is "Apoplex, Cut of the Stone, Falling Sickness, Dead in the Streets, Gout, Head-ach, Jaundice, Lethargy, Leprosie, Lunatick, Overlaid and Starved, Palsie, Rupture, Stone and Strangury, Sciatica, and Suddenly."

96. Barbara Shapiro, *John Wilkins, 1614–1673: An Intellectual Biography* (Berkeley: 1969), 238.

97. The development of this "horizontal" approach and the consequent willingness of natural philosophers to disregard efforts to connect experimental phenomena to mechanist matter theory is the theme of Part 2 of Gaukroger's *Collapse of Mechanism*.

CHAPTER 5

1. "Past Presidents of the Royal Society," Royal Society website, http://royalsociety.org/Past-Presidents, retrieved April 20, 2014.

2. Marina Benjamin, "Medicine, Morality and the Politics of Berkeley's Tar-water," in *The Medical Enlightenment of the Eighteenth Century*," eds. Andrew Cunningham and Roger French (Cambridge: 1990), 165–93.

3. A. Rupert Hall, "Medicine and the Royal Society," in *Medicine in Seventeenth-Century England*, ed. Allen G. Debus (Berkeley: 1974), 421–2; and Richard

Sorrenson, "Towards a History of the Royal Society in the Eighteenth Century," *N&R* (1996) 50, no. 1:29–46, esp. 37–8. See also Lotte Mulligan and Glenn Mulligan, "Reconstructing Restoration Science: Styles of Leadership and Social Composition of the Early Royal Society," *Social Studies of Science* (August 1981) 11, no. 3:327–64. The Mulligans argue that the Secretaries to the Society affected the composition of its membership and attribute an increase of medical members to Sloane.

4. For the importance of "weak ties" in a social network, and the even greater value of networks that join strong (face-to-face) and weak ties, see Nicholas A. Christakis and James H. Fowler, *Connected: The Surprising Power of Our Social Networks and How They Shape Our Lives* (New York: 2009), *passim*, esp. 158–67. For its application to the Royal Society, see David S. Lux and Harold J. Cook, "Closed Circles or Open Networks? Communicating at a Distance during the Scientific Revolution," *History of Science* (1998) 36:179–211. See also Rhodri Hayward, "Emmanuel Mendes Da Costa (1717–1791)," in *Travels of Learning: A Geography of Science in Europe*, eds. Ana Simoes, Ana Carneiro, and Maria Paula Diogo (Dordrecht: 2003), 101–14; the work of Rupert Hall and Marie Boas Hall including M. B. Hall, "The Royal Society and Italy, 1667–1795," *N&R* (1982) 37, no. 1:63–81; and Andrea Rusnock, "Correspondence Networks and the Royal Society, 1700–1750," *British Journal for the History of Science* (1999) 32:155–69. I know of no comparable studies of early Enlightenment medicine, but see Marc Ratcliff, *The Quest for the Invisible* (Farnham, Surrey: 2009), for the pivotal role of communication networks in determining the selection of problems and the acceptance of findings in eighteenth-century microscopy.

5. Quoted by Lotte Mulligan and Glenn Mulligan, "Reconstructing Restoration Science," 333, n. 55. For Lister's theory that poisonous insects (or possibly venomous animals) originated attacks of smallpox and syphilis that ultimately became contagious, see Anna Marie Roos, *Web of Nature: Martin Lister (1639–1712): The First Arachnologist* (Leiden: 2011), 346–50.

6. Mulligan and Mulligan, "Reconstructing Restoration Science," 334. See also Michael Hunter, *Science and Society in Restoration England* (Cambridge: 1981), 44.

7. Sorrenson, "Royal Society." See also Charles Weld, *A History of the Royal Society*, vol. 1 (London: 1848), chapters 13–16.

8. Sloane caused great offence when he published his work on Jamaican plants with the imprimatur of the Royal Society above that of the College of Physicians. Harold Cook, "Stubbe and the Virtuosi-Physicians," in *The Medical Revolution of the Seventeenth Century*, eds. Roger French and Andrew Wear (Cambridge: 1989), 246–71, on 265. His election as President of the Society was opposed by its Tory members, including John Byrom. See also Sorrenson, "Royal Society," 34.

9. Jurin (1684–1750), a mathematician, graduated MD from Cambridge in 1716, after serving as a schoolmaster in Newcastle. The following year, he became a Candidate of the College of Physicians, in 1719, a Fellow, and in 1750, the President. He became an FRS in 1717 and served as the Secretary of the Society from 1721 to 1727.

10. A. Rupert Hall, "Medicine and the Royal Society," in *Medicine in Seventeenth-Century England*, ed. Alan Debus (Berkeley: 1974), 421–52.

11. Cecil Wall, H. Charles Cameron, and E. Ashworth Underwood, eds., *A History of the Worshipful Society of Apothecaries of London, 1617–1815*, vol. 1 (London: 1963), 113.

12. George Clark, *A History of the Royal College of Physicians of London* (Oxford: 1964), 1:456.

13. Andrea Rusnock, ed., *The Correspondence of James Jurin (1684–1750)* (Amsterdam: 1996), 215: Jurin to Deidier, London, December 30, 1723. On Deidier, see chapter 9 and Raymond Williamson, "The Plague of Marseilles and the Experiments of Professor Anton Deidier on its Transmission," *Medical History* (1958) 2, no. 4:237–52. Deidier had sent his papers to Dr. John Woodward, who brought them to the Society.

14. *Dissertatio Medica de Morbis Veneris cui Adjungitur Dissertatio Medico-Chirurgica de Tumoribus* (London: 1724). Deidier argued in this work that venereal disease was due to microscopic worms. See below. Palmer's assistant from December 24, 1724, to the fall of 1725 was Benjamin Franklin. J. A. Leo Lemay, *Benjamin Franklin: A Documentary History* (1997), online at http://www.english.udel.edu/lemay/franklin/printer.html, retrieved April 17, 2010.

15. Rusnock, *Correspondence of Jurin*, 293, Jurin to Deidier, London, April 3, 1725.

16. Sorrenson, "Royal Society," Table 3, 37. Sorrenson includes pharmacy and vital statistics (bills of mortality) in this count. See also Susan Lawrence, *Charitable Knowledge* (Cambridge: 1996), *passim*, esp. 233–4.

17. See also Roy Porter, "The Early Royal Society and the Spread of Medical Knowledge," in French and Wear, *Medical Revolution*, 272–93.

18. Dobell, *Antony Van Leeuwenhoek and His 'Little Animals,' Being an Account of the Father of Protozoology and Bacteriology* (New York: 1958), 25–6.

19. Howard Gest, "The Discovery of Microorganisms," *N&R* (2004) 58:187–201.

20. Dobell, *Leeuwenhoek*, 26–7.

21. Also referred to as Stefan or Steven Hamm or Ham. Catherine Wilson, *The Invisible World: Early Modern Philosophy and the Invention of the Microscope* (Princeton: 1977), 132.

22. As noted above, Pierre Borel had associated a "snail-like infection" with venereal disease in 1655, and in the 1680s, Blankaart referred to worms in the semen as a cause of venereal disease.

23. L. C. Palm, "Leeuwenhoek and Other Dutch Correspondents of the Royal Society," *N&R* (1989) 43:191–207.

24. James Jurin replaced Halley as editor of the *Transactions* in 1720 and the correspondence resumed.

25. Charles F. Mullett, "The Cattle Distemper in Mid-Eighteenth-Century England," *Agricultural History* (1946) 20:144–65, on 145. See also "An Abstract of a Letter from Dr. Wincler Chief Physitian of the Prince Palatine . . . to Dr. Fred. Slare . . . " *Philosophical Transactions* (1683) 13:93–5. His name also appears as Winkler and Winckler.

26. Wilkinson, "Veterinary Cross-currents in the History of Ideas on Infectious Disease," *Journal of the Royal Society of Medicine* (November 1980) 73, no. 11:818–27, on 821. Benjamin Marten, *A New Theory of Consumptions: More Especially of a Phthisis or Consumption of the Lungs* (London: 1720), 62. Cotton Mather also refers to it in "The Angel of Bethesda" and might have seen the letter in the *Philosophical Transactions*, but his reference seems to come from Marten. See Cotton Mather, "The Angel of Bethesda: An Essay upon the Common Maladies of Mankind," ed. Gordon W. Jones (Barre, MA: 1972), 44.

27. Marie Boas Hall, "Frederick Slare, F.R.S. (1648–1727)," *N&R* (1992) 46:23–41. See also Daniel L. Brunner, *Halle Pietists in England: Anthony William Boehm and the Society for Promoting Christian Knowledge* (Göttingen: 1993),

and William LeFanu, *Nehemiah Grew, M.D., F.R.S.: A Study and Bibliography of his Writings* (Detroit: 1990), 33–4.

28. William Poole, "A Fragment of the Library of Theodore Haak (1605–1690)," *British Library Journal* (2007), article 6, online at http://www.bl.uk /eblij/2007articles/article6.html.

29. Boehm wrote his influential *Pietas Hallensis* (1705) in Slare's home. Brunner, *Halle Pietists in England*, 83. For Pietist influences on Cotton Mather, see below, chapter 8. Slare hoped to rationalize the pharmacopeia. His *Experiments . . . upon . . . Bezoar-Stones* (London: 1715) successfully debunked a favorite remedy of early modern physicians. A. C. Wootton, *Chronicles of Pharmacy* (London: 1910), 1:15–16. He experimented on phosphorus; chemists thought this newly discovered glowing and inflammable substance distilled from urine might represent the "vital spirit." He was also interested in vital statistics; he published "An Extract of All Persons, That Did, in 1695, in Franckfort on the Maine, Consummate Matrimony, Receive Baptism, and Were Buried," in the *Philosophical Transactions* (1695–7) 19:559–60.

30. After incorporating his degree at Oxford in 1680, he became a Fellow of the Royal College of Physicians in 1685. Although he was one of 13 Dissident physicians who would complain about the College in 1702/3, he later became a leader in the College.

31. M. F. Ashley Montagu, *Edward Tyson, M.D., F.R.S., 1650–1708, and the Rise of Human and Comparative Anatomy in England . . .* (Philadelphia: 1943), 146. In 1690/1 Tyson, Slare, Waller, and Hooke were named to assist Halley in editing the revived *Philosophical Transactions*; Montagu, *Tyson*, 202. Tyson and Slare had a long and cordial relationship.

32. Frederick Slare, "An Experiment . . . in which a Surprizing Change of Colour . . . was Exhibited . . . by the Admission of Air Only," *Philosophical Transactions* (October 1693) 17:898–908, online from the Royal Society website at http://rstl .royalsocietypublishing.org/content/17/192-206/898.full.pdf+html (or search under "Slate"). See also Hall, "Slare," 34.

33. The entry for Slare in the (old) *Dictionary of National Biography* by Philip Joseph Hartog concluded that he "occupies a unique position between that of the earlier physicians, who often neglected clinical observations for fantastic interpretations of chemical and physiological experiments, and the almost exclusively clinical school of Sydenham." Its replacement, by Lawrence M. Principe, emphasizes Slare's chemical work and does not mention Sydenham, "Frederick Slare (1646–1727)," *ODNB* (Oxford: 2004), online at http://www.Oxforddnb.com/view /article/35715. The original entry can be found in the "Archive" to this entry. See chapter 7 for Wagstaffe's attack on Thomas Dover.

34. Hall, "Slare," 36–7.

35. Mark Ratcliff, *Quest for the Invisible*, 33–50. See also John Farley, *The Spontaneous Generation Controversy from Descartes to Oparin* (Baltimore: 1977).

36. Marcia Ramos-e-Silva, "Giovan Cosimo Bonomo (1663–1696): Discoverer of the Etiology of Scabies," *International Journal of Dermatology* (1998) 38, no. 8:625–30, online from the History of Dermatology Society at http://www.dermato.med .br/hds/bibliography/1998giovan-cosimo-bonomo.htm; and Reuben Friedman, "Bonomo and Cestoni: The Dispute Concerning Their Identity, the Authorship of the 'Letter to Redi,' and the Credit for the Discovery of the Acarian Origin of Scabies," in *Abraham Levinson Anniversary Volume: Studies in Pediatrics and Medical History*, ed. Solomon R. Kagan (New York: 1949), 279–96.

37. V. H. Houliston, "Sleepers Awake: Thomas Moffet's Challenge to the College of Physicians of London, 1584," *Medical History* (1989) 33:245–6.

38. Charles Singer, "Notes on the Early History of Microscopy," *Proceedings of the Royal Society of Medicine, Section of the History of Medicine* (1914), 7:247–79, on 269. Mayerne's preface described his own observations of itch mites, "being with a needle prickt forth from their trenches near the pool of water which they have made in the skin." Marjorie Nicholson writes that this may be the "earliest English reference to microscopical observation." "The Microscope and the English Imagination," *Smith College Studies in Modern Languages* (1935) 16, no. 4:1–92, on 8.

39. Singer, "Early History of Microscopy" reproduces the sketch on 274 but dismisses Hauptmann as "a credulous writer, whose ingenuity was accustomed to outrun his judgment."

40. This was reprinted with the illustration as "*Observationes Medicae Duae, Prima De Crinonibus . . . Secunda de Sironibus*," in Michael Ettmüller, *Operum Omnium Medico-Physicorum*, 2nd ed. (Lyons: 1690), 1:348–50. This work has separately paginated items; the "*Observationes*" are at the end of the volume. For the republication of Ettmüller's plate by Andry and then by M.A.C.D. see chapter 7. An English translation appeared as *Ettmullerus Abridg'd* (London: 1699), but it omitted a discussion of pathogenic worms on 213 of the Latin edition. For Ettmüller's priority over Redi, see C. E. Kellett, "Sir Thomas Browne and the Disease Called the Morgellons," *Annals of Medical History* (1935), n. s. 7:467–79.

41. Richard Mead, trans., "An Abstract of Part of a Letter from Dr. Bonomo to Signior Redi Containing Some Observations Concerning the Worms of Humane Bodies," *Philosophical Transactions* (1702/3) 23:1296–99. Italics omitted. This was reprinted in *The Medical Works of Richard Mead, M.D. . . . with . . . the Life of the Author* (London: 1762) with editions in Edinburgh (1763) and Dublin (1767). I have not consulted Mead's *Opera Omnia*, which first appeared in Latin (Gottingen: 1748) with editions in Paris (1751, 1757) and Naples (1752). The letter was originally published in Italian by Redi in *Osservazioni Intorno a Pellicelli del Corpo Umano Fatte dal Dottor Gio: Cosimo Bonomo . . .* (Florence: 1687). Ramos-e-Silva, "Bonomo," reproduces Bonomo's drawings of the mite, the first page of his original letter, and the cover of Redi's book.

42. Luigi Belloni, "Le '*Contagium Vivum*' avant Pasteur" (Paris: 1961), 11.

43. Daniele Ghesquier, "A Gallic Affair: The Case of the Missing Itch-Mite in French Medicine in the Early Nineteenth Century," *Medical History* (1999) 43, no. 1:26–54.

44. Ghesquier, "Missing Itch-Mite," 54.

45. Ray, *Wisdom of God*, 18. Page numbers are from the 7th London ed. (1717) on John McKeown's John Ray website, online at http://www.jri.org.uk/ray /wisdom/wisdom_of_god.pdf. Textual mistakes due to scanning errors have been silently corrected.

46. Ray, *Wisdom*, part 2, 195.

47. Ray, *Wisdom*, part 2, 298–9. See also Michael Hunter, *Science and the Shape of Orthodoxy* (Woodbridge, Suffolk: 1995), 15.

48. Ray, *Wisdom*, part 2, 299.

49. Ray, *Wisdom*, part 2, 309.

50. Ray, *Wisdom*, part 2, 319. Farley, *Spontaneous Generation*, writes that the evidence of parasitic worms in the bodies of mammals was the most successful argument for spontaneous generation.

51. Ray, *Wisdom*, part 2, 301, and 307–10. Like Sloane, Lister and Robinson had joined the College of Physicians in 1687 under the charter of James II. For Lister's theories of contagion, including the claim that smallpox originated in the bite of a venomous insect and then spread through a contagious venom, see Roos, *Martin Lister*, 345–52.

52. For the efforts of Louis Joblot in France and of Malpighi, Vallisneri, Redi, Cestoni, and others in Italy to refute spontaneism, see Marc J. Ratcliff, *Quest for the Invisible*, 42–8.

53. Anna Marie Roos, *The Salt of the Earth: Natural Philosophy, Medicine, and Chymistry in England, 1650–1750* (Leiden: 2007), 87–107, reviews Grew's work on botany and chemistry but does not mention his Nonconformity. See also Brian Garrett, "Vitalism and Teleology in the Natural Philosophy of Nehemiah Grew (1641–1712)," *British Journal of the History of Science* (2003) 36:63–81.

54. Ray says he was inspired by Andrea Cesalpino, the sixteenth-century author who first classified plants according to their fruits and seeds. John Ray, *Historia Plantarum Generalis* (1686) quoted in Ernst Mayr, *The Growth of Biological Thought: Diversity, Evolution and Inheritance* (Cambridge, MA: 2003), 256–7. Mayr cites a translation by Edmund Silk that appeared in Barbara G. Beddall, "Historical Notes on Avian Classification," *Systematic Biology* (1957) 6, no. 3:129–36. John S. Wilkins reproduced the original passage in "the first biological species concept" (May 10, 2009) on his blog Evolving Thoughts, online at http://scienceblogs.com/evolvingthoughts/2009/05/The_first_biological_species_c.php. For the evolution of Mayr's own idea of biological species, see K. de Queiroz, "Ernst Mayr and the Modern Concept of Species," *Proceedings of the National Academy of Sciences* (May 3, 2005), 102 supp. 1: 6600–7, online from PubMed at www.ncbi.nlm.nih.gov/pubmed/.

55. A. J. Cain, "John Ray on 'Accidents,'" *Archives of Natural History* (1996) 23, no. 3:343–68, on 362–3.

56. Anita Guerrini, "Tyson, Edward (1651–1708)," *ODNB* (Oxford: 2004), online ed. January 2008 at http://www.oxforddnb.com/view/article/27961; Charles Newman, "Edward Tyson," *British Medical Journal* (1975) 4:96–7. Newman sees Tyson as "a difficult man, with a chip on his shoulder, always against the government."

57. Harold Cook, *Trials of an Ordinary Doctor: Johannes Groenevelt in Seventeenth-Century London* (Baltimore: 1994), 15.

58. Harold Cook, *Trials*, 14–15.

59. Harold Cook, *Trials*, 15. Like Grew, Groenevelt had been a student of Sylvius in Leyden and was loosely associated with the followers of Sydenham in London; see chapter 7 for his association with John and Benjamin Marten. The College Censors were responsible for professional standards and discipline.

60. Montagu, *Tyson*, 82. Tyson had just become an FRS. A few months later, he received an MD (Cambridge), and in 1683 he became an FRCP.

61. Montagu, *Tyson*, 196–7. Hydatids are the larval form of tapeworms.

62. Montagu, *Tyson*, 212.

63. Nicholas Andry de Boisregard, *An Account of the Breeding of Worms in Human Bodies* (London: 1701), Maxims IX-XXVII, 194–9.

64. *Boerhaave's Aphorisms Concerning the Knowledge and Cure of Disease*, trans. J. DeLacoste (London: 1725), online from ECCO. For Boerhaave's publications, see Gerrit Arie Lindeboom, Bibliographia Boerhaaviana: *List of Publications Written or Provided by Hermann Boerhaave* (Leiden: 1959).

65. The disease that evoked Slare's comments in 1683 may also have been rinderpest.
66. See Wilkinson, "Rinderpest and Mainstream Infectious Disease Concepts in the Eighteenth Century," *Medical History* (1984) 28:129–50, on 130. A shorter version is in same, *Animals and Disease, An Introduction to the History of Comparative Medicine* (Cambridge, UK: 1992). See also C. A. Spinage, *Cattle Plague: A History* (New York: 2003), and Charles F. Mullett, "The Cattle Distemper in Mid-Eighteenth Century England," *Agricultural History* (1946) 20:144–65.
67. Spinage, *Cattle Plague*, 5.
68. Rinderpest attacks domestic ruminants, including cattle, sheep, goats, and deer as well as many wild ungulates. It was endemic on the Eurasian steppes and especially devastating in Africa because of the huge herds of ungulates there. However, to Europeans, it appeared to be a disease of domestic cattle. Spinage, *Cattle Plague*, 9–12.
69. See Spinage, *Cattle Plague*, part 4, "Cures and Remedies," for the huge array of remedies that were tried. Setons were threads passed through the skin to encourage drainage.
70. Wilkinson, "Rinderpest," 132. The following narrative closely parallels Wilkinson's work.
71. *De Contagiosa Epidemia* . . . (Lipsiae: 1713), *Della Contagiosa Epidemia, la quale Inoltrossi ne'Buoi nel Territoriao di Padoa . . . Dissertazione detta nella Università de Padoa il Giorno X. Novembre 1711, e nella Volgare Favela Translata da Don Bartolommeo Dadiali . . .* (Bologna: 1748).
72. Wilkinson, "Rinderpest," 132, quoting *De Contagiosa Epidemia, quae in Patavino Agro, & Tota Fere Veneta Ditione in Boves Irrepsit* (Padua: 1712).
73. Wilkinson, "Rinderpest," 132.
74. Wilkinson, "Rinderpest," 132.
75. Wilkinson, "Rinderpest," 132.
76. Wilkinson, "Rinderpest," 133; Giovanni Maria Lancisi, *Dissertatio Historica de Bovilla Peste* . . . (Rome: 1715).
77. William Brownrigg, *Considerations on the Means of Preventing the Communication of Pestilential Contagion* (London: 1771), 18–19. A multitiered military cordon was imposed in Bari, Naples, in 1690. Tom Koch, *Cartographies of Disease: Maps, Mapping and Medicine* (Redlands, CA: 2005), 19–23.
78. William Brownrigg, *Considerations*, 18. The Elector of Hanover at that time, Georg Ludwig, Duke of Brunswick-Luneberg, became King George I of England, Scotland, and Ireland in 1714. It is not clear that the Hamburg epidemic was plague. In *A Short Discourse concerning Pestilential Contagion* (London: 1720), 8–9, Richard Mead refers to an epidemic disease "called the *Dunkirk Fever*, as being brought from that Place, where it was indeed a Malignant Disease attended with a Diarrhoea, Vomiting, etc. and probably had its Original from the Pestilential Distemper, which some time before broke out at *Dantzick* and *Hamburgh*; But with us was much more mild."
79. Antonio Vallisneri, Sr., variously spelled Valisneri, Valisnieri, and Vallisnieri. For Vallisneri's work on parasitology, see F. N. Egerton, "A History of the Ecological Sciences, Part 30: Invertebrate Zoology and Parasitology during the 1700s," *Bulletin of the Ecological Society of America* (October 2008) 89, no. 4:413–15, and Jonathan Israel, *The Radical Enlightenment: Philosophy and the Making of Modernity 1650–1750* (Oxford: 2001), 143, 148, 150, and 678–9. Israel describes Vallisneri on 143. He became an FRS in 1703, followed by Lancisi in 1706. Marie Boas Hall, "The Royal Society and Italy 1667–1795," *N&R* (August 1982) 37, no. 1:63–81, Table 1, 73.

80. Jonathan Israel, *Radical Enlightenment*, 143. See also 148, 150.

81. *Considerationi ed Esperienze Intorno alla Generazione de'Vermi Ordinari del Corpo Umano* (Padua: 1710). Tyson had commented that tapeworms could bud and reproduce asexually. For Vallisneri's interest in Tyson's observations of worms, see Montagu, *Tyson*, 155.

82. The letters were compiled, edited, and published by Tomaso Piantanida with permission from Cogrossi but not Vallisneri (Milan: 1714). They were reprinted in facsimile with English translation by Dorothy M. Schullian and an introduction by Luigi Belloni for the Sixth International Congress of Microbiology (Rome: 1953).

83. Wilkinson, "Rinderpest," 135.

84. Cogrossi, *New Theory* (1953), 15–23; cf. Wilkinson, "Rinderpest," 136. The mention of *two* insects suggests sexual reproduction.

85. Cogrossi, *New Theory*, 30.

86. Israel, *Radical Enlightenment*, 679, notes that Vallisneri was an animist who believed that "all organic bodies, that are born, that grow . . . have a soul, as we do, and it would not be such a mortal sin . . . to believe that all plants have one."

87. Luigi Belloni, introduction to Cogrossi, *New Theory*, xx–xxi.

88. Belloni, introduction to Cogrossi, *New Theory*, xxii.

89. Belloni, introduction to Cogrossi, *New Theory*, xxiv.

90. Belloni, introduction to Cogrossi, *New Theory*, xxviii.

91. Belloni, introduction to Cogrossi, *New Theory*, xxviii.

92. Belloni, introduction to Cogrossi, *New Theory*, xxix.

93. Belloni, introduction to Cogrossi, *New Theory*, xxxi. Richard Mead owned a run of this journal. See *Bibliotheca Meadiana . . . Apud Samuel M. Baker* (London: 1754 and 1755), 83 #959. He also bought Lancisi, *De Bovilla peste* (1713) and the works of Vallisneri. Belloni notes that later reprints of Vallisneri's letter to Cogrossi included this supplement, accompanied by a Latin poem summarizing the idea of *contagium vivum* by Orazio Borgondio (Horatio Borgondi) SJ of the *Collegio Romano*. Borgondio (1679–1741), a friend of Vallisneri, was a teacher and mentor of the Croatian physicist and mathematician Roger Joseph Boscovich, SJ and FRS, whose atomic theories inspired the monist matter theory of Joseph Priestley. For more on this circle, see Yasmin Annabel Haskell, *Loyola's Bees: Ideology and Industry in Jesuit Latin Didactic Poetry* (Oxford: 2003).

94. Wilkinson, "Rinderpest," 137, quoting Lancisi, *Dissertatio Historica de Bovilla Peste* (Rome: 1715). See also Belloni, introduction to Cogrossi, *New Theory*, xxxi and xlv, n. 24.

95. Lise Wilkinson, "Rinderpest," 138.

96. Cogrossi, *New Theory*, 14.

97. Viruses are much smaller than bacteria. They can only function by invading host cells and so bridge the divide between living and nonliving entities. They can be visualized with electron microscopes.

98. Wilkinson, "Rinderpest," 136. In 1661, a year of bitter conflict between Galenists and innovators in Bologna, the physicist Giovanni Alfonso Borelli and the physician Marcello Malpighi exchanged letters on an epidemic at Pisa. Borelli wrote to Malpighi in Bologna: "I concur with you in thinking that the poisonous seeds imbibed with the air in the guise of a ferment can segregate this quantity of bile, which then causes the fever." Malpighi's reply has been lost. Malpighi served as Vallisneri's mentor and worked closely with Lancisi in Rome in 1692. Howard B. Adelmann, *Marcello Malpighi and the Evolution of Embryology* (Ithaca, NY: 1966) I:199, 484, and 634.

99. Wilkinson, "Rinderpest," 137, referring to Lancisi, *De Noxiis Paludum Effluviis* (on the noxious effluvia from marshes) (Rome: 1717). Lancisi is credited with introducing the word "malaria" for intermittent fevers. See Eliana Ferroni, Tom Jefferson, and Gabriel Gachelin, "Angelo Celli and Research on the Prevention of Malaria in Italy a Century Ago," *Journal of the Royal Society of Medicine* (2012) 105:35–40, doi: 10.1258/jrsm.2011.11k049.

100. Wilkinson, "Rinderpest," 139.

101. Hall, "Royal Society and Italy."

102. These are listed by Belloni, introduction to Cogrossi, *New Theory*, xxxii–xxxiii, and xlvi–xlix, nn. 27–34.

103. The dates of Bates's birth and death are unknown. The Royal Society lists him as a Freemason and member of Lodge no. 20 at the Dolphin, Tower Street, Seven Dials, in 1723. Royal Society, Sackler Archive, online at https://royalsociety.org/library/collections/biographical-records/.

104. Thomas Bates, "A Brief Account of the Contagious Disease Which Raged among the Milch Cowes near London, in the Year 1714. And of the Methods that Were Taken for Suppressing It," *Philosophical Transactions* (1718) 30:872–82. It is not clear that Bates knew of the Italian discussion because Lancisi's Latin treatise appeared in 1715. Spinage notes in *Cattle Plague*, 725, that a copy of Bates's ms. letter dated September 24, 1714, is in the Hertfordshire County Archives, D/EP/F125.

105. Bates, quoted in Wilkinson, "Rinderpest," 140.

106. Mullett, "Cattle Distemper," 146.

107. Spinage, *Cattle Plague*, 133.

108. Mullett, "Cattle Distemper," 145.

109. Wilkinson, "Rinderpest," 130.

110. Wilkinson, "Rinderpest," 131.

111. Marc Ratcliff, *Quest for the Invisible*, 125–6.

112. William Eamon, "With the Rules of Life and an Enema: Leonardo Fioravanti's Medical Primitivism," in *Renaissance and Revolution*, eds. J. V. Field and Frank A. J. L. James (Cambridge: 1993), 29–44.

113. Edward G. Ruestow, *The Microscope in the Dutch Republic* (Cambridge: 1996), 56–60. Ruestow cites a number of Catholic authors who shared this view. The division appears to be between authors such as Kircher and Buffon, who retained Aristotle's belief in spontaneous generation, and those such as Leeuwenhoek, Redi, Ray, and Réaumur who did not.

114. This revived the earlier objection that Thomas Erastus had made to Paracelsianism that replacing a physiological model of disease as an imbalance with an ontological model of disease as an entity amounted to Manicheanism. See for example, Peter Niebyl, "Sennert, Van Helmont, and Medical Ontology," *Bulletin of the History of Medicine* (1971) 45:115–37, Harold Cook, "Good Advice and Little Medicine: The Professional Authority of Early Modern English Physicians," *Journal of British Studies* (1994) 33:1–31, on 12, and Jole Shackelford, *A Philosophical Path for Paracelsian Medicine* (Copenhagen: 2004), 216–33.

115. The distinction between living and nonliving entities, relatively firm in the late nineteenth and early twentieth centuries, is again breaking down following the discovery of viruses, viroids, virusoids, prions, and plasmids.

116. William Bynum, "The Great Chain of Being after Forty Years: An Appraisal," *History of Science* (1975) 13:1–28, online from the NASA Astrophysics Data System (ADSABS) at http://adsabs.harvard.edu/full/1975hissc..13....1b. The word "biology" was first used by Thomas Beddoes in 1799.

117. Ray originally thought that a material "specific essence" was being transmitted within species. Stephen Gaukroger, *The Collapse of Mechanism and the Rise of Sensibility: Science and the Shaping of Modernity, 1660–1760* (Oxford: 2010), 193, following Phillip Sloan, argues that Ray abandoned the concept of "species essentialism" by the 1690s under Locke's influence. This, however, is a claim about Ray's epistemology—he agreed with Locke that we cannot know about essences—not about his theory of generation. Ray conceded that all human-made classifications are artificial because they cannot capture essences, but presumably a perfect "natural" classification would avoid this problem. See Sloan, "John Locke, John Ray, and the Problem of the Natural System," *Journal of the History of Biology* (1972) 5, no. 1:1–53, on 46, esp. n. 121.

118. Gaukroger, *Collapse*, 206–16, summarizes the development of this approach in chemistry through the creation of Geoffroy's table of chemical relationships, and discusses Geoffroy's relationship with Locke, the Royal Society, and Hans Sloane, who nominated him for Fellowship in 1703. For the Huguenot and Paracelsian tradition at Montpellier and at the *Jardin du Roi*, where Geoffroy worked, see 211–12.

119. Despite its constant use, taxonomists do not agree on the definition of species. See John Wilkins, "Species, Kinds and Evolution," *Reports of the National Center for Science Education* (2006) 26, no. 4:36–45, online at http://ncse.com/rncse/26/4/species-kinds-evolution, retrieved May 2, 2012, and Richard A. Richards, *The Species Problem: A Philosophical Analysis* (Cambridge: 2010).

120. See also Walter Pagel, *Joan Baptista Van Helmont*, 35–7.

CHAPTER 6

1. James Hamilton, the eldest son of the Reverend Hans Hamilton, minister of Dunlop in Ayrshire, came to Ireland as a secret agent for James VI of Scotland (later James I of England). Eric St. John Brooks, *Sir Hans Sloane: The Great Collector and His Circle* (London: 1954), 18.

2. Brooks, *Sloane*, chapter 2, "The Sloanes and the Baillies," 23–33.

3. Brooks, *Sloane*, 30.

4. William Sloane married Jane Hamilton of Killyleagh, the heiress of the Hamilton estates at Clanbrassil. Stanley A. Hawkins, "Sir Hans Sloane (1669–1735) [*sic*]: His Life and Legacy," *Ulster Medical Journal* (2010) 79, no. 1:25–9, online at http://www.ums.ac.uk/umjo79/079(1)025.pdf.

5. Hawkins, "Sloane," 26.

6. Brooks, *Sloane*, 35, emphasizes Sloane's "humble circumstances," but in his entry for "Sloane, Sir Hans" in the *Catalog of the Scientific Community*, Richard Westfall commented, "I do not see how to deny that he grew up in relatively prosperous circumstances," http://galileo.rice.edu/Catalog/NewFiles/sloane.html.

7. Boyle brought Stahl from Strasburg to Oxford in about 1659. His students at Oxford included John Locke, Christopher Wren, and many other distinguished British scientists. Guy Meynell, "Locke, Boyle and Peter Stahl," *N&R* (1995) 49, no. 2:185–92.

8. Brooks, *Sloane*, 40–1. The relationship between Sloane and Staphorst was long-lived. In 1692, Sloane and Tancred Robinson commissioned Staphorst to translate Leonhart Rauwolf's *Itinerary* from German to English, with notes by Ray. Brooks, *Sloane*, 123; Stanley R. Friesen, "Sir Hans Sloane: The Spirit of Inquiry," *The Pharos of Alpha Omega Alpha* (1985) 48, no. 2:19–24, on 20.

9. Brooks, *Sloane*, 41–2.

10. Tournefort (c. 1655–1708) was an early supporter of iatromechanism. See Brockliss and Jones, *The Medical World of Early Modern France* (Oxford: 1997), 419. Chirac had recently been appointed professor at Montpellier; he later became physician to the King of France.

11. Tournefort himself had studied botany with Magnol at Montpellier before obtaining his MD at the University of Orange. Pierre Magnol (1638–1715) came from a long line of apothecaries. He was denied a professorship of medicine at Montpellier in 1667 because he was a Huguenot. He converted in 1685, becoming professor in 1694 and director of the botanical garden in 1696. He corresponded with such English botanists as Ray, Sherard, and Petiver. Richard Westfall, *Catalogue of the Scientific Community*, online from the Galileo project at http://galileo.rice.edu/Catalog/NewFiles/magnol.html. Robinson, meanwhile, was in Italy, where he met Malpighi and discussed Ray and Martin Lister with him. On April 18, 1684, he wrote to Ray about the fire that had destroyed Malpighi's house and many of his notes that February. Howard Adelmann, *Marcello Malpighi and the Evolution of Embryology* (Ithaca, NY: 1966), 1:470.

12. This was located in the principality of Orange near Avignon. It belonged to the House of Orange-Nassau and so was Protestant.

13. *De plantis libri XVI.* See also Scott Atran, *Cognitive Foundations of Natural History: Towards an Anthropology of Science* (Cambridge: 1996).

14. The full title was *Prodromus Historia Generalis Plantarum in quo Familia Plantarum per Tabulas Disponitur* (an introduction to the general history of plants in which the families of plants are arranged in tables) (Montpellier: 1689). See also A. C. Crombie and Michael Hoskin, "The Scientific Movement and the Diffusion of Scientific Ideas, 1688–1751," in *The New Cambridge Modern History*, ed. J. S. Bromley (Cambridge: 1971), 1:37–71, on 57–60.

15. Tournefort grouped his species into "sections" and "classes" in addition to genera, arguing that plants should be grouped from their parts of fructification only, whereas Ray favored keeping plants together when they appeared to form a "natural" group in other ways. See also Edwin Lankester, ed., *Memorials of John Ray: Consisting of His Life by Dr. Derham, Biographical and Critical Notices* (London: 1846), 101–2, and A. J. Cain, "John Ray on 'Accidents,'" *Archives of Natural History* (1996) 23, no. 3:343–69.

16. Friesen, "Sloane," 20.

17. Friesen, "Sloane," 20.

18. Brooks, *Sloane*, 45

19. See chapter 8 below.

20. William Eamon, "'With the Rules of Life and an Enema': Leonardo Fioravanti's Medical Primitivism," in *Renaissance and Revolution*, eds. J. V. Field and Frank A. J. L. James (Cambridge, UK: 1993), 29.

21. Brooks, *Sloane*, 66.

22. Brooks, *Sloane*, 78.

23. The first volume of Sloane's *Voyage to the Islands Madera, Barbados, Nieves, St. Christophers and Jamaica; with the Natural History of the Herbs and Trees, Four-footed Beasts, Fishes, Birds, Insects, Reptiles, etc. of the Last of Those Islands* (London: 1707). The second volume finally appeared in 1725. See Friesen, "Sloane," 21.

24. Chocolate is a New World plant cultivated by Europeans in the tropics after 1519. It was sold in London by 1657 and discussed in a book, *The Indian Nectar* (1662) by the radical Protestant Henry Stubbe, who praised its medicinal qualities. Sloane recommended cocoa for children and consumptives, in keeping

with his interest in botanical exploration, pharmaceutical reform, and the use of simpler botanical remedies. James Delbourgo, "Sir Hans Sloane's Milk Chocolate and the Whole History of the Cacao," *Social Text* (Spring 2011) 29, no. 1:71–101, discusses Sloane's botanical and ethnographic interests and his complicity in the imperialist slave and sugar trades, but questions Sloane's personal involvement in the creation of cocoa as a product.

25. Delbourgo, "Milk Chocolate," 75.
26. Arthur MacGregor, "The Life, Character and Career of Sir Hans Sloane," in *Sir Hans Sloane: Collector, Scientist, Antiquary, Founding Father of the British Museum*, ed. Arthur MacGregor (London: 1994), 11–44, on 14.
27. MacGregor, *Sloane*, 15.
28. MacGregor, *Sloane*, 15.
29. Wendy D. Churchill, "Bodily Differences?: Gender, Race and Class in Hans Sloane's Jamaican Medical Practice, 1687–1688," *JHMAS* (2005) 60, no. 4:391–445, on 402. King's parodies, published anonymously, included *The Present State of Physick in the Island of Cajamai to Members of the R.S.* (London: 1710?) and *A Voyage to the Island of Cajamai in America . . . by Jasper Van Slonenbergh* in *Useful Transactions in Philosophy, to be Continued Monthly as They Sell*, Part 3 for May–September 1709 (London: 1708–9). Both are online from ECCO. MacGregor, "Sloane," 39, n. 83, states that the *"Present State"* was part of the *Useful Transactions*, but it seems to be a separate pamphlet despite the "numb. 1" on the title page.
30. In addition to the articles by Delbourgo and Churchill, see for example Kay Dian Kriz, "Curiosities, Commodities and Transplanted Bodies in Hans Sloane's 'Natural History of Jamaica'" *William and Mary Quarterly*, 3rd ser. (January 2000) 57, no. 1:35–78.
31. Churchill, "Differences," 391–445. Churchill claims that "it is clear" that Sloane began his trip as a humoralist and changed his mind once there. I feel it is more likely that he brought his untraditional approach with him.
32. Churchill, "Differences," 399.
33. Many diseases have similar symptoms: without a nosology, there was no reliable way to group symptoms into illnesses. The traditional method was simply to list symptoms from head to toe or, for acute diseases, by the duration of the fevers. See also the discussion of Willis in Chapter 4. Churchill, "Differences," 410–11 contrasts Sloane's unorganized, roughly chronological arrangement with the case histories recounted by his contemporary William Cockburn who grouped cases with the same symptoms: "The entire set of [Cockburn's] cases was itself presented as an adjunct to the more detailed discussion of the manifestation and progression of the illnesses . . . Sloane's . . . focus upon unanalyzed case histories was atypical of his own age."
34. J. Thackray, "Mineral and Fossil Collections," in MacGregor, 123–35, on 123.
35. Brooks, *Sloane*, 134–5. Killyleagh was a Scottish Presbyterian settlement in Ulster, but the case for Sloane's childhood Presbyterianism is not airtight; his maternal grandfather was Archbishop Laud's chaplain.
36. He also held an honorary MD from Oxford but may have received it without subscribing. See Margaret DeLacy, "Timeline for Admission to Fellowship of the College of Physicians 1660–1800," online at http://www.contagionism.org/timeline_for_admission_to_fellow.htm.
37. As President of the Royal Society, Sloane would have had to swear the Oaths of Supremacy and Allegiance and take the Sacrament according to the usage of the Church of England. This requirement evidently prevented Boyle from serving as

President, but John Pringle, a Scottish Presbyterian, became President in 1772. See Michael Hunter, "The Conscience of Robert Boyle," in Field and James, *Renaissance and Revolution*, 147–59, on 155. This rule did not include subscription to the Thirty-Nine Articles.

38. After the publication of Samuel Clarke's *Scripture Doctrine of the Trinity* (London: 1712) and the Salters Hall meeting (1719).

39. Brooks, *Sloane*, 135.

40. Marjorie Caygill, "Sloane's Will and the Establishment of the British Museum," in MacGregor, *Sloane*, 45–6. She adds, "Sloane had a certain sympathy for dissenters," although Anglicans were also among the 48 named Trustees.

41. J. P. Ferguson, *Dr. Samuel Clarke, An Eighteenth-Century Heretic* (Kineton, Warwick: 1976), 215.

42. For the relationship between Sloane and the heterodox language reformer Francis Lodwick, see Felicity Henderson and William Poole, "The Library Lists of Francis Lodwick (1619–1694): An Introduction to Sloane MSS 855 and 859 and a Searchable Transcript," *eBLJ* (2009) Article 1: 1–7, online at http://www.bl.uk/eblj/2009articles/article1.html. On Halley, see Simon Schaffer, "Halley's Atheism," *N&R* (1977) 32:17–40.

43. Edgar Samuel, "Dr. Meyer Schomberg's Attack on the Jews of London, 1746," *Transactions of the Jewish Historical Society of England* (1964) 20:83–111, 87.

44. Kenneth Collins, *Go and Learn: The International Story of Jews and Medicine in Scotland* (Aberdeen: 1988), 37. Sloane had sponsored both Mortimer and Stuart for the Royal Society.

45. Geoffrey Cantor, "Emanuel Mendes da Costa: Constructing a Career in Science," in *From Strangers to Citizens: The Integration of Immigrant Communities in Britain, Ireland and Colonial America, 1550–1750*, eds. Randolph Vigne and Charles Littleton (Brighton: 2001), 230–8, on 230–1. The others mentioned by Da Costa were Martin Folkes, the Duke of Richmond, the Duchess of Portland, and the Bishop of Exeter. He often attended Society meetings as Peter Collinson's guest.

46. Royal Society, Sackler archive, online at http://royalsociety.org/library/collections/biographical-records/, retrieved April 4, 2014.

47. Brooks, *Sloane*, 219.

48. Harold J. Cook, *The Decline of the Old Medical Regime in Stuart London* (Ithaca, NY: 1986), 275, Table 6, lists 70 Fellows in 1695, 65 Fellows in 1704, and 63 Fellows in 1705. Candidates, honorary Fellows, and Licentiates bring the total at the turn of the century to between 126 and 129.

49. Lotte Mulligan and Glenn Mulligan, "Reconstructing Restoration Science: Styles of Leadership and Social Composition of the Early Royal Society," *Social Studies of Science* (August 1981) 11, no. 3:327–64, on 338.

50. Fuller, *Exanthematologia; or, an Attempt to Give a Rational Account of Eruptive Fevers, Especially of Measles and Smallpox* (London: 1730), dedication, xvi. Fuller and Sloane were distantly related.

51. Joseph M. Levine, *Dr. Woodward's Shield: History, Science, and Satire in Augustan England* (Ithaca: 1991), 112 and 321, n. 67, citing a letter from Newton to Woodward that refers to Sloane as "a tricking fellow; nay a villain and rascal for his deceitful and ill usage of you." Cf. Richard Westfall, *Never at Rest: A Biography of Isaac Newton* (Cambridge: 1990), 636.

52. William King, preface to *The Transactioneer*, intro. by Roger D. Lund (Los Angeles: 1988).

53. See Charles Webster, *The Great Instauration: Science, Medicine and Reform, 1626–1660* (New York: 1976), 150–1. Ray (FRS, 1667) was "so entirely scrupulous about Oaths" that he had refused to take the Solemn League and Covenant, but when the Act of Uniformity required a new oath swearing that the Covenant was not binding on those who *had* taken it, he refused it. This disqualified Ray for a Fellowship. William Derham, "Select Remains and Life of Ray," *Memorials of John Ray*, ed. Edwin Lankester (London: 1846), 8–52, on 15–16. Westfall commented, "I think there is no way to avoid calling him a Puritan . . . However, he was not a belligerent Puritan. He fully accepted the Restoration, and Raven claims that he remained in lay communion with the Anglican Church." "Ray [Wray], John" in "Catalogue of the Scientific Community," online from the Galileo Project website at http://galileo.rice.edu/Catalog/NewFiles/ray.html. Ray appears in *The Transactioneer*, 11–12.

54. John Wilkins, *Species: A History of the Idea* (Berkeley: 2009); A. J. Cain, "John Ray on 'Accidents.'" See also Ernst Mayr, *The Growth of Biological Thought: Diversity, Evolution and Inheritance* (Cambridge, MA: 2003), 162–3.

55. Stephen Gaukroger, *The Collapse of Mechanism and the Rise of Sensibility* (Oxford: 2010), 191–6 (Ray) and 206–17 (Geoffroy); Sloane is on 212.

56. Brooks, *Sloane*, 156.

57. Dover acted as physician to the Bristol Corporation of the Poor directed by the Whig John Cary. See Jonathan Barry, "John Cary and the Legacy of Puritan Reform," in *The Practice of Reform in Health, Medicine, and Science, 1500–2000*, eds. Margaret Pelling and Scott Mandelbrote (Aldershot, UK: 2005), 185–206, on 197–8.

58. P. N. Phear, "Thomas Dover, 1662–1742: Physician, Privateering Captain, and Inventor of Dover's Powder," *Journal of the History of Medicine* (1954) 9, no. 2:139–54, on 152.

59. George Clark, *A History of the Royal College of Physicians of London* (Oxford: 1964), 1:531; Phear, "Dover," 153; Kenneth Dewhurst, *The Quicksilver Doctor: The Life and Times of Thomas Dover, Physician and Adventurer* (Bristol: 1957), 136–7.

60. Phear, "Dover," 154; see also Ritchie Calder, *Medicine and Man: The Story of the Art and Science of Healing* (New York: 1958), 22–3.

61. M. Grieve, *A Modern Herbal* (New York: 1971), 2:433; Hermann Prinz, *Dental Materia Medica and Therapeutics*, 5th ed. (St. Louis: 1917), 20–1.

62. Sloane also published an article "Of the Use of Ipecacuanha, for Loosenesses, Translated from a French Paper: with Some Notes," in the *Philosophical Transactions* (January 1, 1698) 20:69–79, but it does not mention Dover. Sloane thought (correctly) that it was a useful medicine, but not effective in every case.

63. Dewhurst, *Quicksilver Doctor*, 152. See also M. R. Lee, "Ipecacuanha: The South American Vomiting Root," *Journal of the Royal College of Physicians of Edinburgh* (2008) 38:355–60, online November 2008 at http://www.rcpe.ac.uk/journal /issue/journal_38_4/lee.pdf. Lee writes that "for 200 years Ipecac was the only effective amoebicide," but its use has now largely been discontinued.

64. Phear, "Dover," 152. See also Philip K. Wilson, *Surgery, Skin and Syphilis: Daniel Turner's London (1667–1741)* (Amsterdam: 1999), 169–70. Turner, a pious Anglican surgeon with an honorary MD from Yale, attacked Dover in *The Ancient Physician's Legacy Impartially Survey'd* (1733) and set off a pamphlet war.

65. Lee, "Ipecacuanha," 356–7. Lee claims that Dover's book "marked a watershed between the old Galenical certainties and the rational enquiry of the Enlightenment."
66. Phear, "Dover," 152.
67. Phear, "Dover," 152.
68. On Coward, see Ann Thomson, *Bodies of Thought: Science, Religion, and the Soul in the Early Enlightenment* (Oxford: 2008), 104–17, 121–34.
69. MacGregor, "Life of Sloane," Appendix 1, "Sloane's Museum at Bloomsbury, as described by Zacharias Conrad von Uffenbach, 1710," in MacGregor, *Sloane*, 30.
70. Furdell, *The Royal Doctors, 1485–1714* (Rochester, NY: 2001), 217.
71. John Cherry, "Medieval and Later Antiquities: Sir Hans Sloane and the Collecting of History," in MacGregor, *Sloane*, n. 75, 220. AIM25 description of Woolhouse's mss, online at http://www.aim25.ac.uk/cgi-bin/vcdf/ detail?coll_id=9719&inst_id=9&nv1=search&nv2=basic; British Library, manuscripts catalogue, "Woolhouse"; and the Royal Society, Sackler archive, "Biographies of Fellows," online at https://royalsociety.org/library/collections/biographical -records/. The Sloane mss. describe him as "physician" to the Hospital for the Blind. Peter John Wallis and Ruth Wallis, *Eighteenth-Century Medics: Subscriptions, Licenses, Apprenticeships* (Newcastle: 1988) lists him as MD, but other sources refer to him as "Mr." Woolhouse.
72. Anita McConnell, "Woolhouse, John Thomas (1666–1734)," *ODNB* (Oxford: 2004), online at http://www.oxforddnb.com/view/article/29954.
73. His offer of seeds was declined. See Matt Goldish, "Newtonian, *Converso* and Deist: The Lives of Jacob (Henrique) De Castro Sarmento," *Science in Context* (1997) 10:651–75, on 666.
74. See chapter 8.
75. De Castro Sarmento and his first wife remarried as Jews at the Bevis Marks Synagogue on March 10, 1721. He became a Licentiate in 1725 and an FRS in 1729, the same year that he became embroiled in a bitter dispute with Dr. Meier Low Schomberg. A close friend of David Nieto, MD, the Chief Rabbi of the Sephardic synagogue, de Castro Sarmento married for a second time to an Anglican in 1756 and renounced Judaism. See chapter 7 for his patent medicine and chapter 8 for his pamphlet on smallpox.
76. See chapter 2, n. 54.
77. MacGregor, "Life of Sloane," 43, n. 183, citing A. E. Gunther, *The Life of the Rev. Thomas Birch, D.D., F.R.S. 1705–1766* (privately printed: 1984), 7. In the 1893 entry for John Martyn the botanist in the [old] D.N.B., George Simonds Boulger wrote that "on the death of Dr. Rutty [Martyn] was unsuccessful in his candidature for the secretaryship of the Royal Society, the successful competitor, Dr. Cromwell Mortimer, being a relative of Sloane." I have not been able to determine this relationship, if any.
78. Julius Groner and Paul F. S. Cornelius, *John Ellis: Merchant, Microscopist, Naturalist, and King's Agent* (Pacific Grove, CA: 1996), 281. John Nichols printed this proposal in *Literary Anecdotes of the Eighteenth Century* (London: 1812) 5:424; now on Google Books. The Huguenot obstetrician Hugh Chamberlen had proposed a similar plan in 1689. See Albert Rosenberg, "The London Dispensary for the Sick Poor," *JHMAS* (1959) 14, no. 1:41–56, on 43.
79. His article of 1743 noting that "the Blood of Animals, when once let out of the Body, [cannot] be kept either fluid or warm by any the most violent agitation" appeared in the *Philosophical Transactions* (1753) 43:473–80.

80. Theodore Brown comments that "similar thoughts had recently been aired in Scotland, but not since Glisson had a prestigious English student of the 'animal economy' publicly doubted the wisdom of mechanism (and never in the pages of the *Philosophical Transactions*)." He sees Mortimer's article as "something of a turning point to physiology and the Royal Society." *The Mechanical Philosophy and the 'Animal Oeconomy'* (PhD Dissertation: Princeton University, 1969; rpt. New York: 1981), 356–7.

81. Dorothea Waley Singer, "Sir John Pringle and his Circle—part I. Life," *Annals of Science* (1949–50) 6:164.

82. A. E. Gunther, *The Founders of Science at the British Museum, 1753–1900* (Suffolk: 1980), 4.

83. A detailed account of Linnaeus's visit can be found in Andrea Wulf, *The Brother Gardeners* (New York: 2009), chapter 3, 48–65. See also B[enjamin] D[aydon] Jackson, "The Visit of Carl Linnaeus to England in 1736," *Svenska Linnésällska-pets Årsskrift* (Yearbook of the Swedish Linnaeus Society) (1926) 9:1–11.

84. R. Hingston Fox, *Dr. John Fothergill and His Friends: Chapters in Eighteenth-Century Life* (London: 1919), 163–4, 185. Bartram was expelled for Unitarianism by his Quaker congregation but continued to practice as a Friend. For Mead's Quaker relatives, see chapter 9.

85. In the 1894 D.N.B. entry for Miller, G. S. Boulger quoted the comment of Richard Pulteney (1730–1801) that Miller was "the only person I ever knew who remembered . . . Mr. Ray. I shall not easily forget the pleasure that enlightened his countenance . . . in speaking of that revered man."

86. Sloane had owned the freehold to the land on which the garden was sited and provided it to the apothecaries. Brooks, *Sloane*, 117.

87. Collinson and Watson were among the trustees named in Sloane's will to superintend the transfer of his collection to the nation. This became the core of the British Museum.

88. Margaret Riley, "The Club at the Temple Coffee House Revisited," *Archives of Natural History* (April 2006) 33:90–100; Raymond Phineas Stearns, "James Petiver, Promoter of Natural Science, c. 1663–1718," *Proceedings of the American Antiquarian Society* (1952) n. s. 62, pt. 2:243–364, on 255. Another botanical group met on Saturday evenings at the Rainbow Coffee House from about 1721 until 1726. Its members included Dillenius, John Martyn, and Philip Miller. D. E. Allen, "John Martyn's Botanical Society: A Biographical Analysis of the Membership," *Proceedings of the Botanical Society of the British Isles* (1967) 6:305–24; Clifford M. Foust, *Rhubarb: The Wondrous Drug* (Princeton: 1992), 97.

89. Dale, a Nonconformist, was the friend, neighbor, biographer, and executor of John Ray. Riley Temple Coffee House argues that Stearns erred in listing Nehemiah Grew as a known member.

90. Riley, "Temple Coffee House," 96.

91. Among the founding Trustees of the Museum was the Presbyterian Lord Willoughby of Parham, who was also President of the heterodox Warrington Academy. See David Philip Miller, "'The 'Hardwicke Circle': The Whig Supremacy and Its Demise in the 18th-Century Royal Society," *N&R* (1998) 52, no. 1:73–91, on 78.

92. Wulf, *The Brother Gardeners*, 8.

93. Penelope Hunting, "Isaac Rand and the Apothecaries' Physic Garden at Chelsea," *Garden History* (Spring 2002) 30, no. 1:1–23. In 1689, many of the garden's plants had been stolen. The apothecaries Samuel Doody, James Petiver, and Isaac Rand also helped preserve the garden before Sloane purchased the Manor of Chelsea in 1713.

94. Wulf, *Brother Gardeners*.
95. One exception was Lord Bute, a fanatical botanist, for whom both Stewartia and Butea were named, the latter after his death. Linnaeus sometimes named noxious weeds for his enemies.
96. For a Dutch parallel, see Harold J. Cook, "The New Philosophy in the Low Countries," in *The Scientific Revolution in National Context*, eds. Roy Porter and Mikulas Teich (Cambridge: 1992), 115–49, on 140–1. I thank Dr. Cook for a prepublication copy of this chapter.
97. Brooks, *Sloane*, 117. Sloane had purchased the Manor of Chelsea in 1713. He also charged the Society five pounds a year for rent. For other strategies used by scientists such as anatomists and astronomers to increase their status, see Mario Biagioli, "Scientific Revolution, Social *Bricolage*, and Etiquette," in Porter and Teich, *Scientific Revolution*, 11–54.
98. For a rare appreciation of Sloane's contribution, see Rafael Chabrán and Simon Varey, "Hernandez in the Netherlands and England," in *Searching for the Secrets of Nature: The Life and Works of Dr. Francisco Hernández*, eds. Simon Varey, Rafael Chabrán, and Dora B. Weiner (Stanford: 2000), 138–50, esp. 143–6 and the introduction, 19–22. Delbourgo views the lavish illustrations as "a symbolic expression of rising British power . . . as well as of Sloane's personal capacity," "Milk Chocolate," 79.
99. Kenneth Dewhurst, "The Correspondence between John Locke and Sir Hans Sloane," *Irish Journal of Medical Science* (May 1960) 5, no. 35:201–12, suggests that they were introduced by William Courten, a naturalist friend of Robinson's.
100. Brooks, *Sloane*, 147; Peter Anstey, *John Locke and Natural Philosophy* (Oxford: 2011), 193–4.
101. Dewhurst, "Correspondence," 208.
102. Theriac, or Venice Treacle, a complex Galenic remedy, contained opium with many other ingredients.
103. Dewhurst, "Correspondence," 209–10. Sloane also treated the continued fevers with Virginia snakeroot, blisters, and plentiful fluids.
104. James II excluded Grew and Sampson as honorary Fellows of the College of Physicians in the same charter that made Sloane and Robinson Fellows. Sampson was a minister in Framlingham and often preached for Obadiah Grew. Ejected in 1662, he obtained an MD and practiced among his co-religionists while collecting material for a history of Puritanism. Richard Westfall says he "played an important role in promoting Grew's scientific career." "Grew, Nehemiah," *Catalogue of the Scientific Community*, online at http://galileo.rice.edu/Catalog /NewFiles/grew.html. See Munk, *Roll*, 1:384–5. See also Gary De Krey, *A Fractured Society: The Politics of London in the First Age of Party, 1688–1715* (Oxford: 1985), 91.
105. Jonathan Israel, *The Radical Enlightenment: Philosophy and the Making of Modernity 1650–1750* (Oxford: 2001), 608–9. This is a big presumption: lectures at Leyden were in Latin; many foreigners used French.
106. William LeFanu, *Nehemiah Grew, M.D., F.R.S.: A Study and Bibliography of His Writings* (Winchester: 1990), 2. See also chapters 2 and 5 above.
107. Edward G. Ruestow, *The Microscope in the Dutch Republic* (Cambridge: 1996), 154, 219–20.
108. For his *Cosmologia Sacra*, which argued for a "vital principle," see chapter 5. Grew's *Musaeum Regalis Societatis* (1680), a catalogue of the holdings of

the Royal Society, was published by subscription as was his *Anatomy of Plants* (1682). Richard Westfall thus was incorrect to describe Edward Lhuyd's *Lithophylacii Britannici Ichnographia* (1699) as one of the earliest instances of scientific publishing by subscription.

109. Theophrastus (371–287) was the first to suggest this. See LeFanu, *Grew*, 26.

110. Richard Pulteney, *Historical and Biographical Sketches of the Progress of Botany in England* (London: 1790) 2:118. Brooks, *Sloane*, 46.

111. Anna Maria Eleanor Roos, *Web of Nature: Martin Lister (1639–1712): The First Arachnologist* (Leiden: 2011). See also J. D. Woodley, "Lister, Martin (bap. 1639, d. 1712)," *ODNB* (Oxford: Sept 2004), online edn. October 2008 at http://www.oxforddnb.com/view/article/16763. Like Roos, John Gascoigne, *Cambridge in the Age of Enlightenment* (Cambridge: 1989), 61–2, considers Lister a Royalist as he received a Fellowship by royal mandate in 1660. Lister obtained only a BA from Cambridge before going to Montpellier and practiced medicine for many years in York without a medical degree. His MD was awarded by Oxford in 1683/4 in recognition of his donations to the Ashmolean Museum. See the *Catalog of the Scientific Community*, "Lister, Martin," online at http://galileo.rice.edu/Catalog/NewFiles/lister.html.

112. Roos, *Web of Nature*, 340–53. Lister believed that patients with syphilis had symptoms that resembled the crests of iguanas and patients with rabies resembled dogs in the way they swallowed and stuck out their tongues.

113. Miles Barker, "'The Motion of their Juice': Science, History and Learning about Plants and Water," paper presented at the Australasian Science Education Research Association, Darwin, Australia, July 9–12, 1998, online at http://www.fed.qut.edu.au/projects/asera/Papers/barker.htm. Lister's theory was presented in "A Further Account Concerning the Existence of Veins in All Kinds of Plants," *Philosophical Transactions* 90:5132–7 and drew on John Ray and Francis Willughby, "Concerning the Motion of the Sap in Trees," *Philosophical Transactions* (1669) 48:963–5. It was later overturned by Stephen Hales.

114. See Levine, *Dr. Woodward's Shield*, 37–40.

115. Harris thought that Robinson had written two pamphlets against Woodward. Harris (?1666–1719) was the inspiration for a pamphlet by Charles Humphreys entitled *The Picture of a High-flying Clergyman* (1716). He was Secretary of the Royal Society in 1709.

116. King parodied Lister's *A Journey to Paris in the Year 1698* in his own *A Journey to London in the Year 1698* (1698).

117. Roger L. Emerson, "Sir Robert Sibbald, Kt., the Royal Society of Scotland and the Origins of the Scottish Enlightenment," *Annals of Science* (1988) 45:41–72, n. 72, on 58. Mulligan and Mulligan, "Reconstructing Restoration Science: Styles of Leadership and Social Composition of the Early Royal Society," *Social Studies of Science* (August 1981) 11, no. 3: 327–64, on 335, argue that "during the last eight years of the century, Sloane attempted to impose his own strong views about the nature of natural philosophy on the Society but succeeded only in creating divisions and encouraging factions."

118. Bentley had published a work denying the authenticity of letters supposedly by a classical tyrant, Phalaris, annoying their editor, Charles Boyle, later Earl of Orrery, and igniting a pamphlet war in which Boyle was backed up by his Tory friends.

119. Fragments of their work, which began in 1713, finally appeared as *The Memoirs of Martinus Scriblerus* in a collection of Pope's prose in 1741. Much of it

was written by Dr. John Arbuthnot. In addition to Pope and Arbuthnot, other members of the group included Jonathan Swift, John Gay, and Thomas Parnell.

120. Roger D. Lund, introduction to the Augustan Reprint Society ed. of William King, *The Transactioneer* (Los Angeles: 1988), vii.

121. See below for the contrasting views of science held by King and Sloane.

122. Richard G. Olson, "Tory-High Church Opposition to Science and Scientism in the Eighteenth Century . . . " in *The Uses of Science in the Age of New-ton*, ed. John G. Burke (Berkeley: 1983), 171–204, on 200. Olson notes that Newtonian mechanics was more acceptable to the Tories.

123. King, *Transactioneer*, 43, italics omitted.

124. The Transactioneer exclaims that his crony Petiver was "a F. of the R.S. indeed! I made him so. 'Tis my way of Rewarding my Friends and Benefactors. We now begin to call it Our Royal Society." King, *Transactioneer*, 33.

125. Gaukroger, *Collapse*, 154.

126. Gaukroger, *Collapse*, 169.

127. Gaukroger, *Collapse*, 173.

128. Hans Sloane, *Voyage to the Islands* . . . (London: 1707), quoted by Harold J. Cook, "Physicians and Natural History," in *Cultures of Natural History*, eds. Nicholas Jardine, James A. Secord, and Emma C. Spary (Cambridge: 1966), 91–105, on 99. In "The New Philosophy in the Low Countries," in Porter and Teich, *Scientific Revolution*, Cook notes that today we value theory but in seventeenth-century Holland the "laborious investigation and collection of the details of nature" surpassed theoretical work in the numbers of people and the expense it involved. This activity also reached much further down the social spectrum. See also Paula Findlen, "Natural History," in *The Cambridge History of Science*, eds. Katharine Park and Lorraine Daston, 3: *Early Modern Science* (Cambridge: 2006), 435–68.

129. John Toland's *Christianity not Mysterious* (1696) argued that reason, not faith, was the foundation of all certainty and was within the grasp of every-one without needing mediation by a learned clergy. Justin Champion, *Republican Learning: John Toland and the Crisis of Christian Culture, 1696–1722* (Manchester: 2003), 80–6. Although the two men had mutual acquaintances, there is no evidence that Sloane had any significant connection to the contro-versial author.

130. T. C. Bond, "Keeping Up with the Latest *Transactions*: The Literary Critique of Scientific Writing in the Hans Sloane Years," *Eighteenth-Century Life* (May 1998) 22:1–17, 2.

131. Martin Martin, "Several Observations in the North Islands of Scotland," *Philo-sophical Transactions* (1695/7) 19:727–9.

132. Mary Jane W. Scott, *James Thomson, Anglo-Scot* (Athens, GA: 1988), 141. Scott notes that the realistic and vivid Scottish poetic language developed by Thom-son and his circle differed from the traditional pastoral imagery used by English contemporaries, one example of the migration of the rhetorical style favored by Sloane's circle into a cultural style. Champion discusses Toland's criticism of Martin's work in *Republican Learning*, 224 and 228–9.

133. P. Stride, "The St Kilda Boat Cough under the Microscope," *Journal of the Royal College of Physicians of Edinburgh* (2008) 38:272–9, online at http://www.rcpe.ac.uk/journal/issue/journal_38_3/stride.pdf.

134. Edward Gray, "An Account of the Epidemic Catarrh, of the Year 1782; Com-piled at the Request of the Society," *Medical Communications* (1784) 1:67–70.

135. Brookes was a rural practitioner in Surrey for a time. He is thought to have been born before 1700. He seems to have had an MD, and he published, compiled, or translated many works on fishing, physic, and natural history. G. T. Bettany, "Brookes, Richard (*fl.* 1721–1763)," rev. Claire L. Nutt, *ODNB* (Oxford: 2004), online at http://www.oxforddnb.com/view/article/3560. See also Richard Wilding, "From the Rise of the Enlightenment to the Beginnings of Romanticism (Robert Plot, Edward Lhwyd, and Richard Brookes, M.D.)," in *History of Paleobotany; Selected Essays*, eds. Alan J. Bowden, Cynthia V. Burek, and Richard Wilding, Geological Society Special Publication 241 (Bath: 2005), 5–12. I assume that the author of *A History of the Most Remarkable Pestilential Disorders that Have Appeared in Europe for Three Hundred Years Last Past* (London: 1721) is the same as the author of *A New and Accurate System of Natural History* (6 vols., London: 1763) and *The General Gazetteer* (London: 1762). He also seems to have been the R. Brookes, MD, who wrote *The General Practice of Physic: Extracted Chiefly from the Writings of the Most Celebrated Practical Physicians . . .* 2 vols. (London: 1751), but he is not the same man as the Reverend Richard Brookes, MA (d. 1737), translator of *The Natural History of Chocolate* (1730) and *The General History of China* (1736) despite a claim to the contrary in the *ODNB*. See also Nicholas Hans, *New Trends in Education*, 154.

136. The author of *Don Ricardo Honeywater Vindicated* (most likely Tobias Smollett) writes that: "It's well known that the Don [Mead] conform'd to the establish'd Religion of his Country after he was Married," (London: 1748, rpt. 1987), ed. Robert Adams Day, 21.

137. Mead and Freind, however, both supported the mathematician Martin Folkes when he ran against Sloane to succeed Newton as President of the Royal Society in 1727. J. S. Rowlinson, "John Freind: Physician, Chemist, Jacobite, and Friend of Voltaire's," *N&R* (May 2007) 61, no. 2:109–27.

138. William MacMichael, *The Gold-Headed Cane* (New York, 1926), 78.

139. For Bradley's views, see chapter 9.

140. Frank Egerton, "Richard Bradley's Illicit Excursion into Medical Practice in 1714," *Medical History* (1970) 14:53–62, on 55.

141. Egerton, "Bradley's Illicit Excursion," 57.

142. Frank Egerton, "Richard Bradley's Relationship with Sir Hans Sloane," *N&R* (1970) 25:59–77, on 61.

143. Boyle, an FRS and suspected Jacobite, had published the spurious "Epistles of Phalaris" that set off the "Battle of the Books." Bradley probably chose Orrery as a dedicatee because of his wealth, but the following year Orrery was imprisoned in the Tower of London.

144. Raymond Williamson, "John Martyn and the Grub-street Journal, with Particular Reference to His Attacks on Richard Bentley, Richard Bradley, and William Cheselden," *Medical History* (1961) 5:361–74, on 363.

145. Pine was an associate of Hogarth's and a governor of the Foundling Hospital. W. Roberts, "R. Bradley, Pioneer Garden Journalist," *Journal of the Royal Horticultural Society* (1939) 64:164–74, on 167.

146. Roberts, "R. Bradley," 169. Three additional numbers were published from April to September 1722.

147. Egerton, "Bradley's Relationship with Sloane," 63–4.

148. Egerton, "Bradley's Relationship with Sloane," 64.

149. Williamson, "John Martyn," 363.

150. William Heberden, Sr. finally found sufficient support to found the botanical garden in 1762.

151. Williamson, "Martyn," 364.

152. D. E. Allen, "Martyn, John (1699–1768)," *ODNB* (Oxford: 2004), online at http://www.oxforddnb.com/view/article/18235.

153. Richard Westfall, *Never at Rest: A Biography of Isaac Newton* (Cambridge: 1990), 209.

154. Sloane also aided Martyn at least with advice. The two met in 1724, and in 1730 Sloane advised him to attend Emmanuel College, Cambridge, but he left without a medical degree. Martyn had been a clerk in the office of his father, a Hamburg merchant. He apparently became interested in botany by participating in excursions offered monthly by Isaac Rand of the Society of Apothecaries. He practiced as an apothecary, but Wallis *Medics* doesn't list any apprenticeship. He founded a Botanical Society at the Rainbow Coffee House in 1721. He did become Professor of Botany in 1732, but ceased to lecture by 1735. See D. E. Allen, "John Martyn's Botanical Society" and Hunting, "Isaac Rand," 12.

155. Egerton, "Bradley's Relationship with Sloane," 69. For Mather, see chapter 8 and Otho Beall, Jr. and Richard Shryock, *Cotton Mather, First Significant Figure in American Medicine* (Baltimore: 1954), 152–3. Mather also corresponded with Petiver, sending him what has been called "the earliest account of plant hybridization yet found," 48.

156. Egerton, "Bradley's Relationship with Sloane," 69.

157. Egerton, "Bradley's Relationship with Sloane," 69.

158. Williamson, "John Martyn," 363.

159. Williamson, "John Martyn"; Richard Foster Jones, *Ancients and Moderns: A Study of the Rise of the Scientific Movement in Seventeenth-Century England* (Berkeley: 1965), 267, 322.

160. Williamson, "John Martyn," 365.

161. Williamson, "John Martyn," 365.

162. Williamson, "John Martyn," 366. Italics omitted.

163. Allen, "Martyn's Botanical Society."

164. Williamson, "John Martyn," 366. The satire is in the *Memoirs of the Society of Grub-Street* (London: 1737) 1:86–9. It was first published in 1730.

165. Williamson, "John Martyn," 367–8. Italics omitted.

166. Egerton, "Bradley's Relationship with Sloane," 72–7.

167. See for example Fielding Garrison, *An Introduction to the History of Medicine* (Philadelphia: 1921), 369: "The names of many English physicians of Queen Anne's time . . . have a literary and social, rather than a scientific interest." Sloane and Mead are among those listed.

168. Maarten Ultee, "Sir Hans Sloane, Scientist," *British Library Journal* (1988) 14:1–20, 1, and see MacGregor, "Sloane," 41, nn. 132–3. MacGregor notes that the essays in his book contradict these claims. In one essay in MacGregor, Marjorie Caygill objects to claims that Sloane "can hardly be credited with founding his own collection."

169. Londa Schiebinger combines feminist and anticolonialist narratives in "Feminist History of Colonial Science," *Hypatia* (Winter 2004) 19, no. 1:233–54. Schiebinger argues that Sloane deliberately suppressed information about the use of botanical abortifacients in the West Indies.

170. James Delbourgo, "Slavery in the Cabinet of Curiosities: Hans Sloane's Atlantic World," online from the British Museum website at http://www

.britishmuseum.org/PDF/Delbourgo%20essay.pdf. Elsewhere, Delbourgo discusses the tangible and intangible personal benefits that accrued to Sloane from slavery. Sloane even had a slave boy in London, presented to him by Alexander Stuart, but the child proved to be too troublesome, and Sloane sent him back to Stuart. See Delbourgo, "Milk Chocolate," 76.

CHAPTER 7

1. This often but not always translates to the disease we call pulmonary tuberculosis.
2. Marten, *A New Theory of Consumptions* (London: 1720), 8. This is now online on the Hathi Trust website at http://catalog.hathitrust.org/Record/009295819. It was first advertised in issue 30 of the *Daily Post* on November 6, 1719, Gale, *Burney Collection Newspapers*. I thank the Wellcome Library for access to this. Marten also cites Gideon Harvey's claim that consumption was contagious (see chapter 2).
3. Singer, "Benjamin Marten, A Neglected Predecessor of Louis Pasteur," *Janus* (1911) 16:81–98, on 82. Singer published large sections from chapter 2 of *A New Theory of Consumptions* verbatim.
4. Marten, *New Theory*, 52 and 54; cf. Singer, "Marten," 83.
5. Marten, *New Theory*, 61, 65, 71–2; Singer, "Marten," 84–5.
6. Marten, *New Theory*, 77; Singer, "Marten," 87–8.
7. Marten, *New Theory*, 78–9; Singer, "Marten," 88.
8. This brief comment was later stressed by Cotton Mather in his summary of Marten.
9. Singer, "Marten," 93.
10. Marten, *New Theory*, 152–64.
11. Marten, *New Theory*, 68.
12. One exception was men practicing as physicians who had MB degrees instead of MDs. They were often referred to as "Doctor." However, these all appear to have been graduates of English universities. American usage was more flexible.
13. Raymond N. Doetsch, "Benjamin Marten and His "New Theory of Consumptions," *Microbiological Reviews* (September 1978) 42, no. 3:521–8. I thank Dr. David Zuck for his assistance on Marten's background.
14. Evidence for the relationship is the will John wrote shortly before his death on January 8, 1737, which left a shilling to "my Brother Benjamin Marten Doctor of Physick," as well as to a sister, Elizabeth Spooner, and to a nephew James Marten, "the only surviving Son of my late Brother James Marten . . . Apothecary." We know of only two cases in which an MD degree was awarded to a Benjamin Marten in the early eighteenth century: one from Aberdeen in 1717 to "Benjamin Martin *anglus*" and one from Leyden, issued to a "Benjamin Martin, *anglo-britannus*" in 1728. An announcement dated January 14, 1718, in the (London) *Post Man and the Historical Account* states that "The University of Aberdeen has conferr'd the Degree of a Doctor in Physick on Benjamin Marten, Physician, near Red Lyon Square," making the identification of the Aberdeen doctor with our author more likely. Gale, Burney Collection, issue 16542. Harold Cooke identifies the *Post Man* as a Whig newspaper, *Trials of an Ordinary Doctor: Johannes Groenevelt in Seventeenth-Century London* (Baltimore: 1994), 164. I thank Michael Moss of the Archives Centre, University of Glasgow, and Jane Pirie of the University of Aberdeen Library for assistance on Marten's MD. I also thank Ruth Wallis of the Project for Historical Biobibliography.

15. Wallis, *Medics* s.v. and F. J. G. Robinson and P. J. Wallis, *Book Subscription Lists: A Revised Guide* (Newcastle Upon Tyne: 1975).

16. David Zuck, personal communication, June 15, 1996. If he had been in practice for twenty years following an apprenticeship and duty in Ireland, he must have begun in 1691. This gives an approximate birth date of 1670. See also Roy Porter, "'Laying Aside Any Private Advantage': John Marten and Venereal Disease," in *The Secret Malady: Venereal Disease in Eighteenth-Century Britain and France*, ed. Linda E. Merians (Lexington, KY: 1996), 51–67, on 52.

17. Lithotomy is the operation for removing bladder stones.

18. For a complete account, see Harold Cook, *Trials*, 121.

19. Cantharides is made by grinding dried blister beetles. The powder irritates the skin, and when it is ingested it irritates the bladder and urethra. It was often used as an aphrodisiac.

20. (London: Jeffrey Wale and John Isted, 1706). I thank Harold Cook for the full title citation and for additional information.

21. Edward Tyson was a strong supporter of the apothecaries and used his position as Censor of the College of Physicians to exercise his influence in their favor. See above, chapter 2, and M. F. Ashley Montagu, *Edward Tyson, M. D. F.R.S., 1650–1708* (Philadelphia: 1943), 214. For Tyson's very persistent defense of Groenevelt, which annoyed his College colleagues, see above and Cook, *Trials*, 14–16, 146–7, and 154.

22. The earliest edition I have found is the fifth (London: 1707).

23. Harold Cook, "Marten, John (fl. 1692–1737)," *ODNB* (Oxford: 2004), online at http://www.oxforddnb.com/view/article/56721.

24. Marten was indicted and acquitted for obscenity in 1709. See *The Post Boy*, November 29, 1709, issue 2270, Gale, Burney Collection.

25. Spinke identified Marten with the author of the book *The Charitable Surgeon* written by "T.C., Surgeon" and published by Edmund Curll. Curll later printed an advertisement denying that they were the same man. For the rest of the story, see Paul Baines and Pat Rogers, *Edmund Curll, Bookseller* (Oxford: 2007), 36–7 and 114.

26. Raymond Doetsch, "Benjamin Marten's '*New Theory of Consumptions*,'" *Microbiological Reviews* (September 1978) 42:521–8, on 526. A "Thomas Spooner, medic" of Lemon Street, London, published "*A Short Account of the Itch . . .*" under the pseudonym of "T.S." in 1714 (London: T. Child) and was a frequent advertiser of various remedies. This short work and Spooner's *A Compendious Treatise of the Diseases of the Skin* went through many editions. See Philip K. Wilson, *Surgery, Skin and Syphilis, Daniel Turner's London (1667–1741)*, (*Clio Medica* 54) (Amsterdam: 1999), 63, and Wallis, *Medics*, under "Spooner."

27. David Zuck, personal communication, May 31, 1996. See also Doetsch, "Marten's '*New Theory*,'" 526.

28. Doetsch, "Marten's '*New Theory*,'" 526.

29. Doetsch, "Marten's '*New Theory*,'" 526.

30. David Zuck, personal communication, May 31, 1996.

31. John also tells us that he kept an apothecary's shop in Barnaby Street, London, practiced in a house in Hatton Garden, and had married a wealthy woman. Hatton Garden is very close to Theobald's Row (now Theobald's Road), Benjamin Marten's address. Porter, "John Marten and Venereal Disease," 52; John Rocque, *London, Westminster and Southwark*, 1746, online at http://www.motco.com /map/81002/. Barnaby Street was in Bridge Ward south of the Thames, near St. Thomas's Hospital.

32. James G. Donat, "The Rev. John Wesley's Extractions from Dr. Tissot: A Methodist Imprimatur," *History of Science* (2001) 19:285–98. See also Michael Stolberg, "Self-pollution, Moral Reform and the Venereal Trade: Notes on the Sources and Historical Context of *Onania* (1716)," *Journal of the History of Sexuality* (January–April 2000) 9, no. 1–2:37–61.

33. See Donat, "Wesley's Extractions," 287–8.

34. Stolberg, 53, n. 80, quotes a contemporary source on John Marten's opulence.

35. In the winter of 1710–11, Groenevelt was apparently in debtors' prison. Cook, *Trials*, 196.

36. Harold Cook, private communication, June 6, 1996, quoting Groenevelt, *A Compleat Treatise of the Stone and Gravel* (1710).

37. Same, and see Cook, *Trials*, 196–8.

38. Francois dele Boë Sylvius, *Opera Medica*, Amsterdam, 1679. See "Tuberculosis" in *Morton's Medical Bibliography*, ed. Jeremy M. Norman (Aldershot, UK: 1991), 366.

39. Cook, *Trials*, 202 and 262, n. 74.

40. *Fundamenta* was reprinted at Brussels in 1687 and 1693, at Lyons in 1692, and at Brussels in 1737.

41. *Fundamenta Medicinae Scriptoribus, tam inter Antiquos quam Recentiores, Praestantioribus Deprompta . . . Cui Subnectitur Appendix, Praescribendi Methodum in Quibusdem Morbis Exhibens* (London: J. Wyat and J. Woodward, 1714); *Fundamenta Medicinae, Scriptoribus . . . Autore Joanne Groenveld* [sic] . . . *Secundum Dictata D. Zypaei* (London: 1715).

42. It is not clear to me that Groenevelt intended to claim to be more than the editor and author of the appendix; indeed, it is not certain that Groenevelt was involved in the edition. The publisher may have appropriated his name, a not unprecedented event in this period. He died in 1716.

43. *The Grounds of Physick, Containing so Much of Philosophy, Anatomy, Chimistry, and the Mechanical Construction of a Humane Body as is Necessary to the Accomplishment of a Physitian . . .* (London: 1715). References are to this edition. A second English translation of Groenevelt's 1715 Latin text appeared in 1753 under the title *The Rudiments of Physick* (Sherborne and London). See also Harold Cook, "Physick and Natural History in Seventeenth-Century England," in *Revolution and Continuity: Essays in the History and Philosophy of Early Modern Science*, eds. Peter Barker and Robert Ariew (Washington, DC: 1991), 63–80, on 75.

44. *Grounds of Physick*, 153.

45. *Grounds of Physick*, 154–5.

46. *Grounds of Physick*, 154.

47. *Grounds of Physick*, Appendix, 292.

48. Cook, *Trials*, 121.

49. *Munk's Roll*, 1:428. The full title was *Processus Integri in Morbis Fere Omnibus Curandis* (complete method of curing almost all diseases) (London: 1695), online from EEBO. Crell dedicated his poem to his "*Amico Integerrimo, Medico Expertissimo*" (to his very upstanding friend and most expert doctor).

50. Cook, *Trials*, 196.

51. Cook, *Trials*, 196.

52. Cook, *Trials*, 197, 203.

53. Richard Mead, *De Imperio Solis ac Lunae in Corpora Humana, & Morbis inde Oriundis* (1704), trans. by Thomas Stack as *A Treatise concerning the Influence of*

the Sun and Moon upon Human Bodies and the Diseases Thereby Produced (London: 1748), 61–2. Mead spells his name "Groenvelt."

54. Montagu, *Edward Tyson*, 82, and see chapters 2 and 5 above.

55. See chapter 2.

56. Cook, *Trials*, 197–8. Groenevelt, a member of the church by 1674, may also have known the merchant and free-thinker Francis Lodwick (1619–1694), who came from a prominent Flemish family active in the church. Lodwick was elected FRS in 1681. I have not found evidence of a direct connection.

57. Cook, *Trials*, 22.

58. The Fire of London had destroyed the old library, but a new one opened in 1688. By the end of the eighteenth century, only Fellows (not Licentiates) enjoyed access. Perhaps Marten consulted its books with help from a friendly Fellow such as Tyson.

59. Wallis, *Medics*, "Martin, Benjamin, medic"; F. J. G. Robinson and P. J. Wallis, *Book Subscription Lists*; P. J. Wallis and Ruth Wallis, *Book Subscription Lists Extended Supplement to the Revised Guide* (Newcastle: 1996). The books were on music and history.

60. Roy Porter comments that John Marten "followed the practice of many quack authors in giving his address in his books," "John Marten" in Merians, *Secret Malady*, 66. John's address, however, is much more explicit than Benjamin's: "From my House in *Hatton-Garden*, on the Left-Hand beyond the Chappel, turning in from *Holborn*, *John Marten*, Surgeon, being writ over the door."

61. Marten, *New Theory* 2nd ed. (London: 1722), preface, xi. I thank the Countway Library, Harvard University, for a photocopy of this rare preface.

62. Porter, "John Marten," in Merians, *Secret Malady* offers an analysis of John Marten's rhetoric. "Dr. Marten's" drops for Gleets were still sold on Fleet Street in 1748. See Roy Porter, "Lay Medical Knowledge in the Eighteenth Century: The Evidence of the *Gentleman's Magazine*," *Medical History* (1985) 29:138–68, Appendix: Table of Proprietary Medicines (*Pharmacopoeia Empirica*) from the *Gentleman's Magazine* (1748) 18:348–50.

63. Perhaps the best known of these surgeons was William Hunter, who obtained an MD from Glasgow in 1750. See Wilson, *Surgery, Skin and Syphilis*, chapter 8, "Self-Styled Gentleman in London's Middle Class," 191–226, for Daniel Turner's climb.

64. Charles F. Mullett, "Physician vs. Apothecary, 1669–1671," *Scientific Monthly* (1939) 49, no. 6:558–65.

65. Examples can be found in P. S. Brown, "Medicines Advertised in Eighteenth-Century Bath Newspapers," *Medical History* (1976) 20:152–68. John Fothergill's name was stolen for an entire book, *Rules for the Preservation of Health by J. Forthergell* [*sic*]. On alchemical medicine and the creation of proprietary medicines, see also Paul Kléber Monod, *Solomon's Secret Arts* (New Haven: 2013), 125–34. On the marketing of remedies see Roy Porter, *Health for Sale: Quackery in England, 1660–1850* (Manchester: 1989).

66. See Marjorie H. Nicolson, "Ward's 'Pill and Drop' and Men of Letters," *Journal of the History of Ideas* (1968) 29:177–96. Ward's pill contained antimony; his drop included mercury. Antimony, a poison introduced into common medical use by Paracelsus, could reduce fevers.

67. Wilson, *Surgery, Skin and Syphilis*, 170–3.

68. Cockburn, who had left Leyden without a degree in 1693, obtained his License from the College of Physicians in 1694, and an MD from Aberdeen in 1697. He was elected an FRS in 1696. See Charles Creighton, "Cockburn, William

(1669–1739)," rev. Anita Guerrini, *ODNB* (Oxford: 2004), online at http://www.oxforddnb.com/view/article/5777. For Dover's criticism of Cockburn's secrecy, see Philip K. Wilson, "Exposing the Secret Disease: Recognizing and Treating Syphilis in Daniel Turner's London," in Merians, *Secret Malady*, 76.

69. For Dover, see chapter 6 above.

70. Andreas-Holger Maehle, *Drugs on Trial: Experimental Pharmacology and Therapeutic Innovation in the Eighteenth Century* (*Clio Medica* 53) (Amsterdam: 1999), esp. 76–7.

71. Richard Barnett, "Dr Jacob de Castro Sarmento and Sephardim in Medical Practice in 18th Century London," *Transactions of the Jewish Historical Society of England* (1978–80) 27:84–114, on 88. De Castro Sarmento became the physician to the household of successive Portuguese Ambassadors in London and sought to popularize Newtonian science in Portugal through his editions and works in Portuguese. For his connection to Sloane, see above, chapter 6.

72. Alex Sakula, "The Doctors Schomberg and the Royal College of Physicians: An Eighteenth Century Shemozzle," *Journal of Medical Biography* 2 (1994) 113–19, on 115.

73. The author of *A Dictionary of the English Language* (1755), Johnson was awarded honorary doctorates in 1765 by Trinity College Dublin and 1775 by Oxford.

74. Among the authors praising James's Powders was Oliver Goldsmith, whose physician, William Hawes, claimed they killed Goldsmith in 1774.

75. Burton Chance, "Sketches of the Life and Interests of Sir Hans Sloane: Naturalist, Physician, Collector and Benefactor," *Annals of Medical History* (1938) n.s. 10:390–404, on 397. Its most important ingredient was zinc oxide.

76. Brooks, *Sir Hans Sloane*, 85.

77. See Roy Porter, "The Early Royal Society," in *The Medical Revolution of the Seventeenth Century*, eds. Roger French and Andrew Wear (Cambridge: 1989), 272–93, esp. 283–4.

78. See chapter 2 for the members of the Oracle group. For venereology as a "fringe" profession, see W. F. Bynum, "Treating the Wages of Sin: Venereal Disease and Specialism in Eighteenth-Century Britain," in *Medical Fringe and Medical Orthodoxy, 1750–1850*, eds. W. F. Bynum and Roy Porter (London: 1987), 5–28. Venereology was one of the few specialties that focused on a disease widely agreed to be contagious.

79. See Roy Porter, *Health for Sale*.

80. Harold J. Cook, "Pechey, John (*bap.* 1654, *d.* 1718)," *ODNB* (Oxford: 2004), online at http://www.oxforddnb.com/view/article/21737. See also C. J. S. Thompson, *The Quacks of Old London* (New York: 1993), 134–5, and Porter, *Health for Sale*, 5, 80, 190, and 194.

81. Harold J. Cook, "Browne, Richard (1647/8–1693/4?)," *ODNB* (Oxford: 2004) online edn, May 2005 at http://www.oxforddnb.com/view/article/3694, accessed April 26, 2014.

82. M. M. Goldsmith, "Mandeville, Bernard (*bap.* 1670, *d.* 1733)," *ODNB* (Oxford: 2004), online at http://www.oxforddnb.com/view/article/17926. See also same, *Private Vices, Public Benefits: Bernard Mandeville's Social and Political Thought* (Bambridge: 1985).

83. Cook, *Trials*, 200.

84. See also Wilson, *Surgery, Skin and Syphilis*, 161: "being a 'quack' or a 'guardian of orthodoxy' depended as much upon the way a practitioner administered treatments as upon his qualifications to practice."

85. This excludes works on sex or less circumspect books on gynecology. See for example C. A. Baragar, "John Wesley and Medicine," *Annals of Medical History* (1928) 10:59–63, and C. J. Lawrence, "William Buchan, Medicine Laid Open," *Medical History* (1975) 19:20–35.

86. Singer was not the first to stumble on Marten's work and be transfixed. In "The Germ Theory," in the *American Monthly Microscopical Journal* (September 1891) 12:205–6, "E.H. Griffith" (probably Ezra Hollace Griffith, a Fellow of the Royal Microscopical Society) described a first edition he found "while visiting an old Curiosity Shop at the foot of Pike's Peak, Colorado." He added that he did not recall anyone who knew the "Germ Theory" had been propounded before 1865 and quoted chunks of the book. Griffith's own article was reprinted twice in local medical publications and then apparently disappeared from scholarly knowledge until being swept up by a Google scanner.

87. A search through the Burney Collection of Newspapers yields about a hundred advertisements in such newspapers as *The Postman, The Daily Post, The Weekly Journal, The British Journal, The London Journal, The Whitehall Evening Post,* and *The Public Ledger* for either the first (1720) or second edition (1722). The last advertisement listed was in *The Whitehall Evening Post*, June 3, 1727, issue 1362.

88. Beall and Shryock, *Cotton Mather*, 87–91, and "Angel," chapter 7; "Conjecturalies, or, some Touches upon, A New Theory of many Diseases," 149–54. This is also in Cotton Mather, "The Angel of Bethesda," ed. Gordon W. Jones, 43–8. Bradley's *New Improvements* first appeared in 1717 and had several editions.

89. Benjamin Franklin, ms. letter to Samuel Mather, Passy, May 12, 1784, Massachusetts Historical Society, online at http://www.masshist.org/database/533. See also I. Bernard Cohen, *Benjamin Franklin's Science* (Cambridge, MA: 1996), 174. The visit took place in 1724, shortly before Franklin left for England, where Mandeville would befriend him.

90. Doetsch, 5. Barry's work was published in Dublin in 1726 and in London in 1727.

91. Edward Barry, *A Treatise on Consumptions* (London: 1727), 273.

92. Sir Edward Barry was born in Cork and held MDs from Leyden (1719), Dublin (1740), and Oxford (by incorporation, 1761). He became a Fellow of the Royal Society in 1733, President of the Irish College of Physicians in 1749, and Professor of Physic in the University of Dublin, 1745–1761, before moving to London. He became a baronet in 1775 and died in 1776. Jean Loudon, "Barry, Sir Edward, first Baronet (1696–1776)," *ODNB* (Oxford: 2004), online at http://www.oxforddnb.com/view/article/1554.

93. A first edition was recently listed for sale by Mr. Nigel Phillips, with an armorial bookplate with the name "Jolliffe." I thank Mr. Phillips for information concerning this copy. I welcome additional information about other copies. My second edition has no evidence of earlier owners.

94. Orrery was the instigator of the "Battle of the Books" between "Ancients" and "Moderns" at Oxford. See chapter 6.

95. The copy Mather used has not been traced. A copy at Johns Hopkins has a bookplate of "I Baker Holroyd Esq., Sheffield Place, Sussex." The copy in the National Library of Medicine came from Sir John Anstruther of that Ilk, Baronet (1753–1811), who had been educated at Glasgow under Professor John Millar. He was the son of another Sir John Anstruther, Baronet. A copy in the Royal Faculty of Physicians and Surgeons of Glasgow came from the

Greenock Medical and Surgical Association (1818–1851), and a label says it came from John Speirs, MD, in 1820. The copy in the University of Illinois at Chicago has a signature of "Alexander Wolcott," who was presumably one of two Alexander Wolcotts, one the son of a West Country surgeon, the other a Yale graduate and Connecticut doctor. A copy at the University of British Columbia names Thomas Hindmarsh, John Hindmarsh, and a Mary or Ann Hindmarsh in order. A second edition in the College of Physicians, Philadelphia, bears a bookplate of Lord Arundell of Wardour, of a Cornish Catholic family, and was donated by Dr. S. Weir Mitchell, a Fellow of the College. A copy at Reading has early signatures of Wm. Johnston and Geor. Johnston.

Many librarians contributed to this information. In particular I thank Toby Appel of the Yale University Medical Library; Sarah Bakewell of the Wellcome Institute Library; Jonathan Battey of the Newcastle University Library; James Beaton of the Library of the Royal College of Physicians and Surgeons of Glasgow; Lois Black of the New York Academy of Medicine; Joanne Chaison of the American Antiquarian Society; Kevin Crawford of the College of Physicians of Philadelphia; Geoffrey Davenport, Archivist of the Royal College of Physicians, London; Patricia Erwin of the Mayo Medical Library in Rochester, NY; Maria Falconer of the Wangensteen Library, University of Minnesota; Barbara Francis of the Health Sciences Library, University of Florida; Mervyn Griffiths of the Library of the Medical Society of London; David Knott of the Reading University Library; Janet McMullin of Christ Church Library, Oxford; Doris Mahoney of the University of Michigan Library; Ed Mormon, then of the Welch Library, Johns Hopkins University; Mary O'Doherty, Mercer Library, Royal College of Surgeons, Dublin; Helen Peden of the British Library; Lee Perry of the University of British Columbia Library; Elizabeth Tunis of the National Library of Medicine; Colleen Weum of the University of Washington in Seattle; Janet Wright of the Branford Price Millar Library, Portland State University; and the University Library at Göttingen. I also thank John Millburn for information on Benjamin Martin the optician. Above all I thank Dr. David Zuck, who was generous with his time in London and assistance with primary materials.

96. Private communication from Mervyn Griffiths, Medical Society of London, February 13, 1997.

97. *A Natural and Medicinal History of Worms Bred in the Bodies of Men and Other Animals; Taken from the Authorities, and Observations of all Authors . . . from Hippocrates to this Time* . . . trans. by Joseph Browne (London: 1721), online from Google Books. This first appeared as *Historia Naturalis et Medica Latorum Lubricorum* (Geneva: 1715).

98. Le Clerc, *History of Worms*, 278.

99. Published by Alexis-Xavier-René Mesnier and sold also in Brussels. The work is on Google Books with all the drawings and the sequel.

100. The "M." was read by some as an abbreviation for Monsieur, thus, the imprimatur was issued to Monsieur A. C. D.

101. "Preface," ii.

102. *Suite du Système d'un Médecin Anglois, sur la Guérison des Maladies* (sequel of the system of an English doctor, on the cure of illnesses, by which is indicated the sorts of vegetables and minerals which are unfailing poisons for killing the different species of small animals that cause our diseases) (Paris: 1727), 4.

103. William Bulloch refers to this work in his *History of Bacteriology* (New York: 1979), 32, and reproduces drawings of some of the insects on 33.

104. M.A.C.D., *Système*, 5.

105. *De la Génération des Vers dans les Corps de l'Homme* (Paris: 1700), trans. as *An Account of the Breeding of Worms in Human Bodies* (London: 1701). The Google version of the 1720 French edition does not show the plate. For Andry, see chapter 5.

106. M.A.C.D., *Système*, 5. Michael Ettmüller (1644–1683) should be distinguished from his son, Michael-Ernest Ettmüller (1673–1733), who published many of his father's works. Ettmüller's plate is reproduced in C. E. Kellett, "Sir Thomas Browne and the Disease called the Morgellons," *Annals of Medical History*, n.s. (1935) 76:467–79.

107. Jean Astruc, *De Morbis Venereis*, trans. as *A Treatise of Venereal Diseases, in Nine Books . . . from the Last Latin Edition Printed at Paris*, vol. 1 (London: 1754), online from ECCO. The first English translation, which I have not seen, appeared in 1737. Some library catalogues describe the author of the *Système* as "Robert Boyle" or state that the *Système* is mistakenly attributed to the chemist Robert Boyle, but Astruc does not give a first name.

108. See also *Suite du Système*, 6.

109. Astruc, *De Morbis Venereis*, 127–8. See also Catherine Wilson, *The Invisible World: Early Modern Philosophy and the Invention of the Microscope* (Princeton: 1995), 170–1.

110. The best account seems to be H. F. A. Peypers, "*Un Ancien Pseudo-Précurseur de Pasteur ou Le 'Système d'un Médecin Anglois sur la Cause de Toutes les Maladies,' (1726),*" *Janus* (1896/7) 1:57–66, 121–31, 251–62.

111. Catherine Wilson, *Invisible World*, 171, suggests that the author may have been convinced his idea was right even though he had no way to demonstrate it.

112. In addition to the widely read work of Astruc above, see for example, Herman Boerhaave, *Methodus Studii Medici*, ed. Albert Haller (Amsterdam: 1751), 2:653, "This work scarcely deserves mention, because the author fraudulently demonstrated with a microscope insects hidden in the blood to credulous [people], to which [insects] he attributed all human diseases and the cure through their being devoured by other insects. The fraud was clearly exposed by Valisneri vol. III p. 217 and [by] anonymous in *a Letter to Mr. Astruc* in the *Journal des Savants*, 1740," [my trans.]. See also Catherine Wilson, *Invisible World*, 223–4. The "*Lettre à M. Astruc*" is in the *Journal des Sçavans*, (March 1740), 375–85, online from Google.

113. Remi Kohler, "Nicolas Andry de Bois-Regard (Lyon 1658–Paris 1742): The Inventor of the Word 'Orthopedics' and the Father of Parasitology," *Journal of Childhood Orthopedics* (August 2010) 4, no. 4:349–55, published online April 15, 2010, doi: 10.1007/s11832-010-0255-9.

114. This author should be distinguished from the later Pierre Desault who became famous as a Parisian surgeon. The elder Desault's book was entitled "*Dissertation sur les Maladies Vénériennes . . . Avec Deux Dissertations, l'Une sur la Rage, l'Autre sur la Phtisie . . .*" (Bordeaux: 1733, Paris: 1738). It was granted an imprimatur by Andry and was translated by John Andree as *A Treatise on the Venereal Distemper . . . with Two Dissertations: The First on Madness from the Bite of Mad Creatures; The Second on Consumptions . . .* (London: 1738). Andree's translation, which I have used, was dedicated to Daniel Turner, see 10–12. It is online from ECCO.

115. Guerin published three volumes of work by Desault, with the *Maladies Vénéri-ennes* forming vol. 1. The Sackler Archive of the Royal Society lists "An abstract of the 2ⁿᵈ volume of Dr. Peter Desault's *Dissertations of Physick*, printed at Paris in 1735" (1737), a thesis on gout that was translated and discussed before the Society by Dr. Claude Amyand (d. 1740). Amyand was surgeon to George II and, with Sloane, presided over the first smallpox inoculation. See "Amyand" under "Fellows" online at https://royalsociety.org/library/collections/.

116. Hector Grasset, "*La Théorie Parasitaire et la Phtisie Pulmonaire au XVIII Siècle,*" *France Médicale*, November 17, 1899. This article was translated by Thomas C. Minor as "The Parasitic Theory and Pulmonary Phthisis in the Eighteenth Century," in two parts in the *Cincinnati Lancet-Clinic* 93, n.s. 44: (January 6, 1900), 22–6 and (January 13, 1900), 38–43. I quoted p. 38 of Minor's translation. I have not been able to locate the original French version. The translation contains several misspellings of proper nouns; for example Nyander appears as both "Nysander" and "Nylander." See also Peypers, "*Un Ancien Pseudo-Précurseur de Pasteur.*"

117. Desault, *Dissertation*, trans. Andree, 289.

118. Desault, *Dissertation*, trans. Andree, 294.

119. Desault, *Dissertation*, trans. Andree, 302.

120. Desault, *Dissertation*, trans. Andree, 297.

121. Sloane suffered a disabling stroke at the age of 79 in 1739. Though he didn't retire as President of the Royal Society until 1741, his activity and influence were waning by 1740. For Andree's investigation of hemlock, see Susan Lawrence, *Charitable Knowledge: Hospital Pupils and Practitioners in Eighteenth-Century London* (Cambridge: 1996), 214–6, 246–7.

122. He was also a surgeon to the workhouse in St. Clement's Dane about 1780, Kevin P. Siena, *Venereal Disease, Hospitals and the Urban Poor: London's 'Foul Wards' 1600–1800* (Rochester: 2004), 26.

123. The younger Andree obtained an MD in 1798. Older accounts of the Andrees confuse their publications. Wikipedia, "John Andree (physician)" following older biographies, erroneously attributes to John Andree, Sr. a treatise on "Inoculation impartially considered in a Letter to Sir E. Wilmot, Bart," (London: 1765). This, however, together with a translation of "Baron Van Swietin" [*sic*] *Account of the Most Common Diseases Incident to Armies* . . . (Dublin: 1766) with which it was apparently published in one volume, were by John Andrew, MD, a Devon physician and supporter of inoculation.

124. Grasset, 41, says "[?Jean] Bouillet, in 1730, pointed out various verminous diseases prevailing at Beziers," and adds "The treatise of Moreali," (i.e., *Des Fièvres Malignes et Contagieuses Produites par des Vers* [Modena: 1739]).

125. "Of the Pestilential Fever and Plague," *Academical Lectures on Fevers . . . Read in the Royal College at Paris* (London: 1747), 179–320.

126. In addition to Andree Sr., who lived until 1785, Benjamin Franklin saw both the debates over inoculation in the 1720s and the debates over contagionism at mid-century.

127. Between 1700 and 1738, 746 English-speaking students matriculated at Leyden. Of those, 55 became members of the College of Physicians: 28 became Fellows, 14 became Licentiates, and 12 Extra-Licentiates. Four of the Leyden-educated Fellows became President of the College. Forty-five became FRS. E. Ashworth Underwood, *Boerhaave's Men at Leyden and After* (Edinburgh: 1977), 20, 135, and 149.

CHAPTER 8

1. Genevieve Miller, *The Adoption of Inoculation for Smallpox in England and France* (Philadelphia: 1957), 30. See also Donald R. Hopkins, *Princes and Peasants: Smallpox in History* (Chicago and London: 1983).

2. Miller, *Inoculation*, 33. See also chapter 4, "Smallpox" in Charles Creighton, *A History of Epidemics in Britain*, eds. D. E. C. Eversley, E. Ashworth Underwood, and Linda Ovenall, vol. 2: *From the Extinction of the Plague to the Present Time* (London: 1965), 434–622.

3. "Kahn says that Vollgnad and Schultz had reported it in Poland in the 1670s. The Danish anatomist Bartholin described it in 1666 and 1673, it was reported from Scotland in 1715, and Greek peasants were also already familiar with the practice." Hopkins, *Princes and Peasants*, 46. See also Miller, *Inoculation*, 43, and Creighton, *Epidemics*, 464–5 and 471–4.

4. Quoted in Miller, *Inoculation*, 28. The verse might be read as "take the variola to their blood vessels" (*illorum venis variolas mitte*).

5. Andrea Rusnock, *The Correspondence of James Jurin (1684–1750): Physician and Secretary to the Royal Society* (Amsterdam: 1996), 130.

6. *A Treatise upon the Small Pox in Two Parts* (London, 1723).

7. Miller, *Inoculation*, 43; Creighton, *Epidemics*, 474.

8. Hopkins, *Princes and Peasants*, 46; Miller, *Inoculation*, 48.

9. Clopton Havers, FRS (MD Utrecht, 1685), was a Licentiate. The son of a Dissenting clergyman ejected in 1662, he was tutored by Edward Tyson's brother-in-law, Richard Morton, MD, who was also an ejected clergyman. A gifted anatomist, Havers died of a fever in 1702. Jessie Dobson, "Pioneers of Osteogeny: Clopton Havers," *The Journal of Bone and Joint Surgery* (November 1952) 34B, no. 4:701–7, online at http://www.bjj.boneandjoint.org.uk/content/34-B/4/702.html, and Robert L. Martensen, "Havers, Clopton (1657–1702)," *ODNB* (Oxford: 2004), online at http://www.oxforddnb.com/view/article/12633.

10. Secondary sources give various dates for this conversation. Mather's letter to Woodward, dated July 12, 1716, says only that it was "many months" before he heard about inoculation from other sources. *Selected Letters of Cotton Mather*, ed. Kenneth Silverman (Baton Rouge, LA: 1971), 214. See also Miller, *Inoculation*, 52; Hopkins, *Princes and Peasants*, 46, 174, and 248–9. Scholars debate the location of Onesimus's home; Hopkins, 174, suggested that it was eastern Upper Volta, now Burkina Faso.

11. Also known as Emmanuele or Emanuel Timonis, Timonius, or Timoni (1670–1718).

12. A. P. Waterson and Lise Wilkinson, *An Introduction to the History of Virology* (Cambridge: 1978), 202; Miller, *Inoculation*, 52; see also Creighton, *Epidemics*.

13. Miller, *Inoculation*, 57

14. Miller, *Inoculation*, 58.

15. Miller, *Inoculation*, 59.

16. E. St. John Brooks, *Sir Hans Sloane: The Great Collector and His Circle* (London: 1954), 89. A member of the Temple Coffee House Botany Club, Sherard was also related to the botanist James Petiver, Sloane's intimate friend. Raymond Phineas Stearns, "James Petiver, Promoter of Natural Science, c. 1663–1718," *Proceedings of the American Antiquarian Society* (1952) n.s., 62 pt. 2: 243–364, 246, n. 7.

17. "New and safe method of instigating smallpox by transplantation."

18. Miller, *Inoculation*, 60–1.
19. Miller, *Inoculation*, 51; Brooks, *Sloane*, 89; Waterson and Wilkinson, *History of Virology*, 202.
20. Miller, *Inoculation*, 64.
21. Miller, *Inoculation*, 51.
22. Miller, *Inoculation*, 68–9.
23. Maitland had already inoculated Montagu's son in Constantinople on March 18, 1718. Isobel Grundy, *Lady Mary Wortley Montagu: Comet of the Enlightenment* (Oxford: 1999), 162.
24. Miller, *Inoculation*, 72–3.
25. See Arthur M. Silverstein and Genevieve Miller, "The Royal Experiment on Immunity: 1721–1722," *Cellular Immunology* (1981) 63:437–47, 441. See also Grundy, *Lady Mary*, 88
26. Jonathan Andrews and Andrew Scull, *Undertaker of the Mind* (Berkeley: 2001), 96, describe her as "a considerable patroness of religious radicals." See also Audrey T. Carpenter, *John Theophilus Desaguliers: A Natural Philosopher, Engineer and Freemason in Newtonian England* (London and New York: 2011), 33, 203–4, 206.
27. Hamilton was a Scot educated at Leyden: MD Rheims 1686, LRCP 1688, FRCP 1703, FRS 1708. The German Steigherthal, FRS, was MD Utrecht and physician to George I, who brought him to England. He became an honorary FRCP in 1714.
28. Miller, *Inoculation*, 86–8.
29. Miller, *Inoculation*, 82.
30. Miller, *Inoculation*, 83.
31. Miller, *Inoculation*, 68, 78. Miller refers to de Castro Sarmento as "a young Portuguese physician." For his contact with Sloane, see chapter 6; for his patent medicine, see chapter 7.
32. Edgar Samuel, "Sarmento, Jacob de Castro (1692?–1762)," *ODNB* (Oxford: 2004), online at http://www.oxforddnb.com/view/article/24670.
33. Samuel, "Sarmento." Albert M. Hyamson describes him as "the most distinguished English Jew of his day," *The Sephardim of England* (London: 1951), 88 and 106–9. Samuel states he had an MB from Coimbra (1717); Hyamson gives him an MD. He also published several works in Portuguese; I haven't discovered how he acquired English.
34. Jacob de Castro Sarmento, *Materia Medica Physico-Historico-Mechanica* (London: 1736, rpt. 1758). Some authors have erroneously credited de Castro Sarmento with the introduction of quinine (cinchona) into England. Saul Jarcho refers to his work as "a surprising delayed statement," *Quinine's Predecessor, Francesco Torti and the Early History of Cinchona* (Baltimore, MD: 1993), 50 and 252.
35. Miller, *Inoculation*, 87.
36. Miller, *Inoculation*, 88.
37. Miller, *Inoculation*, 97–8. Claude Amyand (d. 1740), the Sergeant Surgeon to the King, was the son of a Huguenot refugee. He helped Jurin collect data concerning inoculation for the Royal Society. Wallis, *Medics*.
38. Rusnock, *James Jurin*, 86–8.
39. This may be a reference to Kircher, who had claimed he saw "animalcules and vermicules" in the pus of smallpox lesions. See chapter 1 above. It is possible that Sloane had read Marten's work.
40. Rusnock, *James Jurin*, 99–101.

41. Andrea Rusnock, *The Correspondence of James Jurin (1684–1750): Physician and Secretary to the Royal Society* (Amsterdam: 1996), 110–11, Leeuwenhoek to Jurin, Delft, July 7, 1722.

42. John T. Desaguliers, "Account of the Appearance of the Matter of the Small Pox seen thro' a Microscope, 13 June, 1723," Royal Society, Sackler Archive, online at https://collections.royalsociety.org/DServe.exe?dsqIni=Dserve.ini&d sqApp=Archive&dsqDb=Catalog&dsqCmd=Show.tcl&dsqSearch=RefNo==%2 7RBO/11/96%27, accessed April 27, 2014. See also Miller, *Inoculation*, 253, and Carpenter, *Desaguliers*, 235. This paper did not appear in print.

43. The concept of a *living* contagion may not be applicable to viruses. See Luis P. Villarreal, "Are Viruses Alive?" *Scientific American* (December 2004), online at http://www.scientificamerican.com/article.cfm?id=are-viruses-alive-2004.

44. Marc J. Ratcliff, *The Quest for the Invisible: Microscopy in the Enlightenment* (Farnham, Surrey: 2009), see e.g., 65.

45. Silverman, *Letters of Cotton Mather*, 214. See also Hopkins, *Smallpox*, 248–9. Mather was nominated and approved for a Fellowship of the Royal Society in 1713. Owing to an oversight, his actual election took place in 1723. Royal Society, Sackler Archive, online at http://www2.royalsociety.org/DServe/dserve .exe, retrieved April 11, 2010.

46. Miller, *Inoculation*, 101.

47. There are numerous accounts of the events in Boston. See for example Roger Zelt, "Smallpox Inoculation in Boston, 1721–1722," *Synthesis* (1977) 4, no. 1:3–14; John Blake, "The Inoculation Controversy in Boston 1721–1722," *New England Quarterly* (1952) 25:489–506; Maxine van De Wetering, "A Reconsideration of the Inoculation Controversy," *New England Quarterly* (1985) 58, no. 1:46–67; David P. Harper, "Angelical Conjunction: Religion, Reason and Inoculation in Boston, 1721–1722," *Pharos* of Alpha Omega Alpha (Winter 2000) 63, no. 1:37–41; C. Edward Wilson, "The Boston Inoculation Controversy: A Revisionist Interpretation," *Journalism History* (1980) 7, no. 1:16–19, 40; and chapter 7, "The Advent of Preventive Medicine," in *Cotton Mather: First Significant Figure in American Medicine*, eds. Otho T. Beall Jr. and Richard H. Shryock (New York: 1979), 93–126.

48. Miller, *Inoculation*, 94.

49. Miller, *Inoculation*, 95. Benjamin Colman, *Some Observations on Receiving the Small-Pox by Ingrafting or Inoculating* (Boston: 1721, London and Dublin: 1722), online from the National Library of Medicine at https://archive.org /details/2546057R.nlm.nih.gov. In light of later practices, it is interesting that Colman stresses the small size of the incisions used for inoculation: "the least you ca[n] well imagine and but Skin deep."

50. Miller, *Inoculation*, 96–7.

51. "Inoculation may have been little more than a focal point for the factional strife which was developing in Boston during the early eighteenth century," Zelt, "Inoculation," 7.

52. Zelt, "Inoculation," 10.

53. Beall and Shryock, *Mather*, 112.

54. Blake, "Inoculation Controversy," 494.

55. Louise A. Breen, "Cotton Mather, the 'Angelical Ministry,' and Inoculation," *Journal of the History of Medicine* (1991) 46:333–57.

56. Beall and Shryock, *Mather*, 105.

57. Beall and Shryock, *Mather*, 108.

58. Miller, *Inoculation*, 93. See also Harper, "Angelical Conjunction," 40–1.
59. Boston, May 21, 1723, in Rusnock, *James Jurin*, 153.
60. Beall and Shryock, *Mather*, 14.
61. See Johanna Geyer-Kordesch, "Passions and the Ghost in the Machine: Or What Not to Ask About Science in Seventeenth- and Eighteenth-Century Germany," in *The Medical Revolution of the Seventeenth Century*, eds. Roger French and Andrew Wear (Cambridge: 1989), 145–63, on 157.
62. For Slare's interest in this work, see chapter 5.
63. Beall and Shryock, *Mather*, 17.
64. Beall and Shryock, *Mather*, 20.
65. Beall and Shryock, *Mather*, 44.
66. Mather knew that plants reproduced sexually. The letter is reprinted in Conway Zirkle, *The Beginnings of Plant Hybridization* (Philadelphia: 1935), 104–6, online from the Hathi Trust. See also Beall and Shryock, *Mather*, 48.
67. Mather was a student at Harvard between 1674 and 1678. Alchemical theses were still being presented then. William R. Newman, *Gehennical Fire: The Lives of George Starkey, An American Alchemist in the Scientific Revolution* (Chicago: 2003), 14–51.
68. Mather to Dr. John Woodward, July 12, 1716, in Silverman, *Letters of Cotton Mather*, 214.
69. Oliver Wendell Holmes rediscovered the ms. for the "Angel" in 1869. Selections appeared in Beall and Shryock, *Mather*. Gordon Jones published the complete text (Barre, MA: 1972).
70. Beall and Shryock, *Mather*, 67, 138.
71. Beall and Shryock, *Mather*, 69.
72. Beall and Shryock, *Mather*, 69.
73. *Bonifacius* appeared in 1710, at the time of Mather's first contacts with Halle. It greatly influenced Benjamin Franklin, who, unlike his brother, became a strong advocate of inoculation. Beall and Shryock, *Mather*, 16, 94–5.
74. See chapter 7 and Beall and Shryock, *Mather*, 154.
75. Benjamin Colman, *Some Observations*, 14.
76. Sloane to Richardson, "Letter 67," *Extracts from the Literary and Scientific Correspondence of Richard Richardson M.D.*, ed. Dawson Turner (Yarmouth: 1835), 171. I thank the Hunt Institute for Botanical Documentation for providing access to a copy of this work, which is now available online from the Internet Archive.
77. Edmund Massey, *A Sermon against the Dangerous and Sinfull Practice of Inoculation. Preach'd at St. Andrew's Holborn, on Sunday, July the 8th, 1722* (London: 1722), 11.
78. Edmund Massey, *A Letter to Mr. Maitland, in Vindication of a Sermon against Inoculation* (Norwich: 1722), 14, online from Google.
79. Peter Razzell, *The Conquest of Smallpox* (Firle, Sussex: 1977), 95–6 noted that the religious objection was even stronger in Calvinist Scotland because it contradicted absolute predestination.
80. For Blackmore's *Discourse on the Plague*, see chapter 9.
81. Thomas Noxon Toomey, "Sir Richard Blackmore, M.D.," *Annals of Medical History* (1922) 4:180–8; Harry Solomon, *Sir Richard Blackmore* (Boston: 1980).
82. Cook, *Trials of an Ordinary Doctor: Johannes Groenevelt in Seventeenth-Century London* (Baltimore: 1994), 168.
83. Gregori Flavio, "Blackmore, Sir Richard (1654–1729)," *ODNB* (Oxford: 2004), online ed., January 2009 at http://www.oxforddnb.com/view/article/2528.

Blackmore also wrote a number of prose works against Arians and served as a vice-president of the Society for the Propagation of the Gospel in America.

84. Richard Blackmore, *A Treatise upon the Small-Pox, in Two Parts* (London: 1723), preface, ix.

85. Blackmore, *Treatise upon the Small-Pox*, preface, xvii.

86. Miller, *Inoculation*, 104. A friend of Swift and Arbuthnot, Wagstaffe had tried to prosecute Thomas Dover for malpractice and may have written the parody *A Letter from the Facetious Dr. Andrew Tripe*, attacking the "modern" practice of Dr. John Woodward. See Levine, *Dr. Woodward's Shield* (Ithaca, NY: 1991), 14. St. Bartholomew's was a bastion of Toryism.

87. As Nathaniel Hodges noted in 1665, reinfections even occurred among plague survivors. Charles F. Mullett, "The English Plague Scare of 1720–23," *Osiris* (1936) 11:487–91. See also the English abstract of C. Huygelen, "[Attempts to inoculate against plague in the eighteenth and nineteenth centuries]," *Verhandelingen-Koninklijke Academie voor Geneeskunde van België* (1999) 61, no. 2:385–409, online from PubMed, National Library of Medicine at http://www .ncbi.nlm.nih.gov/pubmed/10379211.

88. *A Letter to the Reverend Mr. Massey, Occasion'd by his Late Wonderful Sermon Against Inoculation* (London: 1722), 5–6. This has often been attributed to Charles Maitland, but it is not in his style.

89. Fracastoro (following Arab sources) had distinguished smallpox from measles. William Heberden published a more complete differentiation in *Medical Transactions* in 1768.

90. *Mr. Maitland's Account of Inoculating the Small Pox, the Second Edition, to which is added . . . a Word to the Reverend Mr. Massey . . .* (London: 1723), 16–17.

91. The success of Sloane and Steigerthal's experiment of sending an inoculated convict to share a bed with a boy ill with smallpox also helped this argument.

92. Miller, *Inoculation*, 242.

93. Abu Becr Mohammed Ibn Zacariya Al-Razi (Rhazes), *A Treatise of the Smallpox and Measles*, trans. William Alexander Greenhill (London: 1848). This first appeared in Latin in a corrupt edition in 1498; Mead's Latin edition, prepared from an Arabic manuscript, was the first to include Rhazes's distinction between measles and smallpox. The first English translation (from Mead's 1747 ed.) appeared in 1747. See chapter 1 above for Plater's version of this theory.

94. This was William Oliver the elder (1659–1716, MD Rheims), who narrowly escaped after Monmouth's Rebellion, studied medicine in exile abroad, and served as a naval surgeon and physician.

95. William Oliver, *A Practical Essay on Fevers. Containing Remarks on the Hot and Cool Methods of their Cure* (London: 1704), 192. One outcome of Oliver's adoption of a seed theory of disease was his belief in three distinct "species" of smallpox—distinct, middle, and confluent—each determined by its seed. Experience with inoculation would disprove this belief.

96. Benjamin Marten, *A New Theory of Consumptions* (London: 1720), 65.

97. Beall and Shryock, *Mather*, 160.

98. Lise Wilkinson, "The Development of the Virus Concept . . . 5: Smallpox and the Evolution of Ideas on Acute (Viral) Infections," *Medical History* (1979) 3:1–28, on 11, attributes the final defeat of this idea to two treatises by the Italian physician Angelo Gatti in 1764 and 1767.

99. J. Johnson Abraham, *Lettsom, His Life, Times, Friends and Descendents* (London: 1933), chapter 11, 185–204.

100. Thomas Percival, Essay 2, "On the Proportional Mortality of the Small Pox and Measles," in same, *Philosophical, Medical, and Experimental Essays* (London: 1776), 87–108, online from Google.

101. Richard Shryock, "Germ Theories in Medicine Prior to 1870," *Clio Medica* (1972) 7:81–109. Erasmus Darwin also inoculated for measles and found the same thing.

102. Mullett, "Cattle Distemper," 162.

103. Mullett, "Cattle Distemper," 163, and see Daniel Peter Layard, "A Discourse on the Usefulness of Inoculation of the Horned Cattle to Prevent the Contagious Distemper among Them," *Philosophical Transactions 1683–1775* (1757–58) 50:528–38. There were reports that the Turks had tried inoculation against plague as well as smallpox. See Larry Stewart, "The Edge of Utility: Slaves and Smallpox in the Early Eighteenth Century," *Medical History* (1985) 29:54–70, on 69. See also the English abstract of C. Huygelen, "Attempts to Inoculate against Plague." An English doctor, "Mr. White," attempted to inoculate himself and four assistants with plague in 1801; all five died within days.

104. This is still debatable. Razzell believed inoculation worked by attenuating the virus, but Harper, "Angelical Conjunction," 40, writes that "Mather believed that the difference in outcome between cases of smallpox acquired by inoculation and by natural infection was due not to attenuation but to the location of the initial infection. This is correct according to the current hypothesis." Natural smallpox was acquired by inhalation, permitting it to multiply in the respiratory system and migrate to lymph nodes. Inoculation introduced the virus into the bloodstream, allowing the body to muster a defense before it damaged internal organs.

105. According to most accounts, John Hunter injected himself with venereal disease, possibly causing his fatal heart disease, but there is some question as to whether he injected himself or someone else. In the nineteenth century, French physicians also experimented with inoculation for syphilis. Deborah Hayden, *Pox: Genius, Madness, and the Mysteries of Syphilis* (New York: 2003), 29–31, notes that "inoculation experiments involved felons and prostitutes, the most likely subjects, but also servants and even children and infants . . . Doctors began to infect . . . everything living: themselves, their students, chimpanzees, monkeys, horses, rabbits, cats, and rats." She adds that Philippe Ricord, a French syphilologist, inoculated 2500 people with gonorrhea between 1835 and 1838 and Albert Neisser injected a group of prostitutes as young as ten years old with syphilis serum in 1895.

106. Miller, *Inoculation*, 259.

107. Miller, *Inoculation*, 265.

108. For the different ways that other premodern civilizations have conceptualized the causes of epidemic diseases, see Lawrence I. Conrad and Dominik Wujastyk, eds., *Contagion: Perspectives from Pre-modern Societies* (Burlington, VT: 2000). On smallpox, see especially Chia-Feng Chang, "Dispersing the Foetal Toxin of the Body: Conceptions of Smallpox Aetiology in Pre-Modern China," in this volume, 23–38.

109. Not everyone was willing to do so. Mather's opponent William Douglass "pointed out that no one should accept all the quaint things published in the *Philosophical Transactions*, that Mather's sources of information—accounts from the Levant and from untutored Negroes—were at best questionable." Blake, "Inoculation Controversy," 503–4.

110. Miller, *Inoculation*, 83.
111. Miller, *Inoculation*, 90–1.
112. Letter 67: Hans Sloane to Dr. Richardson in Turner, *Extracts*, 170–1 and note.
113. Wallis, *Medics*; Miller, *Inoculation*, 117.
114. Miller, *Inoculation*, 128.
115. Miller, *Inoculation*, 128. Pierce Dod (MD Oxford) became an FRS in 1730 and criticized inoculation in a work of 1746.
116. Rusnock, *Jurin*, 313.
117. Rusnock, *Jurin*, 314.
118. Adrian Wilson, "The Politics of Medical Improvement in Early Hanoverian London," in *The Medical Enlightenment of the Eighteenth Century*, eds. Andrew Cunningham and Roger French (Cambridge: 1990), 29. Wilson's "exclusively" is a bit too strong; as we have seen, Richard Blackmore, an Anglican Whig, opposed inoculation.
119. Wilson, "Politics of Medical Improvement," 33. Levine, *Dr. Woodward's Shield*, 9–10.
120. The collapse of the South Sea Bubble in 1720/1 also created a turbulent political environment. James Craggs, Jr., Secretary of State and son of the Postmaster General widely blamed for the disaster, died of smallpox the day the first report on the Bubble was delivered to Parliament.
121. In Edinburgh, Browne, the "Sydenhamian," had favored venesection and laxatives for smallpox; "Newtonian" opponents used stronger purges and sudorifics.
122. Levine, *Dr. Woodward's Shield*, 20–1.
123. But not always. John Arbuthnot (1667–1735, MD St. Andrews, 1696) was from a Scottish Episcopalian family. He was a close friend of Swift and a member of the "Brothers Club" of Tory wits and of the Scriblerians. He wrote the bulk of *The Memoirs of Martinus Scriblerus*, but he became a strong supporter of inoculation. In this case, and a few others, the FRS trumped the FRCP. For the collaboration of Arbuthnot and Sloane and the Royal African Company, see Stewart, "The Edge of Utility," 63.
124. Wilson, "Politics," 30 and 38.
125. He appears to be unrelated to John Wagstaffe (1633–1677), a skeptical writer who denied the existence of witchcraft.
126. Norman Moore, "Wagstaffe, William (1683/4–1725)," rev. Jean Loudon, *ODNB* (Oxford: 2004), online at http://www.oxforddnb.com/view/article/28402
127. John Radcliffe, the Tory physician and MP, helped Drake in 1706 when he was tried for libel.
128. James Drake, Anthropologia Nova: *Or, a New System of Anatomy . . . and a Short Rationale of Many Distempers . . .* 3rd ed. (London: 1727) vol. 1, 15.
129. The Tory Newtonian Dr. John Freind, however, opposed inoculation. See J. S. Rowlinson, "John Freind: Physician, Chemist, Jacobite, and Friend of Voltaire's," *N&R* (2007) 61, no. 2:109–27, doi: 10.1098/rsnr.2006.0175.
130. Andrea E. Rusnock, *Vital Accounts* (Cambridge: 2002), 44–70: "For Jurin to quantify the efficacy of inoculation, he had to develop categories to enumerate," 59. Many French physicians rejected Jurin's data on the grounds that smallpox in France was not the same as in England and inoculated smallpox was not comparable to naturally acquired smallpox, 87.
131. Rusnock, *Vital Accounts*, 114. Although the project was unsuccessful at the time, scientists have gone back to the weather reports of Jurin's correspondents,

and especially that of Nicolaas Cruquius, for information about climate change. See A. F. V. van Engelen and H. A. M. Geuirts, "Nicholaus Cruquius (1678–1754) and His Meteorological Observations" (De Bilt: 1985), *Koinklijk Nederlands Meterologisch Institut*, online at http://www.knmi.nl/bibliotheek/knmipubmetnummer/knmipub165_IV.pdf.

CHAPTER 9

1. Jacob L. Kool, "Risk of Person-to-Person Transmission of Pneumonic Plague," *Clinical Infectious Diseases* (2005) 40:1166–72, notes that pneumonic plague infects only at a distance of a meter or less.
2. N. C. Stenseth et al., "Plague, Past, Present, and Future," *PLOS Medicine* (January 15, 2008) 5(1): e3. doi: 10: 1371/journal.pmed.0050003, argue that theories of the epidemiology and ecology of *Y. pestis* have been oversimplified and that more likely modes of transmission than the rat-flea-human interaction include other mammals, birds of prey, and possibly human fleas. They note that several thousand people still die of plague each year. The experimental transmission of plague to rabbits by human body lice has been confirmed by S. Ayyadurai, F. Sebbane, et al., "Body lice, *Yersinia pestis* Orientalis, and Black Death," [letter], *Emerging Infectious Diseases* (May 2010) 16:892–3, doi: 10.3201/eid1605.091280.
3. This question has now been resolved by genetic testing of bodies from plague graveyards. See links to the articles by Achtman, Ayyadurai, Haensch, and Schuenemann under "articles" in "Contagionism in History," online at http://www.contagionism.org/.
4. The resemblance between the black rash of plague and the petechiae of typhus contributed to a theory that plague was a more severe form of typhus. See Ann G. Carmichael, "Plague Legislation in the Italian Renaissance," *Bulletin of the History of Medicine* (1981) 57:508–25.
5. The cordon was moved as the plague spread to nearby towns. For the wall, see Claude Tronchon and Thierry Sabot, "*Le Mur de la Peste en Provence*," *Histoire-Genealogie* (May 10, 2007), online at http://www.histoire-genealogie.com/spip.php?article1249.
6. Laurence Brockliss and Colin Jones, *The Medical World of Early Modern France* (Oxford: 1997), 352.
7. Brockliss and Jones, *Medical World of Early Modern France*, 352.
8. François Chicoyneau (1672–1752) was the son of the Chancellor of Montpellier, Michel Chicoyneau, and the son-in-law of Pierre Chirac. He became famous for using mercury "frictions" for syphilis. Julian Martin, "Sauvages' Nosology: Medical Enlightenment in Montpellier," in *The Medical Enlightenment of the Eighteenth Century*, eds. Andrew Cunningham and Roger French (Cambridge: 1990), 113–14; and Junko Therese Takeda, *Between Crown and Commerce: Marseille and the Early Modern Mediterranean* (Baltimore: 2011), 114.
9. Raymond Williamson, "The Plague of Marseilles and the Experiments of Professor Anton Deidier on Its Transmission," *Medical History* (1958) 2:237–52, on 238.
10. Williamson, "Marseilles," 237.
11. Williamson, "Marseilles," 238.
12. The lectures, entitled *Chimie Raisonnée* (Lyon), had an introduction by Pestalossi, who also wrote on the Plague of Marseilles. See Allen G. Debus, "The Paracelsians in Eighteenth-Century France," in *Transformation and Tradition in*

the Sciences: Essays in Honor of I. Bernard Cohen, ed. Everett Mendelsohn (Cambridge: 2003), 193–214, on 198.

13. Allen G. Debus, *The French Paracelsians: The Chemical Challenge to Medical and Scientific Tradition in Early Modern France* (Cambridge: 1991), 146, and same, "Chemistry and the Universities in the Seventeenth Century," *Estudos Avancados* (1990) 4:173–96, doi: 10.1590/S0103-40141990000300009.

14. See chapter 5. The Latin title was *Dissertatio Medica de Morbis Venereis, Cui Adjungitur Dissertatio Medico-Chirurgica de Tumoribus.*

15. Antoine Deidier, *"Avis . . . a l'Auteur de Cette Traduction, pour Servir de Préface . . . "* in same, *Deux Dissertations Médicinales et Chirurgicales, l'Une sur la Maladie Vénérienne . . . L'Autre sur la Nature et la Curation des Tumeurs, Traduction . . . Par un Chirurgien de Paris* (Paris: 1725).

16. Antoine Deidier, *"Avis,"* in *Deux Dissertations.* The student author was M. Sicard. A second, augmented edition of the thesis on venereal disease appeared in Montpellier in 1716 under the name of Giovanni Onorato Raiberti (printed by the widow of H. Pech); a third Latin edition appeared in Rome in 1722. The thesis on tumors was printed in Montpellier in 1711 and reprinted in 1715. The French translation of 1725, *Deux Dissertations,* contains additions including a thesis of 1709 on smallpox. The theses on tumors, smallpox, and plague were again reprinted by D'Houry with additions but without the dissertation on venereal diseases in Paris in 1732 under the title *Traité des Tumeurs contre Nature,* 5th ed.; this was reprinted in 1738. The treatise on venereal disease was enlarged and reprinted by D'Houry in 1735 and 1750. European professors in this period dictated theses for their students to defend. They appeared under the students' names. For a discussion of Linnaeus's practice, see Margaret DeLacy, "A Linnaean Thesis Concerning *Contagium Vivum:* The *"Exanthemata Viva"* of John Nyander and Its Place in Contemporary Thought, with a New Translation by A. J. Cain," *Medical History* (1995) 39:159–85, on 165–70.

17. In the 1720s, Hermann Boerhaave referred one of Sloane's patients, Lady Mary Ferrers, to Deidier and she asked Sloane to persuade her husband to send her to Montpellier. Wayne Wild, *Medicine-by-Post* (Amsterdam: 2006), 98–100.

18. Williamson, "Marseilles," 241.

19. Williamson, "Marseilles," 238–9.

20. Ruth Oratz, "The Plague, Changing Notions of Contagion: London 1665—Marseille 1720," *Synthesis* (1977) 4:4–27, on 17.

21. The disease spread to Avignon and a few neighboring towns within Provence. There has been debate about whether *cordons sanitaires* could have been effective, given the fact that the rats were still free to move about. Brockliss and Jones, *Medical World of Early Modern France,* 350–3, argue that the *cordons* of 1720 and the slightly less effective one of 1668/9 saved the rest of France from plague at the expense of increased mortality within the affected area.

22. Williamson, " Marseilles," 240.

23. Williamson, "Marseilles," 241.

24. Williamson, "Marseilles," 245.

25. Williamson, "Marseilles," 245–6.

26. Antoine Deidier, *Consultations et Observations Médicinales* (Paris: 1754) vol. 3, online from Google.

27. Williamson, "Marseilles," 246.

28. Deidier, *"Cinquième Lettre . . . a . . . M. Montresse,"* Grau de Palaccas near Montpellier, July 6, 1721, in Deidier, *Consultations,* 296–7.

29. Williamson, "Marseilles," 247. By "epidemic," Deidier means carried by poi-
    soned air.
30. Deidier, "*Cinquième Lettre*," in *Consultations*, 301, 372–3.
31. Antoine Deidier, *Discours sur la Contagion de la Peste de Marseille* in the *Traité
    des Tumeurs Contre Nature* (Paris: 1732).
32. Williamson, "Marseilles," 250.
33. Williamson, "Marseilles," 250.
34. Antoine Deidier, *Discours . . . Peste* in *Traité des Tumeurs contra Nature*. For
    Pierre Desault's similar view, see chapter 7.
35. Raymond Williamson, "The Germ Theory of Disease. Neglected Precursors
    of Louis Pasteur: Richard Bradley, Benjamin Marten, Jean-Baptiste Goiffon,"
    Annals of Science (1955) 11, no. 1:44–57, on 53–4. He notes Goiffon was said
    to have inspired Antoine de Jussieu, the botanist. See also Humbert Mollière, *Un
    Précurseur Lyonnais des Théories Microbiennes. J.B. Goiffon et la Nature de la Peste*
    (Basel: 1886).
36. Jean-Baptiste Goiffon, "Avertissement," in Jean-Baptiste Bertrand, comp. *Obser-
    vations Faites Sur la Peste Qui Règne à Present à Marseille et dans la Provence*
    (Lyons: 1721). Bertrand was a physician in Marseilles during the plague. His
    own contagionist history of the epidemic, *Relation Historique de Tout Ce Qui
    S'Est Passé à Marseille Pendent La Dernière Peste*, 2nd ed. (Cologne: 1723), was
    translated by Ann Plumptre as *A Historical Relation of the Plague at Marseilles in
    the Year 1720* (London: 1805). This is not the work entitled *Historical Account
    of the Plague at Marseilles* (London: 1721) often attributed to him, nor does
    Bertrand himself offer an animalcular theory.
37. Goiffon, *Sur la Peste*, my translation, paraphrased and rearranged for coherence.
38. Goiffon, *Sur la Peste*, my translation, paraphrased and rearranged for coherence.
39. Williamson, "Germ Theory," 56; Jean-Baptiste Goiffon, *Relations et Dissertation
    sur la Peste du Gevaudan* (Lyons: 1722).
40. Jean Astruc, *Dissertation sur la Peste de Provence* (Montpellier: 1722) (also trans.
    into Latin by Johann Jacob Scheuchzer, Zurich: 1721), and same, *Dissertation
    sur l'Origine des Maladies Épidémiques et Principalement sur l'Origine de la Peste:
    ou l'On Explique les Causes de la Propagation et de la Cessation de Cette Maladie*
    (Montpellier: 1721), and same, *Dissertation sur la Contagion de la Peste, ou l-On
    Prouve que Cette Maladie est Véritablement Contagieuse . . .* (Toulouse: 1724).
    Astruc's lectures were published in English as *Academical Lectures on Fevers .
    . . Read in the Royal College at Paris* (London: 1747). There appears to be no
    French counterpart to this. The anonymous preface describes Astruc as "my old
    master," so it was probably printed from student notes. In the chapter "Of the
    Pestilential Fever and Plague" (249–273) in this work, he attributes plague to a
    "foreign pestilential contagion," 250. See also Jean Astruc, *Mémoires Pour Servir
    a [sic] L'Histoire de la Faculté de Médecine de Montpellier . . . Revus & Publiés by
    [Anne-Charles] Lorry* (Paris: 1767). In this posthumous work, Astruc wrote that
    he had successfully refuted Chicoyneau's claim that plague was not contagious
    (291); Lorry's *éloge* adds that Astruc "had a complete victory, an example that is
    rare in the Republic of Letters," preface, xlii. Astruc was the son of a Huguenot
    minister who had converted to Catholicism.
41. Brockliss and Jones, *Medical World of Early Modern France*, 751, notes that the
    leader of the contagionist camp, Jean Marie Hecquet, was a Jansenist, but they
    view the heated exchanges between contagionists such as Astruc and anticonta-
    gionists such as François Chicoyneau as a "routine academic squabble" of little

importance. Surely disputes about a disease theory that licensed soldiers to shoot civilians on sight were more than merely academic. Alexander Simpson, "Jean Astruc (1664–1766)—Scholar and Critic," *Proceedings of the Royal Society of Medicine* (1915) 8 (Section of the History of Medicine): 59–71, makes no mention of Astruc's writings on plague.

42. Hector Grasset, *"La Théorie Parasitaire et la Phtisie Pulmonaire au XVIIIe Siècle,"* *France Médicale* (November 17, 1899) trans. as "The Parasitic Theory and Pulmonary Phthisis in the Eighteenth Century," by Thomas C. Minor, *Cincinnati Lancet Clinic,* n.s. 44, whole volume 83 (January 6, 1900): 22–6 and (January 13, 1900): 37–43; and H. F. A. Peypers, *"Un Ancien Pseudo-Précurseur de Pasteur ou le Système d'Un Médecin Anglois sur la Cause de Toutes les Maladies (1726),"* *Janus* (1896/7) 1:57–66, 121–31, and 251–62 offer other examples.

43. The spread of the plague beyond Marseilles into nearby towns, however, led several authors to argue that the quarantine had failed and other causes had arrested the disease. See for example George Pye, *A Discourse of the Plague; Wherein Dr. Mead's Notions are Consider'd and Refuted* (London: 1721).

44. Carmichael, "Plague Legislation in the Italian Renaissance," 511. The first "quarantine" in Ragusa was actually a *"trentino"* imposed on incoming ships for thirty days. Inland cities imposed a "trade and travel" quarantine, 512.

45. Charles Mullett, *The Bubonic Plague and England: An Essay in the History of Preventive Medicine* (Lexington, KY: 1956), 17.

46. Mullett, *Bubonic Plague,* 44.

47. Mullett, *Bubonic Plague,* 44.

48. Mullett, *Bubonic Plague,* 44.

49. Mullett, *Bubonic Plague,* 46.

50. Mullett, *Bubonic Plague,* 66.

51. Mullett, *Bubonic Plague,* 71.

52. Mullett, *Bubonic Plague,* 94. See also Paul Slack, *The Impact of Plague in Tudor and Stuart England* (Oxford: 1985). Slack argues that these measures, including the "shutting up" of families at the very beginning of an epidemic, had a reasonable chance of averting epidemics, see 216–18. Daniel Defoe, *A Journal of the Plague Year* (London: 1722, rpt. Harmondsworth: 1966), 169–85 had a less sanguine view.

53. Mullett, *Bubonic Plague,* 97.

54. Slack, *Impact of Plague,* 208.

55. Mullett, *Bubonic Plague,* 88.

56. Mullett, *Bubonic Plague,* 127.

57. The *Discourse* is dedicated to James Craggs, who had brought this request to Mead.

58. Under torture, William Carstares, an agent for William of Orange and the Ninth Earl of Argyll, named Meade as an associate of men involved in the Rye House Plot. Carstares would become Principal of Edinburgh University. Richard L. Greaves, "Meade, Matthew (1628/9–1699)," *ODNB* (Oxford: 2004), online edn., January 2008 at http://www.oxforddnb.com/view/article/18466.

59. Arnold Zuckerman, "Dr. Richard Mead (1673–1754)," (PhD dissertation, University of Illinois: 1965), 8–9; John Waddington, *Congregational History,* vol. 5: 1850–1880 (London: 1880), 388, online from the Internet Archive at https://archive.org/details/congregationalhistory00wadd. See also Joseph J. Green, "Marshes and Meads," *Friends Quarterly Examiner* (1907) 41:477–90. I thank the Multnomah County Library ILL department for finding this rare

article. Another of Richard Mead's uncles, also named Richard, was the grand-father of Mary Mead, who married the politician John Wilkes in 1747. Thus, Dr. Richard Mead and Wilkes were, by marriage, first cousins once removed. Wilkes became wealthy through this marriage because Mary inherited fortunes from her Quaker uncle William and from her father. Zuckerman, "Mead," 3–5. See also T. L. Underwood, "Edward Haistwell, F.R.S." *N&R* (December 1970) 25, no. 2:179–87.

60. Zuckerman, "Mead," 23. Irenicism is a belief in the value of tolerance and con-ciliation. For Boerhaave's view, see Andrew Cunningham, "Medicine to Calm the Mind: Boerhaave's Medical System, and Why It Was Adopted in Edinburgh," in Cunningham and French, *Medical Enlightenment*, 40–66.

61. For Mead's publication in the *Philosophical Transactions* of Bonomo's letter on the itch mite, see chapter 5 above.

62. Ludmilla Jordanova, "Richard Mead's Communities of Belief in Eighteenth-Century London," in *Christianity and Community in the West*, ed. Simon Ditch-field (Aldershot, UK: 2001), 241–59, argues that Mead was not a Deist because he believed in a living faith, practical philanthropy, and medicine as beneficial, but these traits do not conflict with Deism. I don't see evidence in Mead's work that he was a providentialist who believed God directly intervened in human affairs.

63. Matthew Maty, *Authentic Memoirs of the Life of Richard Mead, M.D.*, (London: 1755), online from Google Books. See also the "Life of Richard Mead" in Mead's *Medical Works*; Zuckerman, "Dr. Richard Mead," and same, "Plague and Contagionism in Eighteenth-Century England: The Role of Richard Mead," *Bulletin of the History of Medicine* (2004) 28, no. 2:273–308; and Anita Guerrini, "Mead, Richard (1673–1754)," *ODNB* (Oxford: 2004); online edn., January 2008 at http://www.oxforddnb.com/view/article/18467.

64. He incorporated his Padua degree at Oxford in 1707 and became a Candidate of the College in 1708. Munk, *Roll.*

65. For Mead and Pope, see Marjorie Hope Nicolson and George S. Rousseau, "*This Long Disease, My Life" Alexander Pope and the Sciences* (Princeton: 1968). For Burnet, see chapters 2 and 8 above.

66. The phrase comes from the "Life of Richard Mead," in *The Medical Works of Richard Mead, M.D. with an Account of the Life and Writings of the Author* (Leyden: 1752, London: 1762, Edinburgh: 1763, Dublin: 1767), 1:10. The Leyden edition is cited by the Wellcome Library catalogue but does not appear in Worldcat; I used the Google Books reprint of the Edinburgh edition to find this quotation. I have not determined the author of this "Life" but it was not the same as Matthew Maty's biography of 1755. "Hearty Whig" is repeated in Mac-Michael, *The Gold-headed Cane* (New York: 1926), 71. For Bentley, see chapters 6 and 7 above.

67. W. B. Howie, "Sir Archibald Stevenson, His Ancestry, and the Riot in the College of Physicians in Edinburgh," *Medical History* (1967) 11:269–84, on 283. Mead also helped the Nonjuror Thomas Hearn regain his post at Oxford. Guerrini, "Mead." Sloane rescued the Jacobite botanist Patrick Blair from Newgate after the "Battle of the Books" between Bentley and Charles Boyle (later Lord Orrery), see Richard Foster Jones, *Ancients and Moderns: A Study of the Rise of the Scientific Movement in Seventeenth-Century England* (Berkeley: 1937).

68. For example, when the French refugee Michel Maittaire needed copies of John Toland's scandalous work *The Pantheisticon*, he found two: one in Sloane's library and one in Mead's. Margaret C. Jacob, *The Radical Enlightenment: Pantheists,*

Freemasons and Republicans (London: 1981), 275. Mead owned one of just three surviving copies of Servetus's *Christiansmi Restituto;* most were burned when their author was condemned. Yvonne Hibbott, "Medical Books of the Sixteenth Century," in *Thornton's Medical Books, Libraries and Collectors,* 3rd ed., ed. Alain Besson (Aldershot, UK: 1990), 61.

69. Sloane wrongly believed that Woodward had been the author of *The Transactioneer;* the true author was William King. Woodward accused Sloane and Petiver of spreading rumors about this. In 1710, Woodward had been thrown off the Council of the Royal Society for rudeness to Sloane. See Levine, *Dr. Woodward's Shield,* 85–92, and Eric St. John Brooks, *Sir Hans Sloane: The Great Collector and His Circle* (London: 1954), 109.

70. There are several accounts of the exchange that ended the duel. According to Woodward, he slipped during the duel. When Mead disarmed him, and told Woodward to ask for his life, he replied that he would have asked for his life to avoid taking Mead's physic, but Mead's sword was completely harmless. Levine, *Dr. Woodward's Shield,* 17. A variant of this quip was also attributed to the artist Geoffrey Kneller following a dispute with John Radcliffe over a garden gate between their properties.

71. The "Scriblerians" were a circle of writers associated with the satirist Jonathan Swift and the poet Alexander Pope, who attacked modern learning under the pseudonym of Martinus Scriblerus. See Richard G. Olson, "Tory-High Church Opposition to Science and Scientism in the Eighteenth Century: The Works of John Arbuthnot, Jonathan Swift, and Samuel Johnson," in *The Uses of Science in the Age of Newton,* ed. James G. Burke (Berkeley: 1983), 171–204. Arbuthnot also appears in William R. Le Fanu, "The Lost Half-Century in English Medicine, 1700–1750," *Bulletin of the History of Medicine* (1971) 46, no. 4:319–48. Other articles by this author spell his name "LeFanu."

72. For evidence of Woodward's irascibility, see his letters to James Jurin on February 13 and 15, 1719, which apparently bewildered Jurin. Andrea Rusnock, *The Correspondence of James Jurin, (1684–1750): Physician and Secretary to the Royal Society* (Amsterdam: 1996), 78–82.

73. Slack, *Impact of Plague,* 327.

74. Richard Mead, *A Short Discourse Concerning Pestilential Contagion* (London: 1720), 3, italics omitted. This work saw seven editions in 1720; an eighth enlarged edition appeared in 1723 and a ninth corrected edition in 1744. See the introduction to *The Medical Works of Mead,* vii.

75. Mead, *Short Discourse,* 12.

76. Mead, *Short Discourse,* 14.

77. Mead, *Short Discourse,* 17.

78. Richard Mead, *A Mechanical Account of Poisons, in Several Essays* (1702), 4th ed. (London: 1747), 15.

79. Mead, *Short Discourse,* 28. For British quarantines, see John Booker, *Maritime Quarantine: The British Experience, c. 1650–1900* (Aldershot, UK: 2007).

80. Mead, *Short Discourse,* 43.

81. Mead, *Short Discourse,* 48.

82. This report is described in Slack, *Impact of Plague,* 334. Slack notes on 407 that Charles I's Huguenot physician Theodore de Mayerne had first recommended a "Board of Health" along with special clothing in 1631.

83. Arbuthnot had a degree from St. Andrews, Mead from Padua, and Sloane from the University of Orange. All three were FRS. Two of the three, Mead and

Sloane, came from Dissenting families; Arbuthnot was a Scot; Sloane was Irish. All three also supported inoculation.

84. Slack, *Impact of Plague*, 334. Sloane was an intimate friend of Samuel Pepys, who had lived through the plague of 1665/6. Brooks, *Sloane*, 148.

85. A severe epidemic in the Baltic from 1709 until 1713 had affected Denmark, Sweden, Estonia, Poland, Prussia, and northern Germany. See Karl-Erik Frandsen, *The Last Plague in the Baltic Region, 1709–1713* (Copenhagen: 2010).

86. Charles F. Mullett, "The English Plague Scare of 1720–23," *Osiris* (1936) 11:487–8.

87. Mullett, *Plague Scare*, 488–90.

88. Mullett, *Plague Scare*, 493. For Gibson, see chapter 2 above.

89. Mullett, *Plague Scare*, 504. Colbatch's treatise, *A Scheme for Proper Methods to Be Taken, Should It Please God to Visit Us with the Plague* (London: 1721) is online on the World Health Organization's Historical Collection: "Rare Books on Plague, Smallpox and Epidemiology" at http://www.who.int/library/collections/historical/en/index5.html.

90. Brookes, *Remarkable Pestilential Disorders* (London: 1722), 48, and see chapter 6 above.

91. Brookes, *Remarkable Pestilential Disorders*, 48.

92. Browne's *Practical Treatise* is available online on the WHO Historical Collection website at http://www.who.int/library/collections/historical/en/index5.html, which spells his name "Brown." Browne should be distinguished from Richard Browne, a partner in the Oracle clinic.

93. On Hodges's etiology and its scientific underpinnings, see Robert Frank, *Harvey and the Oxford Physiologists* (Berkeley: 1980), 240–1.

94. (London: 2nd ed., 1722). The first edition, entitled *Dr. Mead's Short Discourse Explain'd, Being a Clearer Account of Pestilential Contagion, and Preventing* (London: 1721), is online from Google.

95. Benjamin Colman, *Some Observations on Receiving the Small-Pox by Ingrafting or Inoculating* (Boston: 1721), 503.

96. Pye, *Discourse*, 2, and see Mullett, *Bubonic Plague*, 277–8, and Slack, *Impact of Plague*, 329. The anticontagionist Charles Maclean found Pye's refutation effective. See his *Results of an Investigation, Respecting Epidemic and Pestilential Diseases* (London: 1817), 362.

97. Mullett, *Bubonic Plague*, 292. This was not Sir John Pringle (1707–1772) later President of the Royal Society. The anonymous *Considerations on the Nature, Causes, Cure and Prevention of Pestilences . . . by The Free-Thinker* (London: 1721) online from Google, also opposed cordons.

98. Richard Blackmore, *A Discourse upon the Plague with a Prefatory Account of Malignant Fevers in Two Parts* (London: 1721). I thank the Bodleian Library, Oxford, for access to the first edition of this work. The second edition is on the World Health Organization Historical Collections website at http://www.who.int/library/collections/historical/en/index5.html. He also argued that the "worms" seen in the bodies of victims were the effect of putrefaction, not the cause. Mullett, *Plague Scare*, 497, commented that "of all the medical writers at this period, none mirrored more completely ordinary enlightened opinion concerning the plague than . . . Blackmore."

99. Blackmore, *Discourse*, 28–9. For Blackmore's political ties, see Adrian Wilson, "The Politics of Medical Improvement," in Cunningham and French, *Medical Enlightenment*, 29–30; for his opposition to inoculation, see chapter 8.

100. Published in London. For Bradley's long relationship with Hans Sloane, see chapter 6.

101. (London: 2nd ed. 1718); see also Melvin Santer, "How it Happened That . . . *New Improvements of Planting and Gardening* . . . Provided the First Report of an Environment That Contained Green Sulfur Photosynthetic Bacteria," *N&R* (2007) 61, no. 3:327–32, online at http://rsnr.royalsocietypublishing.org /content/61/3/327.

102. Williamson, "Germ Theory," 47–8.

103. Bradley, *Plague of Marseilles*, 2nd ed. (London: 1721), 20–21, online from Google.

104. Charles Singer, *The Development of the Doctrine of Contagium Vivum 1500– 1750* (London: 1913), 14; and see Williamson, "Germ Theory," 57. Cf. Petty's unpublished theory, chapter 4.

105. *A New Discovery*, 4–5. The preface was dated August 31, 1721; the publishers were T. Bickerton and J. Wilford.

106. Defoe, *Journal* (Harmondsworth: 1966), 92–3. This passage is sometimes anachronistically attributed to the date of Defoe's subject, the London plague of 1665.

107. Quoted in Williamson, "Germ Theory," 49. The *General Treatise* was published in separate issues 1721?–1724 and then collected into two volumes in 1726. See also Frank N. Egerton, "A History of the Ecological Sciences, Part 20: Richard Bradley, Entrepreneurial Naturalist," *Bulletin of the Ecological Society of America* (April 2006) 87, no. 2:117–27, and "A History of the Ecological Sciences, Part 29: Plant Disease Studies during the 1700s," *Bulletin of the Ecological Society of America* (2008) 89, no. 3:231–44, search under "Egerton" online at http:// www.esajournals.org/.

108. Apart from the cost to merchants and traders, it was estimated in 1721 that more than two hundred thousand soldiers would be needed for a *cordon sanitaire* around London and other large cities. See Mullett, *Bubonic Plague*, 272. This, however, was exaggerated as the Austrians maintained a cordon of more than one thousand miles with eleven thousand men, drawn from a special military territory along the border that contained about one hundred thousand resident soldiers. Gunther E. Rothenberg, "The Austrian Sanitary Cordon and the Control of the Bubonic Plague: 1710–1871," *Journal of the History of Medicine* (January 1973) 28, no. 1:15–23, on 17–19.

109. Mullett, "Plague Scare," 491.

110. Booker, *Maritime Quarantine*, see esp. Appendix 2: 579–81. Fears about the importation of yellow fever and cholera further complicated the discussion in the early nineteenth century.

111. Adrian Wilson, *The Making of Man-Midwifery: Childbirth in England, 1660– 1770* (Cambridge: 1995), 83. On Manningham, see 82–5 and 114.

112. The Scottish William Cockburn, also identified by Wilson as a Whig "Deventerian" opposed to forceps, is described by Anita Guerrini as a Tory later in his life. He was Swift's physician. Guerrini, "Newtonian Matter Theory, Chemistry and Medicine, 1690–1713," (PhD diss. Indiana University, 1983), 36. Manningham's iatromechanism resembles that of other "Tory Newtonians" studied by Guerrini, such as Cheyne, Keil, and Pitcairne.

113. Donald Gray, "Manningham, Thomas (d.1722)," *ODNB* (Oxford: 2004), online at http://www.oxforddnb.com/view/article/17983.

114. Richard Manningham, *The Plague No Contagious Disease* (London: 1744), 8.

115. Manningham, *No Contagious Disease*, 33.

116. For the transfer of the Council of Trent to Bologna because of a typhus epidemic, see chapter 1.

117. Manningham, *No Contagious Disease*, 14.

118. Manningham, *No Contagious Disease*, 9.

119. This view was derived from Boerhaave's disease theory.

120. Lobb (1678–1763, MD Glasgow) had been a Dissenting minister. His father Stephen Lobb, a prominent Independent pastor, was in and out of jail during the Restoration. Gordon Goodwin, "Lobb, Theophilus (1678–1763)," rev. Lynda Stephenson Payne, *ODNB* (Oxford: 2004), online ed., January 2008 at http://www.oxforddnb.com/view/article/16879.

121. Lobb, *Letters Relating to the Plague*, 123.

122. Illustrated in Thomas Bartholin's *Historiarum Anatomicarum* (Copenhagen: 1664–1671), National Library of Medicine, images from the History of Medicine, online at http://ihm.nlm.nih.gov. See also the article in Wikipedia: http://en.wikipedia.org/wiki/Plague_doctor_costume. A search under "plague doctor" in Google Images retrieves many variants with questionable captions.

123. Mackenzie to Dr. Mead, Constantinople, November 23, 1751, in "Extracts of Several Letters . . . concerning the Plague," *Philosophical Transactions* (1753) 47:384–95, on 394. The letters were communicated by Dr. John Clephane. Mackenzie sent a "Further Account" to Clephane, which appears in the same volume on 514–16. In 1764, Mackenzie also sent Sir James Porter "An Account of the Plague at Constantinople," which appeared in the *Philosophical Transactions* (1764) vol. 54 and was reprinted in the *London Magazine* in 1765. His firsthand accounts were a source for authors such as Brownrigg (see below).

124. "Art[icle] IV. An Inquiry concerning the Cause of the Pestilence, and the Diseases in Fleets and Armies . . . " (Edinburgh: 1759), *The Critical Review* (1759) 8:16–28. Trained as a surgeon, Smollett had MD degrees from Geissen and Aberdeen. For Smollett's identity as a Whig, see Robin Fabel, "The Patriotic Briton: Tobias Smollett and English Politics, 1756–1771," *Eighteenth-Century Studies* (Autumn 1974) 8, no. 1:100–14.

125. Joshua Dixon, *The Literary Life of William Brownrigg, M.D.[,] F.R.S* (London: 1801), 65–6.

126. Brownrigg, *Considerations*, 19. Brownrigg (MD Leyden, 1737) founded the Whitehaven Dispensary in 1783. There is no evidence he ever saw anyone with the plague. An innovative chemist whose work inspired Priestley, he was the first to recognize that there were distinct gases, not just different sorts of air. According to Dixon, he was a devout Christian but was accused of "infidelity and irreligion." See also Herbert T. Pratt, "Brownrigg, William (1711–1800)," *ODNB* (2004), online edn., May 2006 at http://www.oxforddnb.com/view/article/3719.

127. Mullett, *Bubonic Plague*, 320. Guthrie, son of an Edinburgh Episcopalian minister, trained as a surgeon, studied with William Cullen and obtained an MD from St. Andrews in 1770. He lived in Russia. He became an FRS in 1782 but was never a member of the London College of Physicians. Eric H. Robinson, "Guthrie, Matthew (1743–1807)," *ODNB* (Oxford: 2004) online at http://www.oxforddnb.com/view/article/40507.

128. Mullett, *Bubonic Plague*, 320. Charles de Mertens, *An Account of the Plague Which Raged at Moscow, in 1771*, trans. Richard Pearson (London: 1799),

online from the Medical Heritage Library at https://archive.org/details
/accountofplaguew00mert.

129. Black, *Historical Sketch*, 243–5.

130. William Black (1749–1829, MD Leyden, 1771) became a Licentiate in 1787.
He was a medical statistician.

131. William Cullen, *First Lines of the Practice of Physic*, 3rd ed. (Edinburgh: 1781),
quoted in Mullett, *Bubonic Plague*, 318–19.

132. *The Works of William Cullen, M.D.*, ed. John Thomson (Edinburgh: 1827)
2:183.

133. John Howard, *Account of the Principal Lazarettos in Europe* (Warrington:
1789), 32.

134. Howard died in Cherson, in the present Ukraine, after riding through a tem-
pest to attend a dying woman and possibly contracting typhus from her. The
most dramatic, though not always accurate, account of his life can be found
in William Hepworth Dixon, *John Howard and the Prison-World of Europe*
(London: 1850), online from the Hathi Trust at http://catalog.hathitrust
.org/Record/008976226. John Aikin wrote a biography of his friend: *A View
of the Character and Public Services of the Late John Howard, Esq. LL.D., F.R.S.*
(London: 1792). See also Leona Baumgartner, "John Howard and the Public
Health Movement," *Bulletin of the History of Medicine* (1937) 5:489–508; John
Ransom, "John Howard on Communicable Diseases," *Bulletin of the History of
Medicine* (1937) 5:131–47; and Charles-Edward Amory Winslow, *The Conquest
of Epidemic Disease: A Chapter in the History of Ideas* (Madison: 1980), 236–8.

135. Howard, *Lazarettos*, 32.

136. Dr. John Jebb died in 1786. The other possible "Dr. Jebb" is Sir Richard Jebb,
1729–1787, a close friend of John Coakley Lettsom. His father, Dr. Samuel
Jebb, had died in 1772. For John Jebb's writings on prison construction and
government, see Anthony Page, *John Jebb and the Enlightenment Origins of
British Radicalism* (Westport, CT: 2003), 231–5.

137. Gilbert Blane, "A Letter . . . to Rufus King, Esq. Minister Plenipotentiary
from the States of America," in *Observations on the Diseases of Seamen*, 3rd ed.
(London: 1799), 608, online from Google.

138. Charles Maclean, *Results of an Investigation respecting Epidemic and Pestilential
Diseases; Including Researches in the Levant, concerning the Plague*, 2 vol. in 1
(London: 1817), 361, online from Google.

139. Erwin Ackerknecht, "Anticontagionism between 1821 and 1867," *Bulletin of
the History of Medicine* (1948) 22:563–98; an abridged version is in *The International Journal of Epidemiology* (2009) 38, no. 1:7–21, doi: 10.1093/ije
/dyn254.

140. Mullett, *Bubonic Plague*, 371.

141. Ackerknecht, "Anticontagionism."

142. Blane was a Licentiate and FRS; he never became a Fellow of the College. Black
wrote: "By some monkish abuse . . . the . . . privileges of the London College
are monopolized by a small club of Physicians calling themselves Fellows, whose
only merit, or pretensions to superiority, consists in having studied Medicine at
Oxford or Cambridge," *Historical Sketch*, 162.

143. Another book will take a closer look at the formation and ideology of this group
of reformers.

144. Richard Mead, *A Treatise concerning the Influence of the Sun and Moon upon Human Bodies, and the Diseases thereby Produced*, trans. Thomas Stack (London: 1748).

145. Theodore Brown, *The Mechanical Philosophy and the 'Animal Oeconomy': A Study in the Development of English Physiology in the Seventeenth and Early Eighteenth Century* (New York: 1981), 291.

146. Roger French, "Sickness and the Soul: Stahl, Hoffmann and Sauvages on Pathology," in Cunningham and French, *Medical Enlightenment*, 88–110, on 104.

147. Richard Mead, *A Mechanical Account of Poisons, in Several Essays*, 4th ed. (London: 1747), 303.

148. Michael W. Flinn, *The European Demographic System, 1500–1820* (Baltimore: 1981), 60–61, argues that local action in the seventeenth century and national government action in the eighteenth century, particularly quarantines in ports and the Hapsburg *cordon sanitaire* across Europe, prevented epidemics of plague in Western Europe. See also Edward A. Eckert, "The Retreat of Plague from Central Europe, 1640–1720: A Geomedical Approach," *Bulletin of the History of Medicine* (Spring 2000) 74, no. 1:1–28; Peter Christensen, "'In These Perilous Times': Plague and Plague Policies in Early Modern Denmark," *Medical History* (2003) 47:413–50; Paul Slack, *The Impact of Plague*, and same, "The Disappearance of Plague: An Alternative View," *Economic History Review* (August 1981) 34, no. 3:469–76; and Rothenberg, "The Austrian Sanitary Cordon."

CONCLUSION

1. Charles-Edward Amory Winslow, *The Conquest of Epidemic Disease: A Chapter in the History of Ideas* (Madison: 1980), 159–61. Catherine Wilson pondered the decline of interest in *contagium vivum* in *The Invisible World* (Princeton: 1995), 172–5, concluding that it resulted from the very ubiquity of microorganisms, improved microscopes that failed to show the entities previously blamed for illness, and resistance to the idea that God created venomous parasites.

2. Winslow, *Conquest*, 175.

3. A. E. Gunther, *The Founders of Science at the British Museum, 1753–1900* (Halesworth, Sussex: 1980).

4. Peter Razzell, *The Conquest of Smallpox* (Firle, Sussex: 1977), 40–2, Donald Hopkins, *Princes and Peasants: Smallpox in History* (Chicago: 1983), 58. See also Charles Creighton, *A History of Epidemics in Britain* (Cambridge: 1894), vol. 2, 504.

BIBLIOGRAPHY

Note: indications of online versions indicate a resource for *a* full-text copy of the relevant version of a work, not necessarily *the* copy, or the only copy used. At the time of writing, some ECCO and EEBO materials are open access.

PRIMARY UNPRINTED SOURCES

17th–18th Century Burney Collection Newspapers. Digitized by Gale from the British Library, access provided by the Wellcome Library.

"Electronic Enlightenment Project." Eighteenth-century correspondence hosted by the Bodleian Libraries of the University of Oxford, access provided by the Wellcome Library. Online at http://www.e-enlightenment.com/.

Letter from Benjamin Franklin to Samuel Mather, Passy, May 12, 1784. Massachusetts Historical Society. Online at http://www.masshist.org/database/533.

Royal Society:

"The Raymond and Beverly Sackler Archive Resource." Biographical Records: https://royalsociety.org/library/collections/biographical-records/.

"Classified Papers." Volume 18ii, "Desaguliers papers" 1714–1737, no. 14. John Theophilus Desaguliers, "Account of the Appearance of Small Pox Seen through the Microscope," read June 23, 1723. Online at https://royalsociety.org/library/collections/scientific-papers/.

Westfall, Richard. "Catalog of the Scientific Community in the 16th and 17th Centuries." Online at Galileo Project, http://galileo.rice.edu/lib/catalog.html.

REFERENCE WORKS

Innes-Smith, Robert William. *English-Speaking Students of Medicine at the University of Leyden.* Edinburgh: Oliver and Boyd, 1932.

Munk, William. *The Roll of the Royal College of Physicians of London.* Vols. 1 and 2. London: Longman, Green, Longman and Roberts, 1861.

The Oxford Dictionary of National Biography, edited by H. C. G. Matthew and Brian Harrison. Oxford University Press, 2004. Online (by subscription) at http://www.oxforddnb.com/public/index.html.

Robinson, F. J. G., and Peter John Wallis. *Book Subscription Lists: A Revised Guide.* Newcastle upon Tyne: Project for Historical Biobibliography, 1975.

Wallis, Peter John and Ruth Wallis. *Book Subscription Lists: Extended Supplement to the Revised Guide.* Newcastle upon Tyne: Project for Historical Biobibliography, 1996.

Wallis, Peter John, Ruth Wallis, and T. D. Whittet. *Eighteenth-Century Medics: Subscriptions, Licenses, Apprenticeships.* 2nd ed. Newcastle upon Tyne: Project for Historical Biobibliography, 1988.

PRIMARY PRINTED SOURCES

An Address to the College of Physicians, and to the Universities of Oxford and Cambridge; Occasion'd by the late Swarms of Scotch and Leyden Physicians. London: M. Cooper, 1747, online from ECCO.

Andry de Boisregard, Nicholas. *An Account of the Breeding of Worms in Human Bodies . . . with letters . . . from N. Hartsoeker . . . and G. Baglivi . . . from the French . . .* London: H. Rhodes and A. Bell, 1701.

Apperley, Thomas. *Observations in Physick, both Rational and Practical. With a Treatise of the Small-Pox.* London: W. Innys and J. Leake, 1731.

Astruc, Jean. *Academical Lectures on Fevers . . . Read in the Royal College at Paris.* London: J. Nourse, 1747.

———*Dissertation sur l'Origine des Maladies Épidémiques et Principalement sur l'Origine de la Peste . . .* Montpellier: Jean Martel, 1721. Online from the Bibliothèque Interuniversitaire de Médecine, http://www.bium.univ-paris5.fr/histmed/medica/cote?77068x01.

———. *A Treatise of the Venereal Disease in Six Books,* translated by William Barrowby. London: W. Innys and R. Manby, 1737.

———. *A Treatise of the Venereal Disease in Nine Books.* Vol. 1. Translated from the last Latin ed. printed at Paris. London: W. Innys and J. Richardson, C. Davis, J. Clarke, R. Manby, and H. S. Cox, 1754. Online from ECCO.

Ayyadurai, Saravanan, Florent Sebbane, Didier Raoult, and Michel Drancourt. "Body Lice, *Yersinia pestis Orientalis,* and Black Death," [Letter]. *Emerging Infectious Diseases* 16, no. 5 (May 2010). Online from the Centers for Disease Control, http://www.cdc.gov.eid/content/16/5/892.htm.

Barry, Edward. *A Treatise on a Consumption of the Lungs.* Dublin: 1726.

Bates, Thomas. "A Brief Account of the Contagious Disease Which Raged among the Milch Cows near London in the Year 1714 and of the Methods That Were Taken for Suppressing It." *Philosophical Transactions* 30 (1718): 872–85.

Bellers, John. *John Bellers: His Life, Times and Writings.* Edited by George Clarke. London: Routledge and Kegan Paul, 1987.

Bertrand, Jean-Baptiste. *Observations Faites sur la Peste Qui Règne à Présent à Marseille et dans la Provence.* Lyons: A. Laurens, 1721.

Birch, Thomas. *The History of the Royal Society of London for Improving of Natural Knowledge from its first Rise.* Vol. 2. London: A. Millar, 1756.

Blackmore, Richard. *A Discourse upon the Plague with a Prefatory Account of Malignant Fevers in Two Parts.* London: John Clark, 1721.

———. *A Treatise upon the Small Pox in Two Parts. Part I. Containing a Description . . . Part Two, a Dissertation upon the Modern Practice of Inoculation.* London: John Clark: 1723.

Blancard, Stephen (Blankaart). *The Physical Dictionary . . .* 4th ed. London: Sam Crouch, 1702.

Blane, Gilbert. *Observations on the Diseases of Seamen.* London: Murray and Highley, 1799.

Boerhaave, Herman. *Boerhaave's Aphorisms: Concerning Knowledge and Cure of Diseases . . . 1715.* Translated by J. Delacoste, MD. London: A. Bettesworth and C. Hitch; W. Innys and R. Manby, 1725. Online from ECCO.

Bonomo, [Giovanni Cosimo]. "An Abstract of Part of a Letter from Dr. Bonomo to Signior Redi Containing Some Observations concerning the Worms of Humane Bodies." Translated by Richard Mead. *Philosophical Transactions* 23 (1702/3): 1296–9. Online from the Royal Society, doi: 10.1098/rstl.1702.0040.

Borello, Petro (Pierre Borel). *Observationum Microscopicarum Centuria*. The Hague: Adriani Vlacq, 1656. Online from the Swiss Electronic Library, http://dx.doi .org/10.3931/e-rara-4463.

Boyle, Robert. *The Philosophical Works of the Honourable Robert Boyle, Esq., to Which Is Prefixed the Life of the Author*. 5 vols. London: A. Millar, 1744.

Bradley, Richard. *The Plague at Marseilles Considered*. London: W. Mears, 1721.

———. *Precautions against Infection . . . on Account of the Dreadful Plague in France*. London: Thomas Corbett, n. d.

Brookes, R[ichard]. *A History of the Most Remarkable Pestilential Distempers that Have Appeared in Europe for Three Hundred Years Last Past*. London: A. Corbett and J. Roberts, 1721.

Browne, Joseph. Antidotaria; *or, A Collection of Antidotes against the Plague, and Other Malignant Diseases. Together with . . . Remarks . . . Shewing the Necessity of a Farther Reformation of Their New London Dispensatory*. London: J. Wilcox, 1721.

———. *A Practical Treatise of the Plague, and All Pestilential Infections . . . with a Prefatory Epistle Addressed to Dr. Mead . . .* London: J. Wilcox, 1721.

Brownrigg, William. *Considerations on the Means of Preventing the Communication of Pestilential Contagion*. London: Lockyer Davis, 1771.

Castle, George. *The Chymical Galenist . . . In Which are Some Reflections upon a Book Intituled,* "Medela Medicinae." London: Henry Twyford and Timothy Twyford, 1667.

The Censor Censured; or, the Antidote Examin'd, Wherein the Designs of Dr. Pitt, and the Dispensary Physicians are Detected. London: Booksellers of London and Westminster, 1704.

Charleton, Walter. *Physiologia Epicuro-Gassendo-Charletoniana*. London: Thomas Heath, 1654. Online from EEBO.

Cheyne, George. *A New Theory of Acute and Slow Continued Fevers*. London: G. Strahan, 1722.

[Chicoyneau, François]. *Traité des Causes, des Accidens et de la Cure de la Peste*. Paris: Pierre-Jean Mariette, 1744. Online at http://www.biusante.parisdescartes.fr /histmed/medica/cote?25040.

Cockburn, William. *The Present Uncertainty in the Knowledge of Medicines, in a Letter to the Physicians in the Commission for Sick and Wounded Seamen*. London: Benjamin Barker, 1703.

———. *Sea Diseases; or, A Treatise of their Nature, Causes and Cure*. 2nd ed. London: Gor. Strahan, 1706.

Cogrossi, Carlo Francesco. *New Theory of the Contagious Disease among Oxen (1714)*. Translated by Dorothy M. Schullian. Rome: Sixth International Congress of Microbiology, 1953.

Colbatch, John. *A Scheme for Proper Methods to Be Taken, Should It Please God to Visit Us with the Plague*. London: J. Roberts and A. Dodd, 1721. Online from the World Health Organization Historical Collection, http://whqlibdoc.who.int /rare-books/a57045.pdf.

A Collection of Yearly Bills of Mortality, from 1657 to 1758 Inclusive. London: A. Millar, 1759.

Colman, Benjamin. *Some Observations on Receiving the Small-Pox by Ingrafting or Inoculating*. Boston: Samuel Gerrish, 1721.

Considerations on the Nature, Causes, Cure, and Prevention of Pestilences: Being a Collection of Papers, Published on that Subject by The Free-Thinker. London: W. Wilkins, 1721.

"Of Contagion, the Chief Cause of a Plague." Unsigned note in T. Lucretius Carus, *Of the Nature of Things*. Translated by Thomas Creech. Vol. 2. London: George Sawbridge, 1714, 776–82, online from Google.

[Coward, William]. *Second Thoughts Concerning Human Soul . . .* 2nd ed. Corrected and enl. London: A. Baldwin, 1704.

Cullen, William. *Synopsis Nosologiae Methodicae*. 1st ed. Edinburgh: n. p., 1769.

———. *The Works of William Cullen, M.D.* Edited by John Thomson. 2 vols. Edinburgh and London: William Blackwood and T. and G. Underwood, 1827.

Defoe, Daniel. *A Journal of the Plague Year* (1722). Harmondsworth, UK: Penguin Books, 1966.

Deidier, M. Antoine. *Chimie Raisonnée*. Lyon: Marcellin Duplain, 1715. Online from Google.

———. *Consultations et Observations Médicinales*. Vol. 3. Paris: Jean-Thomas Herissant, 1754. Online from Google.

———. *Deux Dissertations Médicinales et Chirurgicales, L'Une sur la Maladie Vénérienne . . . L'Autre sur la Nature & la Curation des Tumeurs*. Paris: Charles-Maurice D'Houry, 1725, online from BIU Santé, http://www2.biusante.parisdescartes .fr/livanc/?do=livre&cote=33104x03.

———. *Traité des Tumeurs contre Nature*. 5th ed. Paris: D'Houry, 1732. Online from Google.

Desault, Pierre. *A Treatise of the Venereal Distemper . . . with Two Dissertations: The First on Madness from the Bite of Mad Creatures; the Second on Consumptions*. Translated by John Andree, MD. London: John Clarke, 1738, online from ECCO.

Dr. Mead's Short Discourse Explain'd; or, His Account of Pestilential Contagion, and Preventing, Exploded. 2nd ed. London: W. Boreham, 1722.

Drake, James. *Anthropologia Nova; or, A New System of Anatomy Describing the Animal Oeconomy, and a Short Rationale of Many Distempers . . .* London: W. and J. Innys, 1727.

Ettmülleri, Michaelis (Michael Ettmüller). *Operum Omnium Medico-Phisicorum*. New ed., Vol. 1. Lyon: Thomas Amaulry, 1690. Online from Google.

The Explainer [pseud.]. *Distinct Notions of the Plague, with the Rise and Fall of Pestilential Contagion*. London: 1722. Online from ECCO.

Fernel, Jean. *On the Hidden Causes of Things*. Edited and translated by John M. Forrester. Leiden: Brill, 2005.

Fothergill, John. *Chain of Friendship: Selected Letters of Dr. John Fothergill of London, 1735–1780*. Edited by Betsy C. Corner and Christopher C. Booth. Cambridge: Belknap Press of Harvard University Press, 1971.

———. *The Works of John Fothergill, M.D.* Edited by John Coakley Lettsom. London: Charles Dilly, 1783.

Fracastorii, Hieronymi (Girolamo Fracastoro). *De Contagione et Contagiosis Morbis et Eorum Curatione, Libri III*. Annotated and translated by Wilmer Cave Wright. New York: G. P. Putnam's Sons, 1930.

Fuller, Thomas. *Exanthematologia; or, An Attempt to Give a Rational Account of Eruptive Fevers, Especially of Measles and Smallpox*. London: C. Rivington and S. Austen, 1730.

[Goiffon, Jean Baptiste]. "Avertissement." In Jean-Baptiste Bertrand, *Observations Faites sur la Peste qui Règne à Présent à Marseille et dans la Provence*. Lyon: Andre Laurens, 1721.

———. *Relations et Dissertation sur la Peste du Gevaudin*. Lyon: Pierre Valfray, 1722.

Grew, Nehemiah. *Cosmologia Sacra* . . . London: W. Rogers, S. Smith and B. Walford, 1701.

Groenevelt, John (Johannes or Jan Groeneveldt; John Greenfield). Appendix to *The Grounds of Physick* . . . [by Francisci Zypaei (Franz or François Vanden Zype)]. Translated from the Latin. London: W. Taylor, J. Osborn, J. Pemberton, 1715. Online from Google.

Guide, Philip. *A Kind Warning to a Multitude of Patients, Daily Afflicted with Different Sorts of Fevers.* London: J. Downing, 1710. Online from ECCO.

Harvey, Gideon. *A Discourse of the Plague* . . . *Together with the State of the Present Contagion.* London: Nath. Brooke, 1665. Online from EEBO.

———. *Great Venus Unmasked; or, a More Exact Discovery of the Venereal Evil, or French Disease* . . . 2nd ed. London: Nath. Brook, 1672. Online from EEBO.

———. *Little Venus Unmask'd* . . . 2nd ed. London: William Thackery, 1670. Online from EEBO.

———. *Morbus Anglicus; or, The Anatomy of Consumptions.* London: Nathaniel Brook, 1666. Online from EEBO.

———. *A New Discourse of the Smallpox, and Malignant Fevers, with an Exact Discovery of the Scorvey.* London: James Partridge, 1685. Online from EEBO.

Hodges, Nathaniel. "An Account of the First Rise, Progress, Symptoms and Cure of the Plague, being the Substance of a Letter from Doctor Hodges to a Person of Quality," in *A Collection of Very Valuable and Scarce Pieces Relating to the Last Plague in the Year 1665.* 2nd ed. London: J. Roberts, 1721. Online from EEBO.

———. *Loimologia; or, an Historical Account of the Plague in London in 1665* . . . *to which is Added, An Essay on the Different Causes of Pestilential Diseases* . . . *by John Quincy.* London: E. Bell and J. Osborn, 1721.

Howard, John. *An Account of the Principal Lazarettos in Europe with Various Papers Relative to the Plague* . . . Warrington: W. Eyres, 1789.

"An Inquiry concerning the Cause of the Pestilence, and the Diseases in Fleets and Armies" by a Physician [Alexander Bruce]. (Anonymous review). *Critical Review* 8 (1759): 16–28. Online from Google.

Jurin, James. *The Correspondence of James Jurin (1684–1750): Physician and Secretary to the Royal Society.* Edited by Andrea Rusnock. Amsterdam: Rodopi, 1996.

King, William. *The Original Works of William King.* 3 vols. London: printed for the editor [John Nichols], 1776. Online from the Internet Archive.

[King, William]. *The Present State of Physick in the Island of Cajamai to Members of the R.S.* London: 1710(?). Online from ECCO.

Layard, Daniel-Peter. "A Letter . . . Relative to the Distemper among the Horned Cattle." *Philosophical Transactions* 70 (1780): 536–45.

Le Clerc, D[aniel]. *A Natural and Medicinal History of Worms, Bred in the Bodies of Men and Other Animals.* Translated by J[oseph] Browne. London: J. Wilcox, 1721. Online from the Internet Archive.

A Letter to the Reverend Mr. Massey, Occasion'd by his Late Wonderful Sermon against Inoculation. London: J. Roberts, 1722.

Lobb, Theophilus. *Letters Relating to the Plague and Other Contagious Distempers.* London: James Buckland, 1745. Online from Google.

———. *Medical Practice of Curing Fevers.* London: John Oswald, 1735. Online from the Hathi Trust.

———. *A Treatise of the Small-Pox* . . . London: T. Woodward, C. Davis, 1731.

M.A.C.D. [pseud.]. *Système D'un Médecin Anglois Sur La Cause De Toutes Les Espèces De Maladies* . . . Paris: Alexis-Xavier-René Mesnier, 1726. Online from Google.

Mackenzie, Mordach. "An Account of the Plague at Constantinople: In a Letter . . . to Sir James Porter." *Philosophical Transactions* 54 (1754): 69–82. Online from the Royal Society, doi:10.1098/rstl.1764.0013.

———. "A Further Account of the Late Plague at Constantinople, in a Letter . . . to John Clephane . . . *Philosophical Transactions* 47 (1751): 514–16. Online from the Royal Society, doi:10.1098/rstl.1751.0088.

———. "Extracts of Several Letters . . . Concerning the Plague at Constantinople." *Philosophical Transactions* 47 (1751–52): 384–95. Online from the Royal Society, doi:10.1098/rstl.1751.0064.

Maclean, Charles. *Results of an Investigation Respecting Epidemic and Pestilential Diseases* . . . 2 vols. in 1. London: Thomas and George Underwood, 1817. Online from Google.

Manningham, Richard. *A Discourse concerning the Plague and Pestilential Fevers* . . . London: J. Robinson, 1758.

Marten, Benjamin, *A New Theory of Consumptions* . . . London: R. Knaplock, A. Bell, J. Hooke and C. King, 1720.

Martin, Martin. "Several Observations in the North Islands of Scotland." *Philosophical Transactions* 19 (1695–7): 727–9. Online from the Royal Society, doi: 10.1098/rstl.1695.0134.

Martyn, John. "An Essay towards a New Theory of Physic, in a Discourse Read Before the Grubean Society, by Ephraim Quibus, M.D." (1730), *Memoirs of the Society of Grub-Street* 1 no. 18 (1737): 86–9. Online from Google.

Massey, Edmund. *A Letter to Mr. Maitland, in Vindication of the Sermon against Inoculation.* Norwich: by order of S. Mascoll, 1722. Online from Google.

———. *A Sermon against the Dangerous and Sinfull Practice of Inoculation. Preach'd at St. Andrew's Holborn* . . . *1722.* London: William Meadows, 1722.

Mather, Cotton. "The Angel of Bethesda: An Essay upon the Common Maladies of Mankind." Edited by Gordon W. Jones. Barre, MA: American Antiquarian Society, 1972.

———. *A Letter, about a Good Management under the Distemper of the Measles.* Boston: 1713.

———. *Selected Letters.* Compiled by Kenneth Silverman. Baton Rouge: State University of Louisiana Press, 1971.

Maty, Matthew. *Authentic Memoirs of the Life of Richard Mead.* London: J. Whiston and B. White, 1755. Online from Google.

Maynwaringe, E[verard]. *The History and Mystery of the Venereal Lues.* London: J. M., 1673. Online from EEBO.

———. [E. M.] Ignota Febris. *Fevers Mistaken, in Doctrine and Practice* . . . London: 1691. Online from EEBO.

———. [E. M.] *Inquiries into the General Catalogue of Diseases* . . . London: 1691. Online from EEBO.

Mead, Richard. *A Discourse concerning the Action of the Sun and Moon on Animal Bodies and the Influence Which This May Have in Many Diseases.* London: 1708. Originally published as *De Imperio Solis et Lunae in Corpora Humana [et] Morbis Inde Oriundis.* London: Raphael Smith, 1704. Online from ECCO.

———. *A Mechanical Account of Poisons, In Several Essays* (1702). 4th ed. London: J. Brindley, 1747.

———. *A Short Discourse concerning Pestilential Contagion, and the Methods to Be Used to Prevent It.* London: Sam Buckley and Ralph Smith, 1720.

———. *A Treatise on the Small Pox and Measles* . . . Translated from the Latin by John Theobald. London: 1747.

Molyneux, Thomas. "Dr. Molineux's Historical Account of the Late General Coughs and Colds: With Some Observations on Other Epidemic Distempers." *Philosophical Transactions* 18 (1694), 105–11.

Musgrave, W. "A Letter from Dr. W. Musgrave, F.R.S. to Dr. Hans Sloane S.R.S. concerning Hydatides Voided by Stool." *Philosophical Transactions* 24 (1704–5): 1797–1800.

Nedham, Marchamont [M. N.]. Medela Medicinae. *A Plea for the Free Profession* . . . *of* . . . *Physick* . . . London: Richard Lownds, 1665.

A New Discovery of the Nature of the Plague . . . *with the Remedy. Contrary to the Opinion of Dr. Meade, Dr. Browne, and Others, Who Give for the First Causes of the Plague* . . . *Air, Diet, and Disease.* London: T. Bickerton, n. d. ca. 1721.

Nichols, John. *Literary Anecdotes of the Eighteenth Century* . . . Vols. 4 and 5. London, for the author, 1812. Online from Google.

Oliver, William. *A Practical Essay on Fevers.* London: T. Goodwin, 1704.

Pestalossi, Jérôme-Jean (Jérôme Jean Pestalozzi). *Avis de Precaution contre La Maladie Contagieuse de Marseille.* Lyon: Frères Bruyset, 1721.

Petty, William. *The Collected Works of Sir William Petty.* Edited by Terence Hutchinson. 8 vols. London: Routledge, 1997.

———. *The Economic Writings of Sir William Petty: Together with the Observations upon the Bills of Mortality More Probably by Captain John Graunt.* Edited by Charles Henry Hull. 2 vols. Cambridge: C. J. Clay and Sons, 1899. Online from Google.

———. *Several Essays in Political Arithmetick.* 4th ed. London: D. Browne, J. Shuckburgh, H. Whiston and B. White, 1755. Online from Google.

Phaer, Thomas. *The Boke of Children.* Edited and annotated by Rich Bowers. Tempe: Arizona Center for Medieval and Renaissance Studies, 1999.

J. Phillips. "Letter to Richard Bradley." In Richard Bradley, *Precautions against Infection* . . . *on Account of the Dreadful Plague in France.* London: Thomas Corbett, n. d.

Plater, Felix. *A Golden Practice of Physick: In Five Books, and Three Tomes* . . . London: Peter Cole, 1662. Online from EEBO.

———. (Felicis Plateri). *De Febribus Liber: Genera, Causas, et Curationes Febrium* . . . Frankfurt: Heredes Andrea Wecheli, Claudium Marinum and Ioannem Aubrium, 1597. Online from the Hathi Trust.

Power, Henry. *Experimental Philosophy in Three Books.* London: John Martin and James Allestry, 1664.

Quincy, John. *An Examination of Dr. Woodward's State of Physick and Diseases.* 2nd ed. London: Andrew Bell, William Taylor and John Osborn, 1720.

Ray, John. *The Wisdom of God Manifested in the Works of the Creation* (1691). 7th ed. London: William Innys, 1714. Online from JRI Initiative, http://www.jri.org .uk/ray/wisdom/.

Richardson, Richard. *Extracts from the Literary and Scientific Correspondence of Richard Richardson, M.D., F.R.S.* Edited by Dawson Turner. Yarmouth: Charles Sloman, 1835. Online from the Internet Archive.

Rivierius, Lazarus (Lazare Riviére) and Jean Fernel. *Four Books of* . . . *Lazarus Riverius. . . . unto Which is Added* . . . *Select Medicinal Counsels of John Fernelius.* Translated by Nicholas Culpeper. London: Peter Cole, 1658. Online from EEBO.

Russell, Alexander. *The Natural History of Aleppo*. 2nd ed., 2 vols. London: G. G. and J. Robinson, 1794.

Sennert, Daniel. *De Febribu Libri IV (De Febribus)*. Geneva: Iacobi Stoer, 1647. Online from Google.

———. *The Institutions or Fundamentals of the Whole Art, Both of Physick and Chirurgery, Divided into Five Books* (1611). Translated by N. D. B. P. London: Lodowick Lloyd, 1656.

———. *Doctor D. Sennertus of Agues and Fevers: Their Differences, Signes, and Cures*. Translated by N. D. B. M. late of Trinity Colledge in Cambridge. London: Lodowick Lloyd, 1658. Online from EEBO.

———. *The Sixth Book of Practical Physic: Of Occult or Hidden Diseases in Nine Parts . . . by Daniel Sennertus . . .* Translated by N. Culpeper and Abdiah Cole. London: Peter Cole, 1662. Online from EEBO.

Short, Thomas. *A Comparative History of the Increase and Decrease of Mankind in England and Several Countries Abroad* (1767). Edited by Richard Wall. Farnbrough, UK: Gregg International, 1973.

Simpson, William. *Zymologia Physica; or, A Brief Philosophical Discourse of Fermentation from a New Hypothesis of Acidum and Sulphur*. London: W. Cooper, 1675.

Slare, Frederick. "An Experiment Made before the . . . Royal Society: In Which a Surprizing Change of Colour . . . was Exhibited . . . by the Admission of Air Only: Applied to . . . the Blood of Respiring Animals." *Philosophical Transactions* 17 (October 1693): 898–908.

———. *Experiments and Observations upon Oriental and Other Bezoar-Stones . . . together with Further Discoveries and Remarks*. London: Tim. Goodwin, 1715.

Sloane, Hans. *An Account of a Most Efficacious Medicine for Soreness, Weakness, and Several Other Distempers of the Eyes*. London: Dan Browne, 1745.

———. "Of the Use of Ipecacuanha for Loosnesses, Translated from a French Paper: with Some Notes." *Philosophical Transactions* 20 (1698): 69–79.

———. *Voyage to the Islands Madera, Barbados, Neives, St. Christophers and Jamaica; with the Natural History of . . . the Last of Those Islands*. 2 vols. London: for the author, 1707–25.

Sprackling, Robert. *Medela Ignorantiae; or a Just and Plain Vindication of Hippocrates and Galen from the Groundless Imputations of M[archamont] N[edham]*. London: Robert Crofts, 1665.

Strother, Edward. *Criticon Febrium; or, a Critical Essay on Fevers*. 2nd ed. London: Charles Rivington, 1718.

Sydenham, Thomas. *The Works of Thomas Sydenham, M.D.* Translated from the Latin ed. of Dr. Greenhill, with a Life of the Author by R. G. Latham (London: Sydenham Society, 1848–50). Facsimile rpt. 2 vols. in 1. London: Classics of Medicine Library, 1979.

Trapham, Thomas. *A Discourse of the State of Health in the Island of Jamaica*. London: R. Boulter, 1679. Online from Google.

Turner, Daniel. *A Discourse concerning Fevers . . .* (1727). 3rd ed. London: John Clarke, 1739.

Twysden, John. *Medicina Veterum Vindicata*. London: J. G., 1666. Online from EEBO.

Van Helmont, Joan Baptista. *Oriatrike; or, Physick Refined . . .* Translated by J. C. [John Chandler]. London: Lodowick Loyd [*sic*], 1662.

Van Swieten, Gerard. *The Commentaries upon the Aphorisms of Hermann Boerhaave . . .* Vol. 2. London: Robert Horsfield and Thomas Longman, 1744. Online from ECCO.

Walwyn, William. *Physick for Families; or, the New, Safe and Powerful Way of Physick, upon Constant Proof Established* . . . London: John Starky, 1681.

Wilkins, John. *An Essay towards a Real Character, and a Philosophical Language.* London: S. Gellibrand and for John Martyn, 1668. Online from EEBO.

Willis, Thomas. *Dr. Willis's Practice of Physick, Being All the Medical Works* . . . Translated by S. P. [Samuel Pordage]. London: T. Dring, C. Harper, and J. Leigh, 1681.

Wincler, Dr. "An Abstract of a Letter from Dr. Wincler Chief Physitian of the Prince Palatine, Dat. Dec. 22. 1682 to Dr. Fred. Slare Fellow of the Royal Society, Containing an Account of a Murren in Switzerland." *Philosophical Transactions* 13 (1683): 93–5.

SELECTED BIBLIOGRAPHY OF SECONDARY SOURCES

Ackerknecht, Erwin. "Anticontagionism between 1821 and 1867." *Bulletin of the History of Medicine* 22 (1948): 562–93. Also abridged and with portions translated in the *International Journal of Epidemiology* 38, no. 1 (2009): 7–21, doi 10.1093/ije/dyn254.

Allen, D. E. "John Martyn's Botanical Society: A Biographical Analysis of the Membership." *Proceedings of the Botanical Society of the British Isles* 6, pt. 4 (1967): 305–24.

Anstey, Peter. "John Locke and Helmontian Medicine." *Studies in the History and Philosophy of Science* 25 (2010): 93–117.

———. *John Locke and Natural Philosophy.* Oxford: Oxford University Press, 2011.

Arrizabalaga, Jon, John Henderson, and Roger French. *The Great Pox: The French Disease in Renaissance Europe.* New Haven: Yale University Press, 1997.

Avramov, Iordan. "An Apprenticeship in Scientific Communication: The Early Correspondence of Henry Oldenburg (1656–63)." *N&R* 53, no. 2 (1999): 187–201.

Ayliffe, Graham A. J., and Mary P. English. *Hospital Infection from Miasmas to MRSA.* Cambridge, UK: Cambridge University Press, 2003.

Barker, Peter, and Roger Ariew, eds. *Revolution and Continuity: Essays in the History and Philosophy of Early Modern Science.* Washington: Catholic University of America, 1991.

Barnett, Richard. "Dr. Jacob De Castro Sarmento and Sephardim in Medical Practice in 18th-Century London." *Transactions of the Jewish Historical Society of England* 27 (1978–80): 84–114.

Bates, Donald G. "Sydenham and the Medical Meaning of Method." *Bulletin of the History of Medicine* 51 (1977): 324–38.

———. "Thomas Willis and the Epidemic Fever of 1661: A Commentary." *Bulletin of the History of Medicine* 39, no. 5 (1965): 393–414.

———. "Thomas Willis and the Fevers Literature of the Seventeenth Century." In Bynum and Nutton, *Theories of Fever*, 47–70.

Beall, Otho Jr., and Richard Shryock. *Cotton Mather, First Significant Figure in American Medicine.* Baltimore: Johns Hopkins University Press, 1954.

Beier, Lucinda McCrae. *Sufferers and Healers: The Experience of Illness in Seventeenth-Century England.* London: Routledge and Kegan Paul, 1987.

Belloni, Luigi. *Le 'Contagium Vivum' avant Pasteur.* Conferences du Palais de la Découverte, ser. D., no. 74. Paris: Imprimerie Alenconnaise, 1961.

Bernier, Jacques. "L'Interprétation de la Phtisie Pulmonaire au XVIIIe Siècle." *Canadian Bulletin of Medical History* 22, no. 1 (2005): 35–56.

Birken, William. "The Dissenting Tradition in English Medicine of the Seventeenth and Eighteenth Centuries." *Medical History* 39 (1995): 197–218.

Blake, John B. "The Inoculation Controversy in Boston, 1721–1722." *New England Quarterly* 25 (1952): 489–506.

Bond, T. C. "Keeping up with the Latest *Transactions*: The Literary Critique of Scientific Writing in the Hans Sloane Years." *Eighteenth-Century Life* 22, no. 2 (1998), 1–17.

Booker, John. *Maritime Quarantine: The British Experience, c. 1650–1900.* Aldershot, UK: Ashgate, 2007.

Booth, Christopher C. *John Haygarth, FRS: A Physician of the Enlightenment (1740–1827).* Philadelphia: American Philosophical Society, 2005.

Bradley, James E. *Religion, Revolution and English Radicalism.* Cambridge, UK: Cambridge University Press, 1990.

———. "The Religious Origins of Radical Politics in England, Scotland, and Ireland, 1662–1800." In *Religion and Politics in Enlightenment Europe.* Edited by James Bradley and Dale Van Kley. Notre Dame: University of Notre Dame Press, 2001, 187–253.

Breen, Louise A. "Cotton Mather, the 'Angelical Ministry' and Inoculation." *Journal of the History of Medicine* 46, no. 3 (1991), 333–57.

Breidbach, Olaf, and Michael T. Ghiselin. "Baroque Classification: A Missing Chapter in the History of Systematics." *Annals of the History and Philosophy of Biology* 11 (2006): 1–30.

Brockliss, Laurence, and Colin Jones. *The Medical World of Early Modern France.* Oxford: Clarendon Press: 1997.

Brooke, John Hedley. "Science and Dissent: Some Historiographical Issues." In *Science and Dissent in England, 1688–1945.* Edited by Paul Wood. Aldershot, UK: Ashgate, 2004.

Brooke, John Hedley, and Ian MacLean, eds. *Heterodoxy in Early Modern Science and Religion.* Oxford: Oxford University Press, 2005.

Brooks, Chandler McC., and Paul F. Cranefield, eds. *The Historical Development of Physiological Thought.* New York: Hafner, 1959.

Brooks, Eric St. John. *Sir Hans Sloane: The Great Collector and His Circle.* London: Batchworth Press, 1954.

Brown, Theodore M. "The College of Physicians and the Acceptance of Iatromechanism in England, 1665–1695." *Bulletin of the History of Medicine* 44 (1970): 12–30.

———. *The Mechanical Philosophy and the 'Animal Oeconomy': A Study in the Development of English Physiology in the Seventeenth and Early Eighteenth Century.* New York: Arno Press, 1981.

Brunner, Daniel L. *Halle Pietists in England: Anthony William Boehm and the Society for Promoting Christian Knowledge.* Göttingen: Wandenhoek and Ruprecht, 1993.

Buck, Peter. "People Who Counted: Political Arithmetic in the Eighteenth Century." *Isis* 73: 28–45.

Buer, Mabel C. *Health, Wealth and Population in the Early Days of the Industrial Revolution.* Rpt. New York: Fertig, 1968.

Bulloch, William. *The History of Bacteriology.* Rpt. New York: Dover, 1979.

Buret, F. *Syphilis in the Middle Ages and in Modern Times.* Translated by A. H. Ohmann-Dumesnil. Vols. 2–3 in 1. Philadelphia: F.A. Davis, 1895.

Burke, John G., ed. *The Uses of Science in the Age of Newton*. Berkeley: University of
    California Press, 1983.
Bynum, William F., and Vivian Nutton, eds. *Theories of Fever from Antiquity to the
    Enlightenment. Medical History*, Supplement 1. London: Wellcome Institute for
    the History of Medicine, 1981.
Bynum, William F., and Roy Porter, eds. *Medical Fringe and Medical Orthodoxy,
    1750–1850*. London: Croom Helm, 1987.
———. *William Hunter and the Eighteenth-Century Medical World*. Cambridge, UK:
    Cambridge University Press, 1985.
Cain, Arthur J. "John Ray on 'Accidents.'" *Archives of Natural History* 23 (1996):
    343–69.
———. "Linnaeus's *Ordines Naturales*." *Archives of Natural History* 20 (1993):
    405–15.
———. "Logic and Memory in Linnaeus's System of Taxonomy." *Proceedings of the
    Linnean Society of London* 169 (1958): 144–63.
———. "Thomas Sydenham, John Ray, and Some Contemporaries on Species."
    Archives of Natural History 26 (1999): 55–83.
Cantor, Geoffrey. "Emanuel Mendes da Costa: Constructing a Career in Science." In
    *From Strangers to Citizens: The Integration of Immigrant Communities in Britain,
    Ireland and Colonial America*. Edited by Randolph Vigne and Charles Littleton.
    Brighton, UK: Sussex Academic Press, 2001, 230–8.
Carmichael, Ann G. "Contagion Theory and Contagion Practice in 15th Century
    Milan." *Renaissance Quarterly* 44 (1991): 213–57.
———. "Plague Legislation in the Italian Renaissance." *Bulletin of the History of
    Medicine* 57 (1983): 508–25.
Carpenter, Audrey T. *John Theophilus Desaguliers: A Natural Philosopher, Engineer
    and Freemason in Newtonian England*. London: Continuum, 2011.
Casteel, Eric Greer. "Entrepôt and Backwater: A Cultural History of the Transfer of
    Medical Knowledge from Leiden to Edinburgh, 1690–1740." Ph.D. dissertation,
    Los Angeles: University of California at Los Angeles, 2007.
Caygill, Marjorie. "Sloane's Will and the Establishment of the British Museum." In
    MacGregor, *Hans Sloane*, 45–68.
Champion, Justin A. I., ed. "Epidemic Disease in London" (1993). London: Centre
    for Metropolitan History, Spring 2001. Online at http://www.history.ac.uk/ihr
    /Focus/Medical/epimenu.html.
———. *Republican Learning: John Toland and the Crisis of Christian Culture,
    1695–1722*. Manchester: Manchester University Press, 2003.
Christensen, Peter. "'In these Perilous Times': Plague and Plague Policies in Early
    Modern Denmark." *Medical History* 47 (2003): 413–50.
Churchill, Wendy D. "Bodily Differences?: Gender, Race and Class in Hans Sloane's
    Jamaican Medical Practice, 1687–1688." *JHMAS* 60 (2005): 391–444.
Clark, George. *A History of the Royal College of Physicians of London*. 2 vols. Oxford:
    Clarendon Press for the Royal College of Physicians, 1964–6.
Clericuzio, Antonio. "Chemical Medicine and Paracelsianism in Italy, 1550–1650."
    In Pelling and Mandelbrote, *Practice of Reform*, 43–58.
———. *Elements, Principles and Corpuscles: A Study of Atomism and Chemistry in the
    Seventeenth Century*. Dordrecht: Kluwer, 2001.
———. ed. *The Internal Laboratory: The Chemical Reinterpretation of Medical Spirits
    in England (1650–1680)*. Dordrecht: Kluwer, 1994.

Cohen, Henry. "The Evolution of the Concept of Disease." *Proceedings of the Royal Society of Medicine* 48, no. 3 (1955): 155–60.

Colwell, H. A. "Gideon Harvey: Sidelights on Medical Life from the Restoration to the End of the XVII Century." *Annals of Medical History* 3, no. 3 (1921): 205–37.

Conrad, Lawrence I., and Dominik Wujastyk, eds. *Contagion: Perspectives from Premodern Societies*. Burlington, VT: Ashgate, 2000.

Cook, Harold J. *The Decline of the Old Medical Regime in Stuart London*. Ithaca, NY: Cornell University Press, 1986.

———. *Matters of Exchange. Commerce, Medicine and Science in the Dutch Golden Age*. New Haven: Yale University Press, 2007.

———. "The New Philosophy in the Low Countries." In Porter and Teich, *Scientific Revolution*, 115–49.

———. "Physicians and Natural History." In *Cultures of Natural History*. Edited by Nicholas Jardine, James A. Secord, and Emma C. Spary. Cambridge, UK: Cambridge University Press, 1996, 91–105.

———. "Physicians and the New Philosophy: Henry Stubbe and the Virtuosi-Physicians." In French and Wear, *Medical Revolution*, 246–71.

———. "Physick and Natural History in Seventeenth-Century England," in *Revolution and Continuity: Essays in the History and Philosophy of Early Modern Science*. Edited by R. Ariew and P. Barker. Washington, DC: Catholic University of America Press, 1991, 63–80.

———. "Practical Medicine and the British Armed Forces after the 'Glorious Revolution.'" *Medical History* 34 (1990): 1–26.

———. "The Rose Case Reconsidered: Physicians, Apothecaries and the Law in Augustan England." *JHMAS* 45 (1990): 527–55.

———. "The Society of Chemical Physicians, the New Philosophy, and the Restoration Court." *Bulletin of the History of Medicine* 61 (1987): 61–77.

———. *Trials of an Ordinary Doctor: Johannes Groenevelt in Seventeenth-Century London*. Baltimore: Johns Hopkins University Press, 1994.

Cope, Zachary. *William Cheselden*. London: E. and S. Livingston, 1953.

Coury, Charles. *Grandeur et Declin d'une Maladie, La Tuberculose au Cours des Ages*. Suresnes: Lepetit, 1972.

Creighton, Charles. *A History of Epidemics in Britain*, (1894). 2nd ed., 2 vols. Edinburgh: Frank Cass, 1965.

Cunningham, Andrew. "Sydenham versus Newton: The Edinburgh Fever Dispute of the 1690s between Andrew Brown and Archibald Pitcairne." In Bynum and Nutton, *Theories of Fever*, 71–98.

———. "Thomas Sydenham and the 'Good Old Cause.'" In French and Wear, *Medical Revolution*, 164–90.

Cunningham, Andrew, and Roger French. *The Medical Enlightenment of the Eighteenth Century*. Cambridge, UK: Cambridge University Press, 1990.

Damkaer, David M. *The Copepodologist's Cabinet: A Biographical and Bibliographical History*. Philadelphia: American Philosophical Society, 2002.

Davenport, Geoffrey. "When Did the College Become Royal?" In *The Royal College of Physicians and its Collections: An Illustrated History*. Edited by Geoffrey Davenport, Ian McDonald, and Caroline Moss-Gibbons. London: James and James, 2001, 26–8.

Davis, Audrey B. 'The Circulation of the Blood and Chemical Anatomy." In Debus, *Science, Medicine and Society*. Vol. 2. 25–37.

————. *Circulation Physiology and Medical Chemistry, 1650–1680.* Lawrence, KS: Coronado Press, 1973.

————. "The Virtues of the Cortex in 1680: A Letter from Charles Goodall to Mr. H." *Medical History* 15 (1971), 293–304.

DeKrey, Gary. *A Fractured Society: The Politics of London in the First Age of Party, 1688–1715.* Oxford: Clarendon Press, 1985.

Debus, Allen G. *Chemistry and Medical Debate: Van Helmont to Boerhaave.* Canton, MA: Science History Publications, 2001.

————. "Chemistry and the Universities in the Seventeenth Century." *Estudos Avançados* 4 (1990): 173–96, doi 10.1590/S0103-40141990000300009.

————. *The English Paracelsians.* New York: Franklin Watts, 1966.

————. *Medicine in 17th Century England.* Berkeley: University of California Press, 1974.

————. "Paracelsian Doctrine in English Medicine." In *Chemistry in the Service of Medicine.* Edited by F. N. L. Poynter. London: Pitman Medical Publishing, 1963, 5–26.

————. "The Paracelsians in Eighteenth-Century France." In *Transformation and Tradition in the Sciences: Essays in Honor of I. Bernard Cohen.* Edited by Everett Mendelsohn. Cambridge, UK: Cambridge University Press, 2003, 193–214.

————. *Science and Education in the Seventeenth Century: The Webster-Ward Debate.* London: Macdonald, 1970.

————. ed. *Science, Medicine and Society in the Renaissance, Essays to Honor Walter Pagel.* 2 vols. New York: Science History Publications, 1972.

————. "Some Comments on the Contemporary Helmontian Renaissance." *Ambix* 19 (1972): 145–50.

————. "Thomas Sherley's 'Philosophical Essay' (1672): Helmontian Mechanism as the Basis of a New Philosophy." *Ambix* 27 (1980): 124–35.

Dekker, Rudolf. "'Private Vices, Public Virtues' Revisited: The Dutch Background of Bernard Mandeville." Translated by Gerard T. Moran. *History of European Ideas* 14 (1992): 481–98.

Delbourgo, James. "Sir Hans Sloane's Milk Chocolate and the Whole History of the Cacao." *Social Text* 29, no. 1 (2011): 71–101.

Deutsch, Helen, and Mary Terrall, eds. *Vital Matters: Eighteenth-Century Views of Conception, Life and Death.* Toronto: University of Toronto Press, 2012.

Dewhurst, Kenneth. "The Correspondence between John Locke and Sir Hans Sloane." *Irish Journal of Medical Science* 6th ser. no. 413; 5, no. 35 (May 1960): 201–12.

————. *The Quicksilver Doctor: The Life and Times of Thomas Dover, Physician and Adventurer.* Bristol: John Wright, 1957.

Dobbs, Betty Jo Teeter. *The Janus Faces of Genius: The Role of Alchemy in Newton's Thought.* Cambridge, UK: Cambridge University Press, 1991.

Dobell, Clifford. *Antony Van Leeuwenhoek and His 'Little Animals': Being an Account of the Father of Protozoology and Bacteriology.* New York: Russell and Russell, 1958.

Doetsch, Raymond N. "Benjamin Marten and His '*New Theory of Consumptions.*'" *Microbiological Reviews* 42, no. 3 (1978): 521–8.

Doig, A., J. P. S. Ferguson, I. A. Milne, and R. Passmore, eds. *William Cullen and the Eighteenth-Century Medical World.* Edinburgh: Edinburgh University Press, 1993.

Drinker, Cecil K. "Doctor Smollett." *Annals of Medical History* 7 (1925): 37–47.

Duffin, Jacalyn. "Jodocus Lommius's Little Golden Book and the History of Diagnostic Semeiology." *JHMAS* 61 (2006): 249–287.

———. *Lovers and Livers: Disease Concepts in History*. Toronto: University of Toronto Press, 2005.

Egerton, Frank. "A History of Ecological Sciences." Parts 1–53 (2001–2015). Online from the Ecological Society of America, http://esapubs.org/bulletin /current/history_links_list.htm.

———. "Richard Bradley's Illicit Excursion into Medical Practice in 1714." *Medical History* 14 (1970): 53–62.

———. "Richard Bradley's Relationship with Sir Hans Sloane." *N&R* 25 (1970): 59–77.

———. *The Roots of Ecology from Antiquity to Haeckel*. Berkeley: University of California Press, 2012.

Eisenstein, Elizabeth. *The Printing Press as an Agent of Change*. Cambridge, UK: Cambridge University Press, 1979.

Ellis, Frank H. "The Background of the London Dispensary." *JHMAS* 20 (1965): 197–212.

Elmer, Peter. "Medicine, Religion and the Puritan Revolution." In French and Wear, *Medical Revolution*, 10–46.

———. "Medicine, Witchcraft and the Politics of Healing." In *Medicine and Religion in Enlightenment Europe*. Edited by Ole Peter Grell and Andrew Cunningham. Aldershot, UK: Ashgate, 2007, 223–242.

Emerson, Roger L. "Sir Robert Sibbald, Kt., the Royal Society of Scotland and the Origins of the Scottish Enlightenment." *Annals of Science* 45 (1988): 41–72.

Ereshefsky, Marc. "Species, Taxonomy, and Systematics." In *Philosophy of Biology: An Anthology*. Edited by Alex Rosenberg and Robert Arp. Chichester, UK: Wiley-Blackwell, 2009, 255–72.

Erhard, Jean. "Opinions Médicales en France au XVIIIe Siècle: La Peste et l'Idée de Contagion." *Annales Histoire, Sciences Sociales* 21, no. 1 (1957): 46–59.

Faber, Knud. *Nosography: The Evolution of Clinical Medicine in Modern Times*. New York: Hoeber, 1930.

Farley, John. "Parasites and the Germ Theory of Disease." In *Framing Disease*. Edited by Charles Rosenberg and Janet Golden. New Brunswick, NJ: Rutgers University Press, 1992, 34–49.

Ferguson, J. P. *Dr. Samuel Clarke: An Eighteenth-Century Heretic*. Kineton, UK: Roundwood Press, 1976.

Field, J. F., and Frank A. J. L. James, eds. *Renaissance and Revolution: Humanists, Scholars, Craftsmen and Natural Philosophers in Early Modern Europe*. Cambridge, UK: Cambridge University Press, 1993.

Fletcher, John Edward. *A Study of the Life and Works of Athanasius Kircher, "Germanus Incredibilis."* Edited by Elizabeth Fletcher. Leiden: Brill, 2011.

Flinn, Michael Walter. *The European Demographic System, 1500–1820*. Baltimore: Johns Hopkins University Press, 1981.

Ford, Brian. *Single Lens: The Story of the Simple Microscope*. London: William Heinemann, 1985.

Fox, Christopher, Roy Porter, and Robert Wokler, eds. *Inventing Human Science: Eighteenth-Century Domains*. Berkeley: University of California Press, 1995.

Frank, Robert G., Jr. *Harvey and the Oxford Physiologists: A Study of Scientific Ideas*. Berkeley: University of California Press, 1980.

———. "The Physician as Virtuoso in Seventeenth-Century England." In *English Scientific Virtuosi in the Sixteenth and Seventeenth Centuries*. Edited by Barbara Shapiro and Robert G. Frank, Jr. Los Angeles: William Andrews Clark Memorial Library, 1979, 57–119.

———. "Science, Medicine and the Universities of Early Modern England: Background and Sources." *History of Science* 11 (1973): 194–216, 239–69.

French, Roger, and Andrew Wear, eds. *The Medical Revolution of the Seventeenth Century*. Cambridge, UK: Cambridge University Press, 1989.

———. "Sickness and the Soul: Stahl, Hoffman and Sauvages on Pathology." In Cunningham and French, eds. *Medical Enlightenment*, 88–110.

Friedman, Reuben. "Bonomo and Cestoni: The Dispute Concerning their Identity, the Authorship of the 'Letter to Redi,' and the Credit for the Discovery of the Acarian Origin of Scabies." In *Abraham Levinson Anniversary Volume: Studies in Pediatrics and Medical History . . .* Edited by Solomon R. Kagan. New York: Froben Press, 1948, 279–96.

Friesen, John. "Archibald Pitcairne, David Gregory and the Scottish Origins of English Tory Newtonianism 1688–1715." *History of Science* 41 (2003): 163–91.

Furdell, Elizabeth Lane. *Publishing and Medicine in Early Modern England*. Rochester: Rochester University Press, 2002.

Garrett, Brian. "Vitalism and Teleology in the Natural Philosophy of Nehemiah Grew (1641–1712)." *British Journal for the History of Science* 36 (2003): 63–81.

Garrison, Fielding H. "Fracastorius, Athanasius Kircher and the Germ Theory of Disease." *Science* 31 nos. 796 and 805 (1 April and 3 June, 1910): 500–502 and 857–859.

Gascoigne, John. *Cambridge in the Age of the Enlightenment*. Cambridge, UK: Cambridge University Press, 1989.

Gaukroger, Stephen. *The Collapse of Mechanism and the Rise of Sensibility: Science and the Shaping of Modernity, 1660–1760*. Oxford: Oxford University Press, 2012.

———. *The Emergence of a Scientific Culture: Science and the Shaping of Modernity, 1210–1685*. Oxford: Oxford University Press, 2008.

Gay, Peter. "The Enlightenment as Medicine and as Cure." In *The Age of the Enlightenment: Studies Presented to Theodore Besterman*. Edited by W. H. Barber. Edinburgh: Oliver and Boyd for the University Court of the University of St. Andrews, 1967, 375–86.

Gest, Howard. "The Discovery of Microorganisms by Robert Hooke and Antoni Van Leeuwenhoek, Fellows of the Royal Society." *N&R* 58 (2004): 187–201.

Geyer-Kordesch, Johanna. "Fevers and other Fundamentals: Dutch and German Medical Explanations, c. 1680–1730." In Bynum and Nutton, *Theories of Fever*, 99–120.

Ghesquier, Daniele. "A Gallic Affair: The Case of the Missing Itch-Mite in French Medicine in the Early Nineteenth Century. *Medical History* 43 (1999): 26–54.

Gillespie, Charles C. "Physick and Philosophy: A Study of the Influence of the College of Physicians of London upon the Foundation of the Royal Society." *Journal of Modern History* 19 (1947): 210–25.

Gillespie, Neal C. "Natural History, Natural Philosophy and Social Order: John Ray and the Newtonian Ideology." *Journal of the History of Biology* 20 (1987): 1–49.

Glass, David Victor, ed. *The Development of Population Statistics*. Farnborough, UK: Gregg International, 1973.

Golinski, Jan V. "A Noble Spectacle. Phosphorus and the Public Cultures of Science in the Early Royal Society." *Isis* 80 (1989): 11–39.

Goslings. W. R. O. "Leiden and Edinburgh: The Seed, the Soil and the Climate." In *The Early Years of the Edinburgh Medical School*. Edited by R. G. W. Anderson and A. D. C. Simpson. Edinburgh: Royal Scottish Museum, 1976, 1–18.

Grasset, Hector. "The Parasitic Theory and Pulmonary Phthisis in the Eighteenth Century." Translated by Thomas C. Minor. *Cincinnati Lancet-Clinic*, n. s., 44 whole volume 83 (6 January and 13 January, 1900): 22–6 and 38–43. Online from Google.

Green, Joseph J. "Marshes and Meads." *Friends Quarterly Examiner* 41 (1907): 477–90.

Grell, Ole Peter, and Andrew Cunningham, eds. *Medicine and Religion in Enlightenment Europe*. Aldershot, UK: Ashgate, 2007.

Groner, Julius, and Paul F. S. Cornelius. *John Ellis: Merchant, Microscopist, Naturalist, and King's Agent*. Pacific Grove, CA: Boxwood Press, 1996.

Grundy, Isobel. *Lady Mary Wortley Montagu: Comet of the Enlightenment*. Oxford: Oxford University Press, 1999.

Guerrini, Anita. "Chemistry Teaching at Oxford and Cambridge, circa 1700." In *Alchemy and Chemistry in the 16th and 17th Centuries*. Edited by Piyo Rattansi and Antonio Clericuzio. Dordrecht: Kluwer Academic Publishers, 1994, 183–200.

———. "James Keill, George Cheyne, and Newtonian Physiology, 1690–1740." *Journal of the History of Biology* 18 (1985): 247–66.

———. "Newtonian Matter Theory, Chemistry, and Medicine, 1690–1713." Ph.D. dissertation, Indiana University, 1983.

———. "'A Scotsman on the Make': The Career of Alexander Stuart." In *The Scottish Enlightenment: Essays in Reinterpretation*. Edited by Paul Wood. Rochester, NY: University of Rochester Press, 2000, 157–76.

———. "The Tory Newtonians: Gregory, Pitcairne, and Their Circle," *Journal of British Studies* 25 (1986): 288–311.

Haakonssen, Knud, ed. *Enlightenment and Religion: Rational Dissent in Eighteenth-Century Britain*. Cambridge, UK: Cambridge University Press, 1996.

Hall, Marie Boas. "Frederick Slare, F.R.S. (1628–1727)." *N&R* 46 (1992): 23–41.

———. "Oldenburg, the '*Philosophical Transactions*,' and Technology." In Burke, *Uses of Science*, 21–47.

———. "The Royal Society and Italy, 1667–1795." *N&R* 37 (1982): 63–81.

———. "The Royal Society's Role in the Diffusion of Information in the Seventeenth Century (1)." *N&R* 29 (1975): 173–92.

Hall, A. Rupert. "Medicine and the Royal Society." In Debus, *Medicine in Seventeenth-Century England*, 421–52.

Hamlin, Christopher. "Commentary: Ackerknecht and 'Anticontagionism:' A Tale of Two Dichotomies." *International Journal of Epidemiology* 38 (2009): 22–7. Online at doi: 10.1093/ije/dyn256.

Hans, Nicholas. *New Trends in Education in the Eighteenth Century*. Rpt. London: Routledge and Kegan Paul, 1966.

Hartston, William. "Medical Dispensaries in Eighteenth-Century London." Abridged in *Proceedings of the Royal Society of Medicine*, Section of the History of Medicine 56 (1963): 753–8.

Hawkins, Stanley A. "Sir Hans Sloane (1660–1735): His Life and Legacy." *Ulster Medical Journal* 79 (2010): 25–9. Online from PubMed at http://www.ncbi.nlm.nih.gov/pmc/articles/PMC2938984/.

Headrick, Daniel R. *When Information Came of Age: Technologies of Knowledge in the Age of Reason and Revolution, 1700–1850*. Oxford: Oxford University Press, 2000.

Heimann, Peter M., and J. E. McGuire. "Newtonian Forces and Lockean Powers: Concepts of Matter in Eighteenth-Century Thought." *Historical Studies in the Physical Sciences* 3 (1971): 255–61.

Henriques, Henry Straus Quixano. *The Jews and the English Law.* Oxford: Oxford University Press, 1908. Online from the Internet Archive.

Henry, John. "The Matter of Souls: Medical Theory and Theology." In French and Wear, *Medical Revolution,* 87–113.

———. "Medicine and Pneumatology: Henry More, Richard Baxter, and Francis Glisson's '*Treatise on the Energetic Nature of Substance.*'" *Medical History* 31 (1987): 15–40.

———. "Occult Qualities and the Experimental Philosophy: Active Principles in Pre-Newtonian Matter Theory." *History of Science* 24 (1986): 335–81. Online from the SAO/NASA Astrophysics Data System, http://adsabs.harvard.edu /full/1986HisSc..24..335H.

Hill, Christopher. *Change and Continuity in Seventeenth-Century England.* Cambridge, MA: Harvard University Press, 1975.

———. *The World Turned Upside Down: Radical Ideas During the English Revolution.* New York: Viking Press, 1972.

Hirai, Hiro. *Le Concept de Semence dans les Théories de la Matière à la Renaissance: De Marsile Ficin à Pierre Gassendi.* Turnhout: Brepols, 2005.

———. "Kircher's Chymical Interpretation of the Creation and Spontaneous Generation." In *Chymists and Chymistry.* Edited by Lawrence Principe. New York: Science History Publications, 2007, 77–88.

Holmes, Geoffrey. "Science, Reason and Religion in the Age of Newton." *British Journal for the History of Science* 11 (1978): 164–71.

Hopkins, Donald R. *Princes and Peasants: Smallpox in History.* Chicago. University of Chicago Press, 1983.

Horne, R. A. "Atomism in Ancient Medical History." *Medical History* 7 (1963): 317–29.

Houliston, V. H. "Sleepers Awake: Thomas Moffet's Challenge to the College of Physicians of London, 1584." *Medical History* 33 (1989): 235–46.

Howard-Jones, Norman. "Fracastoro and Henle: A Reappraisal of Their Contribution to the Concept of Communicable Diseases." *Medical History* 21 (1977): 61–8.

Howie, W. B. "Sir Archibald Stevenson, His Ancestry, and the Riot in the College of Physicians at Edinburgh." *Medical History* 11 (1967): 269–84.

Hunter, Michael. *Science and Society in Restoration England.* Cambridge, UK: Cambridge University Press, 1981.

Huygelen, C. "The Immunization of Cattle against Rinderpest in Eighteenth-Century Europe." *Medical History* 41 (1997): 182–96.

Hyamson, Albert M. *The Sephardim of England: A History of the Spanish and Portuguese Jewish Community 1492–1951.* London: Methuen, 1951.

Israel, Jonathan I. *Radical Enlightenment: Philosophy and the Making of Modernity, 1650–1750.* Oxford: Oxford University Press, 2001.

Jacob, James R., and Margaret C. Jacob. "The Anglican Origins of Modern Science: The Metaphysical Foundations of the Whig Constitution." *Isis* 71 (1980): 251–67.

Jewson, Nicholas D. "Medical Knowledge and the Patronage System." *Sociology* 8 (1974): 369–85.

Jones, Richard Foster. *Ancients and Moderns: A Study of the Rise of the Scientific Movement in Seventeenth-Century England.* 2nd ed. Berkeley: University of California Press, 1965.

Kang, Minsoo. "From the Man-Machine to the Automaton-Man: The Enlighten-ment Origins of the Mechanistic Imagery of Humanity." In Deutsch and Terrall, *Vital Matters*, 148–73.

Kargon, Robert Hugh. *Atomism in England from Hariot to Newton.* Oxford: Clar-endon Press, 1966.

Keele, Kenneth D. "The Sydenham-Boyle Theory of Morbific Particles." *Medical History* 18 (1974): 240–8.

Kellett, C. E. "Sir Thomas Browne and the Disease Called the Morgellons." *Annals of Medical History*, n. s., 7 (1915): 467–79.

King, Lester S. *The Medical World of the Eighteenth Century.* Chicago: University of Chicago Press, 1958.

———. *The Philosophy of Medicine: The Early Eighteenth Century.* Cambridge, MA: Harvard University Press, 1978.

———. *The Road to Medical Enlightenment, 1650–1695.* London: McDonald, 1970.

———. "Stahl and Hoffman: A Study in Eighteenth-Century Animism." *Journal of the History of Medicine* 19 (1964): 118–30.

Lankester, Edwin, ed. *Memorials of John Ray: Consisting of His Life by Dr. Derham; Biographical and Critical Notices . . .* London: Ray Society, 1846.

LaTour, Bruno. *Science in Action: How to Follow Scientists and Engineers through Society.* Cambridge, MA: Harvard University Press, 1978.

Lawrence, Susan C. *Charitable Knowledge: Hospital Pupils and Practitioners in Eight-eenth-Century London.* Cambridge, UK: Cambridge University Press, 1996.

Le Fanu (LeFanu), William Richard. "The Lost Half-Century in English Medicine, 1700–1750." *Bulletin of the History of Medicine* 46 (1972): 319–49.

LeFanu, William Richard. *Nehemiah Grew M.D., F.R.S.: A Study and Bibliography of His Writings.* Detroit: Omnigraphics, 1990.

Lee, M. F. "Ipecacuanha: The South American Vomiting Root." *Journal of the Royal College of Physicians of Edinburgh*, 38 (2008): 355–60.

Levine, Joseph M. *Dr. Woodward's Shield.* Ithaca: Cornell University Press, 1991.

Lewis, Rhodri. "The Publication of John Wilkins's '*Essay*' (1668): Some Contextual Considerations." *N&R* 56 (2002): 133–47.

———. *Language, Mind and Nature: Artificial Languages in England from Bacon to Locke.* Cambridge, UK: Cambridge University Press, 2007.

Lindeboom, G. A. *Hermann Boerhaave, the Man and His Work.* London: Methuen, 1968.

Loudon, Irvine. "The Origins and Growth of the Dispensary Movement in Eng-land." *Bulletin of the History of Medicine* 55 (1982): 322–42.

Lüthy, Christoph H. "Daniel Sennert's Slow Conversion from Hylemorphism to Atomism." *Graduate Faculty Philosophy Journal* 35 (2005): 99–121.

Lüthy, Christoph H., John Murdoch, and William Newman, eds. *Late Medieval and Early Modern Corpuscular Matter Theories.* Leiden: Brill, 2001.

Lüthy, Christoph H., and William R. Newman. "Daniel Sennert's Earliest Writings (1599–1600) and Their Debt to Giordano Bruno." *Bruniana and Campanel-liana* 6 (2000): 261–79.

Lux, David, and Harold J. Cook. "Closed Circles or Open Networks? Communicat-ing at a Distance during the Scientific Revolution." *History of Science* 36 (1998): 179–211.

MacGregor, Arthur, ed. *Sir Hans Sloane, Collector, Scientist, Antiquarian, Founding Father of the British Museum,* London: British Museum Publications, 1995.

———. "The Life, Character and Career of Sir Hans Sloane." In MacGregor, *Hans Sloane*, 11–44.

MacMichael, William. *A Brief Sketch of the Progress of Opinion on the Subject of Contagion: With Some Remarks on Quarantine.* London: 1825.

———. *The Gold-Headed Cane.* 2nd ed. London: John Murray, 1926.

Maehle, Andreas-Holger. *Drugs on Trial: Experimental Pharmacology and Therapeutic Innovation in the Eighteenth Century.* Amsterdam: Rodopi, 1999.

Major, Ralph. "Athanasius Kircher." *Annals of Medical History*, 3rd ser., 1 (1939): 105–20.

Marshall, John. *John Locke, Toleration and Early Enlightenment Culture.* Cambridge, UK: Cambridge University Press, 2006.

Mason, S. F. "Science and Religion in Seventeenth-Century England." In *The Intellectual Revolution of the Seventeenth Century.* Edited by Charles Webster. London: Routledge and Kegan Paul, 1974.

Mayr, Ernst. "Species Concepts and Their Application." In *The Philosophy of Biology.* Edited by David L. Hull and Michael Ruse. Amherst, NY: Prometheus Books, 1998.

———. "What Is a Species and What Is Not?" *Philosophy of Science* 63 (1996): 262–77.

McCarl, Mary Rhinelander. "Publishing the Works of Nicholas Culpeper, Astrological Herbalist and Translator of Latin Medical Works in Seventeenth-Century London." *Canadian Bulletin of Medical History* 13 (1996): 225–76.

McGuire, J. E. "Atoms and the Analogy of Nature." *Studies in the History and Philosophy of Science* 1 (1970): 3–58.

McLachlan, H. John. *Socinianism in Seventeenth-Century England.* London: Oxford University Press, 1951.

Mendelsohn, J. Andrew. "Alchemy and Politics in England 1649–1665." *Past and Present* 135 (1992): 30–78.

Merians, Linda E., ed. *The Secret Malady: Venereal Disease in Eighteenth-Century Britain and France.* Lexington: University Press of Kentucky, 1996.

Meynell, G. G. "John Locke and the Preface to Thomas Sydenham's '*Observationes Medicae.*'" *Medical History* 50 (2006): 93–110.

Michael, Emily. "Daniel Sennert on Matter and Form: At the Juncture of the Old and the New." *Early Science and Medicine* 2 (2007): 272–99.

———. "Sennert's Sea Change: Atoms and Causes." In Luthy, Murdoch, and Newman, *Corpuscular Matter Theories*, 331–62.

Miller, Genevieve. *The Adoption of Inoculation for Smallpox in England and France.* Philadelphia: University of Pennsylvania Press, 1957.

Milton, J. R. "Locke, Medicine and the Mechanical Philosophy." *British Journal for the History of Philosophy* 9 (2001): 221–43.

Mollière, Humbert. *Étude d'Histoire Médicale: Un Précurseur Lyonnais des Théories Microbiennes: J.-B. Goiffon et la Nature Animée de la Peste.* Basel: Henri Georg, 1886.

Monod, Paul Kléber. *Solomon's Secret Arts: The Occult in the Age of Enlightenment.* New Haven: Yale University Press, 2013.

Montagu, M. F. Ashley. *Edward Tyson, M.D., F. R. S., 1650–1708, and the Rise of Human and Comparative Anatomy in England: A Study in the History of Science.* Philadelphia: American Philosophical Society, 1943.

Monti, Achille. *The Fundamental Data of Modern Pathology: History, Criticisms, Comparisons, Applications.* Translated from the Italian by John Joseph Eyre. London: New Sydenham Society, 1900.

Mortimer, Ian. "Diocesan Licensing and Medical Practitioners in South-West England, 1660–1780." *Medical History* 48 (2004): 49–68.

Mullett, Charles F. *The Bubonic Plague and England: An Essay in the History of Preventive Medicine.* Lexington: University of Kentucky Press, 1956.

———. "The Cattle Distemper in Mid-Eighteenth Century England." *Agricultural History* 20 (1946): 144–65.

———. "Physician vs. Apothecary, 1669–1671." *The Scientific Monthly* 49 (1939): 558–65.

Mulligan, Lotte, and Glenn Mulligan. "Reconstructing Restoration Science: Styles of Leadership and Social Composition of the Early Royal Society." *Social Studies of Science* 11 (1981): 327–64.

Newman, William R. "The Corpuscular Transmutational Theory of Eirenaeus Philalethes." In Rattansi and Clericuzio, *Alchemy and Chemistry*, 161–82.

———. "The Corpuscular Theory of J. B. Van Helmont and Its Medieval Sources." *Vivarium* 31 (1993): 161–91.

———. *Gehennical Fire: The Lives of George Starkey, an American Alchemist in the Scientific Revolution.* Cambridge, MA: Harvard University Press, 1994.

Nicolson, Marjorie Hope. "The Microscope and English Imagination." *Smith College Studies in Modern Languages* 16 (1935): 1–92

——— "Ward's Pill and Drop and Men of Letters." *Journal of the History of Ideas* 29 (1968): 172–96.

Nicolson, Marjorie Hope and Rousseau, George S. *"This Long Disease, My Life": Alexander Pope and the Sciences.* Princeton: Princeton University Press, 1968.

Niebyl, Peter H. "The English Bloodletting Revolution, or Modern Medicine before 1850." *Bulletin of the History of Medicine* 51 (1977): 464–83.

———. "The Helmontian Thorn." *Bulletin of the History of Medicine* 45 (1971): 570–95.

———. "Science and Metaphor in the Medicine of Restoration England." *Bulletin of the History of Medicine* 47 (1973): 356–74.

———. "Sennert, Van Helmont, and Medical Ontology." *Bulletin of the History of Medicine* 45 (1971): 115–37.

———. "Venesection and the Concept of the Foreign Body." Ph. D. dissertation. New Haven: Yale University, 1969.

Nutton, Vivian. "The Reception of Fracastoro's Theory of Contagion: The Seed That Fell among Thorns?" *Osiris*, 2nd ser., 6 (1990): 196–234.

———. "The Seeds of Disease: An Explanation of Contagion and Infection from the Greeks to the Renaissance." *Medical History* 27 (1983): 1–34.

Olsen, Richard G. *Science and Religion, 1450–1900.* Baltimore: Johns Hopkins University Press, 2004.

———. "Tory High-Church Opposition to Science and Scientism in the Eighteenth Century: The Works of John Arbuthnot, Jonathan Swift, and Samuel Johnson." In Burke, *The Uses of Science*, 191–204.

Oratz, R. "The Plague. Changing Notions of Contagion: London 1665—Marseilles 1720." *Synthesis* 4 (1977): 4–27.

Osler, Margaret J., and Paul Lawrence Farber, eds. *Religion, Science and Worldview.* Cambridge, UK: Cambridge University Press, 1985.

Pagel, Walter. "Harvey and the 'Modern' Concept of Disease." *Bulletin of the History of Medicine* 42 (1968): 496–509.

———. "Humoral Pathology: A Lingering Anachronism." *Bulletin of the History of Medicine* 29 (1955): 299–308.

———. *Joan Baptista Van Helmont.* Cambridge, UK: Cambridge University Press, 1982.

———. *Paracelsus: An Introduction to Philosophical Medicine in the Era of the Renaissance.* 2nd ed. Basel: Karger, 1982.

———. "The Reaction to Aristotle in Seventeenth-Century Biological Thought: Campanella, Van Helmont, Glanvill, Charleton, Harvey, Glisson, Descartes." In Underwood, *Science Medicine and History.* Vol. 2: 489–509.

———. "Religious and Philosophical Aspects of Van Helmont's Science and Medicine." *Bulletin of the History of Medicine,* Supplement 2, 1944: 1–44.

———. "Religious Motives in the Medical Biology of the XVIIth Century." *Bulletin of the History of Medicine* 3 (1935): 97–128, 213–31, 265–312.

———. *The Smiling Spleen: Paracelsianism in Storm and Stress.* Basel: Karger, 1984.

———. "Van Helmont's Concept of Disease—To Be or Not to Be? The Influence of Paracelsus." *Bulletin of the History of Medicine* 46 (1972): 419–54.

Palm, L. C. "Leeuwenhoek and Other Dutch Correspondents of the Royal Society." *N&R* 43 (1989): 191–207.

Partington, James Riddick. *A History of Chemistry.* 4 vols. London: Macmillan, 1961–70.

Pavord, Anna. *Searching for Order.* London: Bloomsbury, 2005.

Pelling, Margaret. *Cholera, Fever and English Medicine, 1825–1865.* Oxford: Oxford University Press, 1978.

Peypers, H. F. A. "Un Ancien Pseudo-Précurseur de Pasteur ou le '*Système d'un Médecin Anglois sur la Cause de Toutes les Maladies*' (1726)." *Janus* 1 (1896–7): 57–66, 121–31, 251–62. Online from Google.

Phear, D. N. "Thomas Dover 1662–1742: Physician, Privateering Captain, and Inventor of Dover's Powder." *Journal of the History of Medicine* 9 (1954): 139–56.

Poole, William. "A Fragment of the Library of Theodore Haak (1605–1690)." *eBLJ* Article 6 (2007): 1–38. Online from the Electronic British Library Journal, http://www.bl.uk/eblj/2007articles/pdf/ebljarticle62007.pdf.

Popkin, Richard H. "Serendipity and the Clark: Spinoza and the Prince of Condé." *Clark Library Newsletter* 10 (1986): 4–7.

———. *The Third Force in Seventeenth-Century Thought.* Leiden: E. J. Brill, 1992.

Porter, Roy. *Health for Sale: Quackery in England, 1660–1850.* Manchester: Manchester University Press, 1989.

———. "The Early Royal Society and the Spread of Medical Science." In French and Wear, *Medical Revolution,* 272–93.

———. "Lay Medical Knowledge in the Eighteenth Century—The Evidence of *The Gentleman's Magazine.*" *Medical History* 29 (1985): 138–68.

———. "'Laying Aside any Private Advantage:' John Marten and Venereal Disease." In Merrians, *Secret Malady,* 51–67.

———. "Medical Science and Human Science in the Enlightenment." In Fox, Porter, and Wokler, *Inventing Human Science,* 53–87.

———. "Was There a Medical Enlightenment in Eighteenth-Century England?" *British Journal for Eighteenth-Century Studies* 5 (1982): 49–63.

Porter, Roy, and Mikulas Teich, eds. *The Scientific Revolution in National Context.* Cambridge, UK: Cambridge University Press, 1992.

Power, D'Arcy. "Medicine." In *Johnson's England.* Edited by A. S. Turberville. Vol. 2. Oxford: Clarendon Press, 1933, 265–86.

Poynter, F. N. L., ed. *Chemistry in the Service of Medicine.* London: Pitman Medical Publishing, 1963.

———. "Nicholas Culpeper and His Books." *JHMAS* 17 (1962): 152–67.

———. "A Seventeenth-Century Medical Controversy: Robert Witty versus William Simpson." In Underwood, *Science, Medicine and History*, Vol. 2: 72–81.

Principe, Lawrence M., ed. *New Narratives in Eighteenth-Century Chemistry*. Dordrecht: Springer, 2007.

Randall, John Herman. *The School of Padua and the Emergence of Modern Science*. Padua: Editrice Antenore, 1961.

Ranger, Terence, and Paul Slack, eds. *Epidemics and Ideas: Essays on the Historical Perception of Pestilence*. Cambridge, UK: Cambridge University Press, 1996.

Ratcliff, Marc J. *The Quest for the Invisible*. Farnham, UK: Ashgate, 2009.

Rather, L. J. "Towards a Philosophical Study of the Idea of Disease." In Brooks and Cranefield, *Development of Physiological Thought*, 351–73.

Rattansi, Piyo M., and Antonio Clericuzio, eds. *Alchemy and Chemistry in the 16th and 17th Centuries*. Dordrecht: Kluwer Academic Publishers, 1994.

Rattansi, Piyo M. "The Helmontian-Galenist Controversy in Restoration England." *Ambix* 12 (1964): 1–23.

———. "Paracelsus and the Puritan Revolution." *Ambix* 11 (1963): 24–32.

———. "The Social Interpretation of Science in the Seventeenth Century." In *Science and Society, 1600–1900*. Edited by Peter Mathias. Cambridge, UK: Cambridge University Press, 1972, 1–32.

Razzell, Peter. *The Conquest of Smallpox*. Firle, UK: Caliban Books, 1977.

Reilly, P. Conor. *Athanasius Kircher S.J.: Master of a Hundred Arts*. Wiesbaden: Edizioni Del Mondo, 1974.

[Rhazes] Abú Bakr Mohammad Ibn Zakaríyá ar-Rází. *A Treatise on the Smallpox and Measles*. Translated by William Alexander Greenhill. London: Sydenham Society, 1848. Online from the Internet Archive.

Richardson, Linda Deer. "The Generation of Disease: Occult Causes and Diseases of the Total Substance." In Wear, French, and Lonie, *Medical Renaissance*, 175–94.

Riley, James. *The Eighteenth-Century Campaign to Avoid Disease*. New York: St. Martin's Press, 1987.

Riley, Margaret. "The Club at the Temple Coffee House Revisited." *Archives of Natural History* 33 (2006): 90–100.

Riley, William A. "Early References to the Relation of Flies to Disease." *Science* 31 no. 790 (1910): 263–4.

———. "Kircher and the Germ Theory of Disease." *Science* 31 no. 800 (1910): 666.

Risse, Guenter. "Clinical Instruction in Hospitals: The Boerhaavian Tradition in Leyden, Edinburgh, Vienna and Pavia." *Clio Medica* 21 (1987–8): 1–19.

———. "Epidemics and Medicine: The Influence of Disease on Medical Thought and Practice. *Bulletin of the History of Medicine* 53 (1979): 505–19.

Roberts, W. "R. Bradley, Pioneer Garden Journalist." *Journal of the Royal Horticultural Society* 64 (1939): 164–74.

Romagnoli, Giovanni. "L'Evoluzione del Concetto della Contagiosità e della Profilassi della Tuberculosi Polminare Attraverso i Secoli, dall'Antichità Fino alla Scoperta del Bacillo de Koch." *Giornale de Batteriologia, Virologia ed Immunologia ed Annali dell'Ospedale Maria Vittoria de Torino* 61 (1968): 233–73.

Romanell, Patrick. *John Locke and Medicine*. Buffalo, NY: Prometheus Books, 1984.

Rook, Arthur. "Medicine at Cambridge, 1660–1760." *Medical History* 13 (1969): 107–22.

Roos, Anna Maria Eleanor. *The Salt of the Earth: Natural Philosophy, Medicine, and Chymistry in England, 1650–1750*. Leiden: Brill, 2007.

———. *Web of Nature: Martin Lister (1639–1712): The First Arachnologist*. Leiden: Brill, 2011.

Rose, Craig. "Politics and the London Royal Hospitals, 1683–92." In *The Hospital in History*. Edited by Lindsay Granshaw and Roy Porter. London: Routledge, 1989, 123–48.

Rosenberg, Albert. "The London Dispensary for the Sick Poor." *JHMAS* 14 (1959): 41–56.

———. *Sir Richard Blackmore: A Poet and Physician of the Augustan Age*. Lincoln: University of Nebraska Press, 1953.

Rosenberg, Charles. "Commentary: Epidemiology in Context." *International Journal of Epidemiology* 38 (2009): 28–30. Online at doi: 10.1093/ije/dyn257.

Rousseau, George. "'Sowing the Wind, and Reaping the Whirlwind': Aspects of Change in Eighteenth-Century Medicine." In *Studies in Change and Revolution: Aspects of English Intellectual History, 1640-1800*. Edited by Paul Korshin. Menston UK: Scholar Press, 1972, 129-159.

Rousseau, George, and Roy Porter, eds. *The Ferment of Knowledge: Studies in the Historiography of Eighteenth-Century Science*. Cambridge, UK: Cambridge University Press, 1980.

Rudrum, Alan. "Research Reports no. VI: Theology and Politics in Seventeenth-Century England." *Clark Newsletter* 15 (1988): 5–7.

Ruestow, Edward G. *The Microscope in the Dutch Republic*. Cambridge, UK: Cambridge University Press, 1996.

Rurah, John. "John Pechey, 1655–1718." *American Journal of the Diseases of Children* 39 (1930): 179–84.

Rusnock, Andrea. "Correspondence Networks and the Royal Society, 1700–1750." *British Journal of the History of Science* 32 (1999): 155–69.

———. *Vital Accounts: Quantifying Health and Population in Eighteenth-Century England and France*. Cambridge, UK: Cambridge University Press, 2002.

Schaffer, Simon. "The Glorious Revolution and Medicine in Britain and the Netherlands." *N&R* 43 (1989): 167–90.

———. "Natural Philosophy." In Rousseau and Porter, *Ferment of Knowledge*, 55–91.

Schofield, R. E. "An Evolutionary Taxonomy of Eighteenth-Century Newtonianisms." *Studies in Eighteenth-Century Culture* 7 (1998): 175–92.

Schwoerer, Lois. *The Ingenious Mr. Henry Care, Restoration Publicist*. Baltimore: Johns Hopkins University Press, 2001.

Scott, Mary Jane W. *James Thomson, Anglo-Scot*. Athens: University of Georgia Press, 1988.

Selwood, Jacob. *Diversity and Difference in Early Modern London*. Farnham, UK: Ashgate, 2010.

Shackelford, Jole. *A Philosophical Path for Paracelsian Medicine: The Ideas, Intellectual Context, and Influence of Petrus Severinus (1540/2–1602)*. Copenhagen: Museum Tusculanum Press, 2004.

Schaffer, Simon. "Natural Philosophy." In Rousseau and Porter, *Ferment of Knowledge*, 55–91.

Shapin, Steven, and Simon Schaffer. *Leviathan and the Air-Pump: Hobbes, Boyle and the Experimental Life*. Princeton: Princeton University Press, 1985.

Shapin, Steven. "Social Uses of Science." In Rousseau and Porter, *Ferment of Knowledge*, 93–139.

Shapiro, Barbara. *John Wilkins, 1614–1672: An Intellectual Biography*. Berkeley: University of California Press, 1969.

Shryock, Richard Harrison. *The Development of Modern Medicine: An Interpretation of the Social and Scientific Factors Involved*, (1936). Madison: University of Wisconsin Press, 1979. Rpt.

———. "Germ Theories in Medicine Prior to 1870." *Clio Medica* 7 (1972): 81–109.

Shuttleton, David E. "'A Modest Examination:' John Arbuthnot and the Scottish Newtonians." *British Journal for Eighteenth-Century Studies* 18 (1995): 47–62.

Siena, Kevin P. "Contagion, Exclusion and the Unique Medical World of the Eighteenth-Century Workhouse." In *Medicine and the Workhouse*. Edited by Jonathan Reinarz and Leonard Schwarz. Rochester: University of Rochester Press, 2013.

———. *Venereal Disease, Hospitals and the Urban Poor: London's "Foul Wards,"* *1600–1800*. Rochester: Rochester University Press, 2004.

Silverstein, Arthur M., and Genevieve Miller. "The Royal Experiment on Immunity: 1721–1722." *Cellular Immunology* 61 (1981): 437–47.

Singer, Charles. "Benjamin Marten, a Neglected Predecessor of Louis Pasteur." *Janus* 16 (1911): 81–9.

———. *The Development of the Doctrine of Contagium Vivum 1500–1750: A Preliminary Sketch*. London: for the author, 1913.

———. "Notes on the Early Development of Microscopy." *Proceedings of the Royal Society of Medicine, Section of the History of Medicine* 7 (1914): 247–79.

Singer, Charles, and Dorothea Singer. "The Scientific Position of Girolamo Fracastoro (1478–1553) with Especial Reference to the Source, Character and Influence of His Theory of Infection." *Annals of Medical History* 1 (1917): 1–29.

Sirota, Brent S. "The Trinitarian Crisis in Church and State: Religious Controversy and the Making of the Postrevolutionary Church of England, 1687–1702." *Journal of British Studies* 52 (2013): 26–54.

Slack, Paul. *The Impact of Plague in Tudor and Stuart England*. Rpt. Oxford: Clarendon Press, 2003.

Slaughter, M. M. *Universal Languages and Scientific Taxonomy in the Seventeenth Century*. Cambridge, UK: Cambridge University Press, 1982.

Snobelen, Stephen D. "To Discourse of God: Isaac Newton's Heterodox Theology and His Natural Philosophy." In *Science and Dissent in England, 1688–1945*. Edited by Paul Wood. Aldershot, UK: Ashgate, 2004, 19–38.

———. "'God of Gods, and Lord of Lords': The Theology of Isaac Newton's General Scholium to the '*Principia.*'" *Osiris* 2nd ser., 16 (2001): 169–208.

———. "Isaac Newton, Heretic: The Strategies of a Nicodemite." *British Journal of the History of Science* 32 (1999): 381–419.

———. "Isaac Newton, Socinianism and the 'One Supreme God.'" In *Socinianism and Arminianism: Antritrinitarians, Calvinists, and Cultural Exchange in Seventeenth-Century Europe*. Edited by Martin Mulsow and Jan Rohls. Leiden: Brill, 2005, 241–98. Online from the Newton Project, http://www.isaac-newton.org/articles.

Solomon, Harry M. *Sir Richard Blackmore*. Boston: Twayne, 1980.

Sowerby, Scott. "Forgetting the Repealers: Religious Toleration and Historical Amnesia in Later Stuart England." *Past and Present* 215 (2012): 85–123.

Spinage, C. A. *Cattle Plague: A History*. New York: Kluwer Academic/Plenum Publishers, 2003.

Stearns, Raymond Phineas. "James Petiver, Promoter of Natural Science, c. 1663–1718." *Proceedings of the American Antiquarian Society*, n. s., 62 (1952): 243–364.

Stern, Alexandra Minna, and Howard Markel. "Commentary: Disease Etiology and Political Ideology: Revisiting Erwin H Ackerknecht's Classic 1948 Essay, 'Anti-contagionism between 1821 and 1867.'" *International Journal of Epidemiology* 38 (2009): 31–3.

Stevenson, Lloyd G. "New Diseases in the Seventeenth Century." *Bulletin of the History of Medicine* 39 (1965): 1–21.

Stewart, Larry. "The Edge of Utility: Slaves and Smallpox in the Early Eighteenth Century." *Medical History* 29 (1985): 54–70.

Stride, P. "The St. Kilda Boat Cough under the Microscope." *Journal of the Royal College of Physicians of Edinburgh* 38 (2008): 272–9.

Sutherland, L. S., and L. G. Mitchell, eds. *The History of the University of Oxford*. Vol. 5: *The Eighteenth Century*. Oxford: Clarendon Press, 1986.

Syfret, R. H. "Some Early Critics of the Royal Society." *N&R* 8 (1950): 20–64.

———. "Some Early Reactions to the Royal Society." *N&R* 7 (1950): 207–58.

Takeda, Junko Therese. *Between Crown and Commerce: Marseille and the Early Modern Mediterranean*. Baltimore: Johns Hopkins University Press, 2011.

Temkin, Owsei. *Galenism: Rise and Decline of a Medical Philosophy*. Ithaca, NY: Cornell University Press, 1973.

———. *The Double Face of Janus*. Baltimore: Johns Hopkins University Press, 1977.

Thomas, Henry. "The Society of Chymical Physitians: An Echo of the Great Plague of London, 1665." In Underwood, *Science, Medicine and History*, Vol. 2: 55–71.

Traill, R. R. "Richard Morton (1637–1698)." *Medical History* 14: 1970, 166–74.

Trevor-Roper, Hugh. *Europe's Physician: The Various Life of Theodore de Mayerne*. Edited by Blair Worden. New Haven: Yale University Press, 2006.

Tröhler, Ulrich. "An Early 18th Century Proposal for Improving Medicine by Tabulating and Analysing Practice." 2007. Online from the James Lind Library, http://www.jameslindlibrary.org/articles.

———. "The Introduction of Numerical Methods to Assess the Effects of Medical Interventions during the 18th Century: A Brief History." 2010. Online from the James Lind Library, http://www.jameslindlibrary.org/articles.

———. *"To Improve the Evidence of Medicine": The 18th Century British Origins of a Critical Approach*. Edinburgh: Royal College of Physicians of Edinburgh, 2000.

———. "Quantification in British Medicine and Surgery 1750–1830, With Special Reference to Its Introduction into Therapeutics." Ph.D. dissertation. London: University College, 1978.

Tronchon, Claude, and Thierry Sabnot. "Le Mur de la Peste in Provence." 2007. Online from Histoire-Généalogie, http://www.histoire-généalogie.com/spip .php?article1249.

Ultee, Maarten. "Sir Hans Sloane, Scientist." *British Library Journal* 14 (1988): 1–20.

Underwood, E. Ashworth. *Boerhaave's Men at Leyden and After*. Edinburgh: Edinburgh University Press, 1977.

———. ed. *Science, Medicine and History: Essays on the Evolution of Scientific Thought and Medical Practice Written in Honour of Charles Singer*. 2 vols. London: Oxford University Press, 1953.

Verwaal, Ruben E. "Hippocrates Meets the Yellow Emperor: On the Reception of Chinese and Japanese Medicine in Early Modern Europe." M.A. thesis. University

of Utrecht Medical Center, 2010. Online from Utrecht University, http://dspace
.library.uu.nl/handle/1874/179050.

Waddington, Ivan. "The Role of the Hospital in the Development of Modern Medicine." *Sociology* 7 (1973): 211–5.

Waddington, John. *Congregational History*. Vol. 5: *1850–1880*. London: Longmans Green, 1880.

Walmsley, Jonathan. "'*Morbus*'—Locke's Early Essay on Disease." *Early Science and Medicine* 5 (2000): 366–93.

Wear, Andrew. *Knowledge and Practice in English Medicine, 1550–1680*. Cambridge, UK: Cambridge University Press, 2000.

———. "Medical Practice in Late Seventeenth- and Early Eighteenth-Century England: Continuity and Union." In French and Wear, *Medical Revolution*, 294–320.

Wear, Andrew, Roger Kenneth French, and Iain Lonie, eds. *The Medical Renaissance of the Sixteenth Century*. New York: Cambridge University Press, 1985.

Webb, Robert K. "The Emergence of Rational Dissent." In Haakonsen, *Enlightenment and Religion*, 12–41.

Webster, Charles. "English Medical Reformers of the Puritan Revolution: A Background to the 'Society of Chymical Physitians.'" *Ambix* 14 (1967): 16–41.

———. *The Great Instauration: Science, Medicine and Reform, 1626–1660*. New York: Holmes and Meyer, 1976.

———. ed. *Health, Medicine and Mortality in the 16th Century*. Cambridge, UK: Cambridge University Press, 1979.

———. ed. *The Intellectual Revolution of the Seventeenth Century*. London: Routledge and Kegan Paul, 1974.

Westfall, Richard. *Never at Rest: A Biography of Isaac Newton*. Cambridge, UK: Cambridge University Press, rpt. 1990.

Wigelsworth, Jeffrey Robert. "'Their Grosser Degrees of Infidelity:' Deists, Politics, Natural Philosophy, and the Power of God in Eighteenth-Century England." Ph.D. Dissertation, University of Saskatchewan, 2005. Online from the University of Saskatchewan http://ecommons.usask.ca/handle/10388/etd-09292005-141239.

Wilbur, Earl Morse. *A History of Unitarianism*. 2 Vols. Cambridge, MA: Harvard University Press, 1945–52.

Wilkins, John. *Species: A History of the Idea*. Berkeley: University of California Press, 2009.

Wilkinson, Lise. *Animals and Disease: An Introduction to the History of Comparative Medicine*. Cambridge, MA: Cambridge University Press, 1992.

———. "The Development of the Virus Concept as Reflected in Corpora of Studies on Individual Pathogens. 5: Smallpox and the Evolution of Ideas on Acute (Viral) Infections." *Medical History* 23 (1979): 1–29.

———. "Rinderpest and Mainstream Infectious Diseases Concepts in the Eighteenth Century." *Medical History* 28 (1984): 129–50.

Williams, George H. *The Radical Reformation*. Philadelphia: Westminster Press, 1962.

Williamson, Raymond. "The Germ Theory of Disease. Neglected Precursors of Louis Pasteur (Richard Bradley, Benjamin Marten, Jean-Baptiste Goiffon)." *Annals of Science* 11 (1955): 44–57.

———. "John Martyn and the 'Grub-Street Journal,' with Particular Reference to His Attacks on Richard Bentley, Richard Bradley, and William Cheselden." *Medical History* 5 (1961): 361–73.

———. "The Plague of Marseilles and the Experiments of Professor Anton Deidier on Its Transmission." *Medical History* 2 (1958): 237–51.

Williamson, Elizabeth. *A Cultural History of Medical Vitalism in Enlightenment Montpellier.* Aldershot, UK: Ashgate, 2003.

Wilson, Adrian. *The Making of Man-Midwifery: Childbirth in England, 1660–1770.* Cambridge, MA: Harvard University Press, 1995.

———. "On the History of Disease-Concepts: The Case of Pleurisy." *History of Science* 38 (2000): 271–319.

Wilson, Adrian, and T. G. Ashplant. "Present-Centered History and the Problem of Historical Knowledge." *Historical Journal* 31 (1988) 253–74.

———. "Whig History and Present-Centered History." *Historical Journal* 31 (1988): 1–16.

Wilson, Catherine. *The Invisible World: Early Modern Philosophy and the Invention of the Microscope.* Princeton: Princeton University Press, 1997.

Wilson, Leonard. "Medical History without the Medicine." *Journal of the History of Medicine* 35 (1980): 5–7.

Wilson, Philip K. *Surgery, Skin and Syphilis: Daniel Turner's London 1657–1741.* Amsterdam: Rodopi, 1999.

Winslow, Charles-Edward Amory. *The Conquest of Epidemic Disease: A Chapter in the History of Ideas* (1943). Madison: University of Wisconsin Press, 1980.

Withey, Alun. *Physick and the Family: Health, Medicine and Care in Wales, 1600–1750.* Manchester: Manchester University Press, 2011.

Wolff, Jacob. *The Science of Cancerous Disease from Earliest Times to the Present.* Translated by Barbara Ayoub. Canton, MA: Science History Publications, 1989.

Wood, Paul, ed. *Science and Dissent in England, 1688–1945.* Aldershot, UK: Ashgate, 2004.

Woodward, John. *To Do the Sick No Harm: A Study of the British Voluntary Hospital System to 1875.* London: Routledge and Kegan Paul, 1974.

Wootton, David. *Bad Medicine: Doctors Doing Harm Since Hippocrates.* Oxford: Oxford University Press, 2007.

Wulf, Andrea. *The Brother Gardeners: Botany, Empire, and the Birth of an Obsession.* New York: Knopf, 2009.

Yolton, John. *Thinking Matter: Materialism in Eighteenth-Century Britain.* Minneapolis: University of Minnesota Press, 1984.

Zelt, Roger P. "Smallpox Inoculation in Boston, 1721–1722." *Synthesis* 4 (1977): 3–14.

Zuckerman, Arnold. "Dr. Richard Mead (1673–1754): A Biographical Study." Ph.D. dissertation, University of Illinois, 1965.

———. "Plague and Contagionism in Eighteenth-Century England: The Role of Richard Mead." *Bulletin of the History of Medicine* 28 (2004): 273–308.

INDEX

Page numbers in *italics* refer to figures.

The manufacturer's authorised representative in the EU is Springer Nature Customer Service Centre GmbH, Europaplatz 3, 69115 Heidelberg, Germany. If you have any concerns regarding our products, please contact ProductSafety@springernature.com

Printed and bound by CPI Group (UK) Ltd, Croydon, CR0 4YY

23/04/2026

02095587-0020